Eiselt · von Frajer
Operations Research Handbook

Horst A. Eiselt · Helmut von Frajer

Operations Research Handbook
Standard Algorithms and Methods with Examples

Walter de Gruyter · Berlin · New York · 1977

CIP-Kurztitelaufnahme der Deutschen Bibliothek

Eiselt, Horst A.
Operations research handbook: standard algorithms and methods with examples/Horst A. Eiselt; Helmut von Frajer. — 1. Aufl. — Berlin, New York: de Gruyter. 1977.
ISBN 3-11-007055-3

NE: Frajer, Helmut von:

Library of Congress Cataloging in Publication Data

Eiselt, Horst A. 1950 —
   Operations research handbook.
   Bibliography: p.
   1. Operations research — Handbooks, manuals, etc.
I. Frajer, Helmut von, 1948 — joint author.
II. Title.
T 57.6.E37 001.4'24 77-5572
ISBN 3-11-007055-3

© Copyright 1977 by Walter de Gruyter & Co., Berlin 30.
All rights reserved, including those of translation into foreign languages. No part of this book may be reproduced in any form — by photoprint, microfilm, or any other means — nor transmitted nor translated into a machine language without written permission from the publisher. Printing: Karl Gerike, Berlin. — Binding: Lüderitz & Bauer, Buchgewerbe GmbH, Berlin. — Cover design: Armin Wernitz, Berlin. — Printed in Germany.

Preface

In writing this handbook the authors had in mind not only the students of economics, social sciences, and mathematic but also the professionals in all fields engaging in optimization problems and the control of projects.

The most important and most often utilized algorithms for the solution of these problems have been collected and described in a uniform and hopefully understandable manner. With this goal set, each section was divided as follows:

a) Hypotheses: Here the problem is formulated and the prerequisites are explained.

b) Principle: Briefly the general concept is presented.

c) Description: In this section each step of the algorithm or method is explained in a standard format.

d) Example: For each paragraph an example is completely solved. An effort has been made in selecting the examples to find those which illustrate even the special properties of certain algorithms and methods.

The reader should already be familiar with the basic concepts of differentiation, integration, and statistics. A summary of the topics in Linear Algebra that are necessary is provided at the beginning of the book.

Because this book was conceived as a desk reference, the corresponding theory and necessary proofs which are found in the myriad of textbooks in Operations Research have been omitted. For details concerning the material in each chapter, beyond the outright applications, it is essential that the reader becomes familiar with the technical literature.

6 *Preface*

To this purpose a selection of the standard literature and advanced research studies is also included in the bibliography.

The solid arrows in the sketch below indicate the relationships of the chapters of the book to each other, as well as the minimum mathematics suggested for full understanding of each chapter. The dotted arrows show other relationships that exist between the material in each chapter although they will not be expounded upon further.

It was never the intension of the authors to convey through this sketch exactly how the field of Operations Research is structured either in theory or practice. Nevertheless it does illustrate certain general relationships between the various areas of Operations Research.

The authors are very grateful to Prof. Dr. H. Noltemeier, University of Goettingen, for the many invaluable suggestions that he has offered from time to time during the formulation of this book. The authors also thank Cpt. Lee Dewald, who devoted much of his last semester in Goettingen assisting in the translation of the original manuscript; and Mrs. Schillings, who arduously sorted through the translation and typed the final copy. Last but not least the authors acknowledge Walter de Gruyter & Co. as partners in publishing.

Goettingen, in spring 1977   Horst A. Eiselt
                                                   Helmut von Frajer

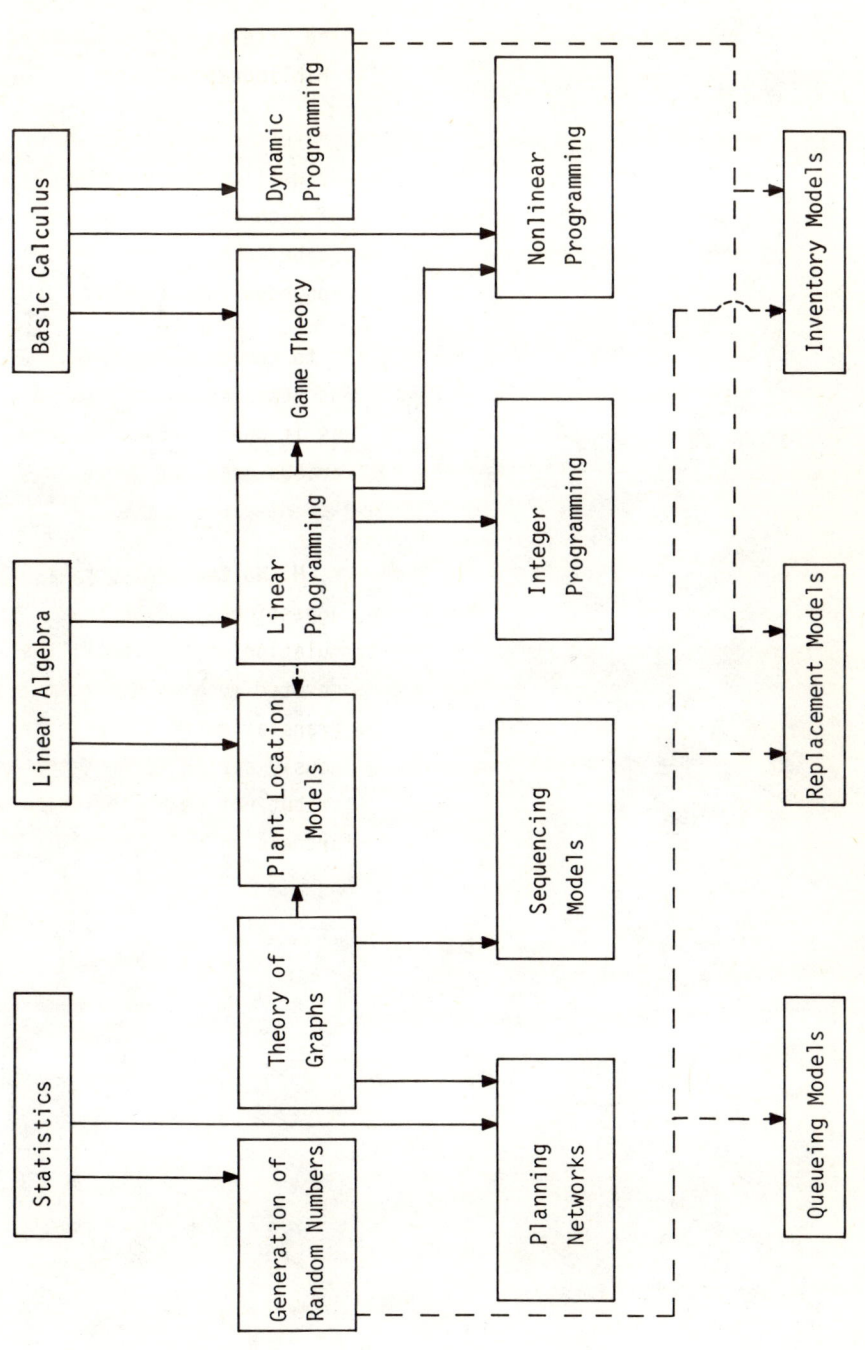

Contents

	Definitions and Symbols	15
0.	Summary of Matrix Algebra and Allied Topics	
0.1	Definitions	17
0.2	Elementary Operations	18
1.	Linear Programming	
1.1	General Methods	
1.1.1	The Primal Simplex-Algorithm	22
1.1.2	The Two-Phase Method	25
1.1.3	The Primal Simplex-Algorithm without Explicit Identity Matrix	30
1.1.4	The Dual Simplex-Algorithm	34
1.1.5	Sensitivity Analisys and Parametric Programming (S.A. and P.P.)	38
1.1.5.1	S.A. and P.P. with Expected Alterations	39
1.1.5.2	S.A. and P.P. with Unexpected Alterations	
1.1.5.2.1	Subsequent Alterations of the Restriction Vector	43
1.1.5.2.2	Subsequent Alterations of Coefficients of the Objective Function	45
1.2	Shortened Methods	
1.2.1	The Transportation Problem	48
1.2.1.1	The Northwest-Corner Rule	49
1.2.1.2	The Row Minimum Method	51
1.2.1.3	The Column Minimum Method	54
1.2.1.4	The Matrix Minimum Method	57
1.2.1.5	The Double Preference Method	60
1.2.1.6	VOGEL's Approximation Method (VAM)	66
1.2.1.7	The Frequency Method	71
1.2.1.8	The Stepping-Stone Method	72
1.2.2	The Hungarian Method (Kuhn)	75

1.2.3	The Decomposition Principle (Dantzig; Wolfe)	81
1.2.4	FLOOD's Technique	89
1.3	Theorems and Rules	
1.3.1	The Dual Problem	91
1.3.2	Theorems of Duality	93
1.3.3	The Lexicographic Selection Rule	94
2.	Integer Programming	
2.1	Cutting Plane Methods	
2.1.1	The GOMORY-I-All Integer Method	96
2.1.2	The GOMORY-II-All Integer Method	100
2.1.3	The GOMORY-III-Mixed Integer Method	103
2.1.4	The GOMORY-III-Mixed Integer Method with Intensified Cuts	106
2.1.5	The Primal Cutting Plane Method (Young; Glover; Ben-Israel; Charnes)	108
2.2	Branch and Bound Methods	
2.2.1	The Method of LAND and DOIG	111
2.2.2	The Method of DAKIN	118
2.2.3	The Method of DRIEBEEK	122
2.2.4	The Additive Algorithm (Balas)	129
2.3	Primal-Dual Methods	
2.3.1	A Partitioning Procedure for Mixed Integer Problems (Benders)	134
3.	Theory of Graphs	
3.0.1	Definitions	143
3.0.2	The Determination of Rank in Graphs	146
3.0.3	The Number of Paths in a Graph	148
3.0.4	The Determination of the Strongly Connected Components of a Graph	149
3.1	Shortest Paths in Graphs	
3.1.1	The Algorithm of DIJKSTRA	151
3.1.2	The Algorithm of DANTZIG	154

3.1.3	The FORD Algorithm I (shortest path(s))	159
3.1.4	The FORD Algorithm II (longest path(s))	160
3.1.5	The Tripel Algorithm	162
3.1.6	The HASSE Algorithm	166
3.1.7	The Cascade Algorithm	168
3.1.8	The Algorithm of LITTLE.............................	169
3.1.9	The Method of EASTMAN	174
3.2	Flows in Networks	
3.2.1	The Algorithm of FORD and FULKERSON	178
3.2.2	The Algorithm of BUSACKER and GOWEN	183
3.2.3	The Method of KLEIN	187
3.2.4	The Out-of-Kilter Algorithm (Ford; Fulkerson)	191
3.3	Shortest Spanning Subtrees of a Graph	
3.3.1	The Method of KRUSKAL	200
3.3.2	The Method of SOLLIN	203
3.3.3	The Method of WOOLSEY	205
3.3.4	The Method of BERGE	207
3.4	Gozinto Graphs	210
3.4.1	The Method of VAZSONYI.............................	210
3.4.2	The Method of TISCHER	212
3.4.3	The Method of FLOYD	213
3.4.4	The Gozinto List Method	214
4.	Planning Networks	
4.0.1	The Critical Path Method (CPM).....................	217
4.0.2	The CPM Project Acceleration	220
4.0.3	The Program Evaluation and Review Technique (PERT)	224
4.0.4	The Metra Potential Method (MPM)	227
4.0.5	The Graphical Evaluation and Review Technique (GERT)	230

12 Contents

5.	Game Theory	
5.1	Non Matrix Games	
5.1.1	The Normal Form	236
5.1.2	NASH's Solution of the Bargaining Problem	241
5.1.3	The Extensive Form	242
5.2	Matrix Games	
5.2.1	A Method for Determining Pure Strategy Pairs for Two-Person Zero-Sum Games	251
5.2.2	A Method for Solving Two-Person Zero-Sum Games with the Simplex-Algorithm	253
5.2.3	An Approximization Method for Two-Person Zero-Sum Games ("learning method"; Gale; Brown)	256
5.2.4	The LEMKE-HOWSON Algorithm for the Solution of Bimatrix Games	260
5.3	Decisions under Uncertainty (games against nature)	264
5.3.1	The Solution of WALD	265
5.3.2	The Solution of HURWICZ	266
5.3.3	The Solution of SAVAGE and NIEHANS	266
5.3.4	The Solution of BAYES	267
5.3.5	The Solution of LAPLACE	268
5.3.6	The Solution of HODGES and LEHMANN	268
6.	Dynamic Programming	
6.0.1	The n-Period Model	270
6.0.2	The Infinite-Period Model (policy iteration routine)	276
7.	Queueing Models	280
7.0.1	The 1-Channel, 1-Stage Model	282
7.0.2	The 1-Channel, r-Phase Model	284
7.0.3	The k-Channel, 1-Stage Model	285
8.	Nonlinear Programming	
8.1	Theorems and Special Methods	
8.1.1	The Theorem of KUHN and TUCKER	288

8.1.2	The Method of LAGRANGE	289
8.1.3	A Method for the Optimization of Nonlinear Separable Objective Functions under Linear Constraints	292
8.2	General Methods	
8.2.1	The Method of WOLFE (short form)	296
8.2.2	The Method of FRANK and WOLFE	302
8.2.3	The Method of BEALE	307
8.2.4	An Algorithm for the Solution of Linear Complementarity Problems (Lemke)	312
8.2.5	The Gradient Projection Method (Rosen)	315
9.	Generation of Random Numbers (simulation)	321
9.0.1	The AWF-Cubes (Graf)	321
9.0.2	The Midsquare Method (J.v.Neumann)	322
9.0.3	A Mixed Congruence Method	323
9.0.4	A Multiplicative Congruence Method	324
10.	Replacement Models	
10.1	Replacement Models with Respect to Gradually Increasing Maintenance Costs	
10.1.1	A Model Disregarding the Rate of Interest	325
10.1.2	A Model Regarding the Rate of Interest	326
10.2	Replacement Models with Respect to Sudden Failure	
10.2.1	A Model Disregarding the Rate of Interest	328
10.2.2	A Model Regarding the Rate of Interest	331
11.	Inventory Models	333
11.0.1	The Classical Inventory Model (Andler)	333
11.0.2	An Inventory Model with Penalties for Undersupplied Demands	334
11.0.3	An Inventory Model with Terms for Delivery	335
11.0.4	An Inventory Model with Damage to Stock	337
11.0.5	An Inventory Model with Rebates (different price intervals)	338

11.0.6	An Inventory Model with Respect to Transportation Capacity	340
12.	Sequencing Models	
12.0.1	JOHNSON's Algorithm for Two Machines	343
12.0.2	JOHNSON's Algorithm for Three Machines (special case)	345
12.0.3	A Heuristic Solution for a Sequencing Problem	347
13.	Plant Location Models	
13.1	Exact Methods	
13.1.1	The Optimal Plant Location in a Transportation Network I	350
13.1.2	The Optimal Plant Location in a Transportation Network II	351
13.1.3	The Optimal Plant Location on a Straight Line	353
13.1.4	The Optimal Plant Location with Respect to Rectangular Transportation Movements	354
13.2	Heuristic Methods	
13.2.1	The Center of Gravity-Method	356
13.2.2	A Solution by Vector Summation	358
13.2.3	An Iterative Method	363
	Appendix	367
	Table 1 : $q^k = (1 + i)^k$	367
	Table 2 : $q^{-k} = (1 + i)^{-k}$	367
	Table 3 : e^{-k}	368
	Table 4 : Random numbers with an equal distribution.............................	370
	Table 5 : Area under the standardized normal distribution function.....................	372
	Bibliography	373

Definitions and Symbols

A : $A_{[m \times n]}$: $[m \times n]$ - dimensional matrix A

I : identity matrix

Θ : null matrix

A^T : transpose of matrix A

A^{-1} : inverse of matrix A

$\mathbb{R}^n$: n-dimensional real euclidian space

If not otherwise defined, a problem P is given as follows:

P : $\left.\begin{matrix}\min\\\max\end{matrix}\right\} \pi = f(x); A \cdot x \gtreqless b; x \geq \Theta$,

where $A:A_{[m \times n]}$; $x:x_{[n \times 1]}$; $b:b_{[m \times 1]}$.

$x \in [a;b] = a \leq x \leq b$, where $b \geq a$: closed interval

$x \in (a;b] = a < x \leq b$, where $b > a$
$x \in [a;b) = a \leq x < b$, where $b > a$ $\Big\}$: half-closed intervals

$x \in (a;b) = a < x < b$, where $b > a$: open interval

$a: = a + b$: valuation

$\exists$: there is ...

$\forall$: for all ...

$\leftrightarrow \equiv$ iff : equivalence relation

$\Rightarrow$: implication

$[a]$: largest integer smaller than a

$<a>$: smallest integer larger than a

$\emptyset$: empty set

$|a|$: absolute value of a

$|M| = |\{m_i\}|$: the number of elements in M

$M_1 \cup M_2$: union of the sets M_1 and M_2

$M_1 \cap M_2$: intersection of the sets M_1 and M_2

16 *Definitions and Symbols*

a ∨ b : a or b (alternative)

a ∧ b : a and b (conjunctive)

$\sum_{j : x_j : bv} a_j$: sum over all elements a_j, for whose indices j x_j is a basic variable

$\dfrac{\partial f(x)}{\partial x}$: partial derivative of f(x) with respect to x

grad f(x) : gradient of f(x) = vector of the partial first derivatives of the function f(x) = total derivative of f(x) with respect to the vector x .

A vector a: = $(a_1;...;a_j;...;a_n)$ is lexicographically positive (in symbols: a ≻ 0) ⇔ $(a_k > 0 \mid k = \min\{j \mid a_j \neq 0\})$.

A vector a is lexicographically greater than a vector b (in symbols: a ≻ b) ⇔ (a-b) ≻ 0 .

The vector $a^{(r)}$ is the lexicographic maximum of a set of vectors $a^{(i)}$ (in symbols: $a^{(r)} = \text{lex max}_i \{a^{(i)}\}$) ⇔ $a^{(r)} \succ a^{(i)}$ ∀ i ≠ r .

(The definition for "lexicographically smaller" and the lexicographic minimum are equivalent) .

k-min: the k-th smallest element of a set for k = 1,2,...

rk(A) : rank of the matrix A

bv : basic variable

nbv : non-basic variable

inf : infimum

sup : supremum

q.u. : quantity unit (i.e. ea.; doz.; gal. etc)
m.u. : monetary unit (i.e. $; ¢ etc)
d.u. : distance unit (i.e. mi.; km.; in.; ft. etc)
t.u. : time unit (i.e. hr.; min.; sec. etc)

0. Summary of Matrix Algebra and Allied Topics

0.1 Definitions

Definition 1: A vector in which all elements belong to one row is called a row vector and written $a:=(a_j):=(a_1,\ldots a_n)$. A column vector is defined in the corresponding manner and written

$$a:=(a_i):=\begin{pmatrix}a_1\\ \vdots \\ a_m\end{pmatrix}.$$

Definition 2: An n-dimensional vector $e^{(i)}:=(e_j)$ is called the i-th unit vector, if

$$e_j := \begin{cases}1, & \text{if } j = i \\ 0 \ \forall \ j = 1,\ldots,n;\ j \neq i\end{cases}$$

Definition 3: An n-dimensional vector $e:=(e_j)$ is called a summing vector, if $e_j = 1 \ \forall \ j = 1,\ldots,n$.

Definition 4: An (m × n) - dimensional matrix A is an ordered set of m·n elements. It can be represented as an m-dimensional column vector, whose elements are n-dimensional row vectors, or as an n-dimensional row vector, whose elements are m-dimensional column vectors.

$$A := (a_{ij}) := \begin{pmatrix}a_{11} & \cdots & a_{1n}\\ \vdots & & \vdots \\ a_{m1} & & a_{mn}\end{pmatrix}$$

Definition 5: Let A^T be the transpose of the matrix A, then $A^T:=(a_{ij})^T:=(a_{ji})$, i.e. row- and column-indices

18 Matrix Algebra

are exchanged. The transpose of a column vector is a row vector and vice versa.

Definition 6: An (m × n) - dimensional matrix A is called quadratic, if m = n .

Definition 7: A quadratic matrix A is called the identity matrix, if
$a_{ii} = 1 \ \forall \ i$
$a_{ij} = 0 \ \forall \ i \neq j$.

Definition 8: A quadratic matrix A is called a diagonal matrix, if
$a_{ii} = \varepsilon_i \ \forall i; \ \varepsilon_i \in \mathbb{R} \ ; \ a_{ij} = 0 \ \forall \ i \neq j$.

Definition 9: A quadratic matrix A is called triangular, if
$a_{ij} \in \mathbb{R} \ \forall i \geq j \ ; \ a_{ij} = 0 \ \forall i < j$.

Definition 10: An (m × n) - dimensional matrix A is called the null matrix, if $a_{ij} = 0 \ \forall \ i,j$.

0.2 Elementary Operations

0.2.1 Vector Addition

The vector c is the sum of two equal-dimensional vectors a and b , if $c_i := a_i + b_i \ \forall \ i$.

0.2.2 Vector Subtraction

The vector c is the difference of two equal-dimensional vectors a and b , if $c_i := a_i - b_i \ \forall \ i$.

0.2.3 Multiplication of a Vector with a Scalar

Let $\varepsilon \in \mathbb{R}$ be a scalar and let a be an n-dimensional vector, then $c := \varepsilon \cdot a$ is an n-dimensional vector, if $c_i := \varepsilon \cdot a_i \ \forall \ i = 1,\ldots,n$.

0.2.4 Inner Vector Product

The inner vector product $\varepsilon \in \mathbb{R}$ of an n-dimensional row vector a^T and an n-dimensional column vector b is defined as:

$$\varepsilon := \sum_{i=1}^{n} a_i \cdot b_i \quad .$$

0.2.5 Dyadic Vector Product

The dyadic vector product C of an m-dimensional column vector a and an n-dimensional row vector b is defined as the $(m \times n)$ - dimensional matrix C, so that $C:=(c_{ij}):=(a_i \cdot b_j) \; \forall \; i,j$.

0.2.6 Matrix Addition

The sum C of two equal-dimensional matrices A and B is defined as: $C:=(c_{ij}):=(a_{ij} + b_{ij}) \; \forall \; i,j$.

0.2.7 Matrix Subtraction

The difference C of two equal-dimensional matrices A and B is defined as: $C:=(c_{ij}):=(a_{ij} - b_{ij}) \; \forall \; i,j$.

0.2.8 Multiplication of a Scalar with a Matrix

Let $\varepsilon \in \mathbb{R}$ be a scalar and A a matrix, then $C:=\varepsilon \cdot A$, if $C:=(c_{ij}):=(\varepsilon \cdot a_{ij}) \; \forall \; i,j$.

0.2.9 Multiplication of a Row Vector with a Matrix

An m-dimensional row vector c is the product of an n-dimensional row vector b and an $(n \times m)$ - dimensional matrix A, if

$$c:=(c_i):=(\sum_{j=1}^{n} b_j \cdot a_{ji}) \; \forall \; i = 1,\ldots,m \quad .$$

0.2.10 Multiplication of a Matrix with a Column Vector

An m-dimensional column vector c is the product of an $(m \times n)$-dimensional matrix A and an n-dimensional vector b, if

$$c := (c_i) := (\sum_{j=1}^{n} a_{ij} \cdot b_j) \; \forall \; i = 1,\ldots,m \; .$$

0.2.11 Matrix Multiplication

An $(m \times k)$ - dimensional matrix C is the product of an $(m \times n)$ - dimensional matrix A and an $(n \times k)$ - dimensional matrix B, if $C := (c_{ij}) := (\sum_{\mu} a_{i\mu} \cdot b_{\mu j}) \; \forall \; i,j$.

0.2.12 Matrix Multiplication by Covering

An $(m \times n)$ - dimensional matrix C is the "cover product" of two $(m \times n)$ - dimensional matrices A and B, if
$$C := (c_{ij}) := (a_{ij} \cdot b_{ij}) \; \forall \; i,j \; .$$

0.2.13 Calculation of the Inverse

As the quotient C of two quadratic, regular $(n \times n)$ - dimensional matrices A and B is not defined, consider the following equation $C := A \cdot B^{-1}$, where B^{-1} is the inverse of the matrix B. B^{-1} is calculated as follows:

Step 1: Transform the $(n \times 2n)$ - dimensional matrix $\tilde{B}$ starting with $\tilde{B} := (B;I)$ and set the index $\tau := 1$.

Step 2: Determine $\tilde{b}_{i\tau}^{*} := \begin{cases} 1, & \text{if } i = \tau \\ 0, & \text{otherwise} \end{cases}$

$$\tilde{b}_{\tau j}^{*} := \frac{\tilde{b}_{\tau j}}{\tilde{b}_{\tau \tau}} \; \forall \; j$$

$$\tilde{b}_{ij}^{*} := \tilde{b}_{ij} - \frac{\tilde{b}_{i\tau} \cdot \tilde{b}_{\tau j}}{\tilde{b}_{\tau \tau}}$$

and set $\tau := \tau + 1$.

Step 3: Is $\tau = n + 1$?

If yes: Stop, now the matrix $\hat{B}:=(I;B^{-1})$. B^{-1} is the inverse of B.

If no : Set $\hat{\tilde{b}}_{ij}:=\hat{\tilde{b}}^*_{ij}$. Go to step 2.

1. Linear Programming

1.1 General Methods

1.1.1 The Primal Simplex-Algorithm

Hypotheses
Given the following problem:
P: max $\pi = c \cdot x$
     $A \cdot x \leq b$
     $x \geq 0$

Note: Each problem of the form min $\pi = c \cdot x$ can be transformed into the maximization problem max $-\pi = -c \cdot x$.

Principle
The algorithm begins with an initial feasible solution $x=(0;\ldots;0)$ determining in each iteration a new basic feasible solution (graphically speaking: moving from one corner of the convex set described by the constraints to another) in which the value of the objective function increases or as a minimum does not decrease. The algorithm terminates in an optimal basic solution or an unbounded solution.

Description
Step 1: Add slack variables y to the system, so that $A \cdot x + y = b$ and set up the initial tableau :

x	y	1
A	I	b
-c	0	π

For the following calculations let: $\tilde{A}=(A,I); \tilde{c}=(-c,0); \tilde{b}=b$.

Step 2: $\exists\ \tilde{c}_j < 0$?
If yes: Go to step 3.
If no : Stop, the current solution is optimal.

Step 3: (selection of the pivot-column s)
Determine $\tilde{c}_s := \min\{\tilde{c}_j \mid \tilde{c}_j < 0\}$.
(In principle any negative element $\tilde{c}_j$ will do.)

Step 4: $\exists\ \tilde{a}_{is} > 0$?
If yes: Go to step 5.
If no : Stop, P has an unbounded solution.

Step 5: (selection of the pivot-row r)
Determine the pivot-row r as follows:

$$\frac{\tilde{b}_r}{\tilde{a}_{rs}} := \min\left\{\frac{\tilde{b}_i}{\tilde{a}_{is}}\ \Big|\ \tilde{a}_{is} > 0\right\}\ .\ \tilde{a}_{rs}\ \text{is pivot-element.}$$

Step 6: (iteration, tableau-transformation)
Compute the new tableau:

a) pivot-row r :

$$\tilde{a}^*_{rj} := \frac{\tilde{a}_{rj}}{\tilde{a}_{rs}}\ \forall j = 1,\ldots,n+m;\quad \tilde{b}^*_r := \frac{\tilde{b}_r}{\tilde{a}_{rs}}$$

b) pivot-column s :

$$\tilde{a}^*_{is} := \begin{cases} 1, & \text{if } i = r \\ 0, & \text{otherwise} \end{cases}\ ;\quad \tilde{c}^*_s := 0$$

c) all other rows and columns :

$$\tilde{a}^*_{ij} := \tilde{a}_{ij} - \frac{\tilde{a}_{is} \cdot \tilde{a}_{rj}}{\tilde{a}_{rs}}\ ;\quad \tilde{c}^*_j := \tilde{c}_j - \frac{\tilde{c}_s \cdot \tilde{a}_{rj}}{\tilde{a}_{rs}}$$

$$\tilde{b}^*_i := \tilde{b}_i - \frac{\tilde{b}_r \cdot \tilde{a}_{is}}{\tilde{a}_{rs}}\ ;\quad \pi^* := \pi - \frac{\tilde{b}_r \cdot \tilde{c}_s}{\tilde{a}_{rs}}\ .$$

Go to step 2 .

24 Linear Programming

Note: A variable x_k or y_k is called a basis variable (bv), if

$$\tilde{a}_{ik} = \begin{cases} 1, & \text{if } i = 1 \\ 0, & \text{otherwise} \end{cases} \quad ; \quad \tilde{c}_k = 0 \; .$$

The actual values are $x_k = b_1$ or $y_k = b_1$. All variables, which are not basis variables, are called nonbasis variables (nbv). Then $x_j = 0 \; \forall \; x_j$, nbv; or $y_j = 0 \; \forall \; y_j$, nbv.

Example

Given the following problem P:

P: max $\pi = 5 \cdot x_1 + 6 \cdot x_2$

(1) $3 \cdot x_1 - 2 \cdot x_1 \leq 9$ (1) $3 \cdot x_1 - 2 \cdot x_2 + y_1 = 9$

(2) $-5 \cdot x_1 + 5 \cdot x_2 \leq 15$ (2) $5 \cdot x_1 + 5 \cdot x_2 + y_2 = 15$

(3) $12 \cdot x_1 + 3 \cdot x_2 \leq 12$ (3) $12 \cdot x_1 + 3 \cdot x_2 + y_3 = 12$

$x_1, x_2 \geq 0$ $x_1, x_2, y_1, y_2, y_3 \geq 0$

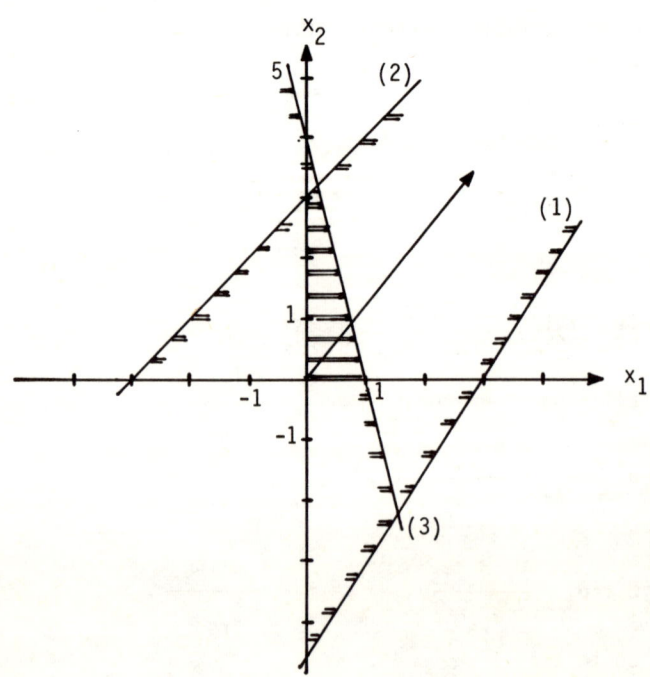

$T^{(1)}$:

x_1	x_2	y_1	y_2	y_3	1
3	-2	1	0	0	9
-5	5	0	1	0	15
12	3	0	0	1	12
-5	-6	0	0	0	0

$T^{(2)}$:

x_1	x_2	y_1	y_2	y_3	1
1	0	1	2/5	0	15
-1	1	0	1/5	0	3
15	0	0	-3/5	1	3
-11	0	0	6/5	0	18

$T^{(3)}$:

x_1	x_2	y_1	y_2	y_3	1
0	0	1	11/25	-1/15	74/5
0	1	0	4/25	1/15	16/5
1	0	0	-1/25	1/15	1/5
0	0	0	19/25	11/15	101/5

Recap of the calculations:

$T^{(1)}$: $x = (0;0)$; $\pi = 0$
$T^{(2)}$: $x = (0;3)$; $\pi = 18$
$T^{(3)}$: $x = \bar{x} = (1/5;16/5)$; $\bar{\pi} = 101/5$.

1.1.2 The Two-Phase Method

<u>Hypotheses</u>

Given the following problem :

P: $\left.\begin{array}{c}\min\\ \max\end{array}\right\}$ $\pi = c \cdot x$

$A \cdot x \gtreqless b \geq \Theta$

$x \geq \Theta$

<u>Principle</u>

The method starts with a non-feasible basic solution (i.e. $x=(0;...;0)$ is not a feasible corner in the given convex set). At the end of the first phase a feasible solution is obtained, if one exists. Beginning with this feasible solution, the second phase determines as in 1.1.1 an optimal basic solution.

26 *Linear Programming*

Description

Phase I (determination of a feasible solution)

Step 1: Transform the restrictions as follows:
$$a \cdot x \leq b \rightarrow a \cdot x + y = b$$
$$a \cdot x = b \rightarrow a \cdot x + z = b$$
$$a \cdot x \geq b \rightarrow a \cdot x - y + z = b$$
where y are slack variables and z are artificial variables. The basis variables in the last k rows are artificial variables. Formulate the initial tableau without the objective function.

Step 2: Determine the coefficients d_j of the artificial objective function as follows:
$$d_j := -\sum_{i=m-k+1}^{m} a_{ij} \quad \forall j, \text{nbv};$$
i.e. calculate the sum of the last k elements in each of the columns belonging to a nbv. The value $\hat{\pi}$ of this objective function is
$$\hat{\pi} := -\sum_{i=m-k+1}^{m} b_i \quad .$$
The complete initial tableau is

	x	y	z	1
		I	0	
A	0		I	b
		-I		
	d		0	$\hat{\pi}$

Step 3: Is $d_j \geq 0 \; \forall j$?
If yes: Go to step 5.
If no : Go to step 4.

Step 4: Select pivot-column and -row and do one iteration as in step 6, Primal Simplex-Algorithm. Go to step 3.

Step 5: Are all artificial variables nbv ?
  If yes: Go to step 7.
  If no : Go to step 6.

Step 6: Are all artificial variables, which are still in the basis, equal to zero ?
  If yes: Eliminate all rows which have "1" in the columns of these variables and then eliminate the corresponding columns. Go to step 8.
  If no : Stop, P has no feasible solution .

Step 7: Eliminate all columns belonging to artificial variables.

Step 8: Replace the artificial objective function in the remaining tableau with the actual objective function, in which the nbv's are retained and the bv's are expressed as linear combinations of nbv's.

Phase II (determination of an optimal solution)

Step 9: Apply the Primal Simplex-Algorithm to the tableau, obtained in step 8, beginning with step 2.
Note: Step 8 can be omitted, when the actual objective function is added to the initial tableau at the beginning of the first phase and the coefficients are calculated in the usual manner, but they shall not determine the pivot-column. At the end of the first phase the artificial objective function must also be eliminated. The second phase begins with the remaining tableau.

The M-Method, described in some publications, is - apart from some formal modifications - identical to the Two-Phase Method.

28 Linear Programming

Example

Given the following problem P:

P: max $\pi = (-6 \cdot x_1 + 6 \cdot x_2)$

(1) $2 \cdot x_1 + 3 \cdot x_2 \leq 6$
(2) $-5 \cdot x_1 + 9 \cdot x_2 = 15$
(3) $-6 \cdot x_1 + 3 \cdot x_2 \geq 3$
$x_1, x_2 \geq 0$

$\rightarrow$

(1) $2 \cdot x_1 + 3 \cdot x_2 + y_1 = 6$
(2) $-5 \cdot x_1 + 9 \cdot x_2 + z_1 = 15$
(3) $-6 \cdot x_1 + 3 \cdot x_2 - y_2 + z_2 = 3$
$x_i, y_i, z_i \geq 0 \; \forall \; i = 1,2$.

In this case, the feasible region includes all points lying on the straight line (2) between Q and R

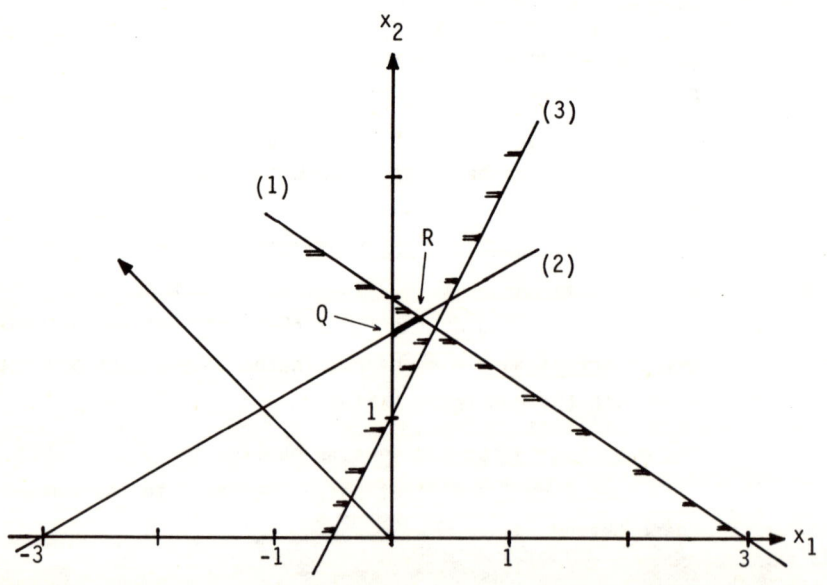

The actual objective function shall be computed simultaneously

Phase I

$T^{(1)}$:

x_1	x_2	y_1	y_2	z_1	z_2	1
2	3	1	0	0	0	6
-5	9	0	0	1	0	15
-6	3	0	-1	0	1	3
11	-12	0	1	0	0	-18
6	-6	0	0	0	0	0

$T^{(2)}$:

x_1	x_2	y_1	y_2	z_1	z_2	1
8	0	1	1	0	-1	3
13	0	0	3	1	-3	6
-2	1	0	-1/3	0	1/3	1
-13	0	0	-3	0	4	-6
-6	0	0	-2	0	2	6

$T^{(3)}$:

x_1	x_2	y_1	y_2	z_1	z_2	1
1	0	1/8	1/8	0	-1/8	3/8
0	0	-13/8	11/8	1	-11/8	9/8
0	1	1/4	-1/12	0	1/12	14/8
0	0	13/8	-11/8	0	19/8	-9/8
0	0	3/4	-5/4	0	5/4	33/4

$T^{(4)}$:

x_1	x_2	y_1	y_2	z_1	z_2	1
1	0	3/11	0	-1/11	0	3/11
0	0	-13/11	1	8/11	-1	9/11
0	1	5/33	0	2/33	0	20/11
0	0	0	0	1	1	0
0	0	-8/11	0	10/11	0	102/11

For the purpose of control the coefficients of the actual objective function are computed one more time according to step 8.

$\pi = (-6 \cdot x_1 + 6 \cdot x_2)$; $x_1 = 3/11 - 3/11\, y_1$; $x_2 = 20/11 - 5/33\, y_1$;
$\to \pi = -6(3/11 - 3/11\, y_1) + 6(20/11 - 5/33\, y_1)$
$\to \pi - 8/11\, y_1 = 102/11$.

Phase II

$T^{(5)}$:

x_1	x_2	y_1	y_2	1
1	0	3/11	0	3/11
0	0	-13/11	1	9/11
0	1	5/33	0	20/11
0	0	-8/11	0	102/11

$T^{(6)}$:

x_1	x_2	y_1	y_2	1
11/3	0	1	0	1
13/3	0	0	1	2
-5/9	1	0	0	5/3
8/3	0	0	0	10

Recap of the calculations:

$T^{(1)}$: $x = (0;0)$; $\pi = 0$
$T^{(2)}$: $x = (0;1)$; $\pi = 6$
$T^{(3)}$: $x = (3/8;14/8)$; $\pi = 33/4$
$T^{(4)}$: $x = (3/11;20/11)$; $\pi = 102/11$
$T^{(5)}$: $= T^{(4)}$
$T^{(6)}$: $\bar{x} = (0;5/3)$; $\bar{\pi} = 10$.

1.1.3 The Primal Simplex-Algorithm without Explicit Identity Matrix

Hypotheses

Given the following problem:

P: $\left.\begin{array}{l}\min\\\max\end{array}\right\}$ $\pi = c \cdot x$

$A \cdot x \gtreqless b \geq \theta$

$x \geq \theta$

Principle

The algorithm is equivalent to 1.1.1, only a different tableau is utilized.

Description

Step 1: Transform the system of constraints as follows:

$$\sum_j a_{ij} \cdot x_j \leq b_i \quad \rightarrow \quad \sum_j a_{ij} \cdot x_j + y_i = b_i$$

$$\sum_j a_{ij} \cdot x_j = b_i \quad \rightarrow \quad \sum_j a_{ij} \cdot x_j + z_i = b_i$$

$$\sum_j a_{ij} \cdot x_j \geq b_i \quad \rightarrow \quad \sum_j a_{ij} \cdot x_j - w_i + z_i = b_i$$

Result: The problem $\hat{P}$.

Step 2: Does $\hat{P}$ contain artificial variables z_i ?
If yes: Go to step 3.
If no : Go to step 4.

Step 3: Determine the vector d , so that

$$d_j := -\sum_{i:z_i, bv} a_{ij} \quad \forall \ j \ .$$

Step 4: Formulate the initial simplex-tableau as follows:

	x	w	1
y		Θ	
	A		b
z		-I	
	d		
	-c		π

Step 5: Determine the pivot-column s , so that
$d_s := \min \{ d_j \mid d_j < 0 \}$, or, if d does not exist,
so that $c_s := \min \{ c_j \mid c_j < 0 \}$.

Step 6: Determine the pivot-row r , so that

$$\frac{b_r}{a_{rs}} := \min \left\{ \frac{b_i}{a_{is}} \mid a_{is} > 0 \right\} .$$

The pivot-element a_{rs} is hereby determined.

Step 7: Exchange the variable in row r for the variable in column s and transform the tableau as follows :

$$a^*_{rs} = \frac{1}{a_{rs}}; \quad a^*_{rj} = \frac{a_{rj}}{a_{rs}} \forall j \neq s; \quad a^*_{is} = -\frac{a_{is}}{a_{rs}} \forall i \neq r$$

$$a^*_{ij} = a_{ij} - \frac{a_{is} \cdot a_{rj}}{a_{rs}} \quad \forall i \neq r; j \neq s;$$

$$b^*_i = b_i - \frac{a_{is} \cdot b_r}{a_{rs}} \quad \forall i \neq r$$

$$b^*_r = \frac{b_r}{a_{rs}} \; ; \; c^*_j = c_j - \frac{a_{rj} \cdot c_s}{a_{rs}} \quad \forall j \neq s; \; c^*_s = -\frac{c_s}{a_{rs}}$$

$$\pi^* = \pi - \frac{b_r \cdot c_s}{a_{rs}} \; ; \; d^*_j = d_j - \frac{a_{rj} \cdot c_s}{a_{rs}} \; ; \; d^*_s = -\frac{d_s}{a_{rs}}$$

If the d-vector does not exist, go to step 10; otherwise go to step 8.

Step 8: Is $d_j \geq 0 \quad j$?
    If yes: Eliminate row d and all columns, belonging to the artificial variables z and go to step 9.
    If no : Go to step 5.

Step 9: Are there still rows in the current tableau, corresponding to the artificial variables z ?
    If yes: Stop, P has no feasible solution.
    If no : Go to step 10.

Step 10: Is $c_j \geq 0 \; \forall j$?
    If yes: Stop, the current solution is optimal.
    If no : Go to step 5.

Example

Given the following problem P:

P: max $\pi = (- 6 \cdot x_1 + 6 \cdot x_2)$

(1) $2 \cdot x_1 + 3 \cdot x_2 \leq 6$
(2) $-5 \cdot x_1 + 9 \cdot x_2 = 15$
(3) $-6 \cdot x_1 + 3 \cdot x_2 \geq 3$
    $x_1, x_2 \geq 0$.

$\rightarrow$

$2 \cdot x_1 + 3 \cdot x_2 + y_1 = 6$
$-5 \cdot x_1 + 9 \cdot x_2 + z_1 = 15$
$-6 \cdot x_1 + 3 \cdot x_2 - w_1 + z_2 = 3$
$x_i, z_i, y_1, w_1 \geq 0 \; \forall i = 1, 2.$

$T^{(1)}$:

	x_1	x_2	w_1	1
y_1	2	3	0	6
z_1	-5	9	0	15
z_2	-6	3	-1	3
1	6	-6	0	0

$T^{(2)}$:

	x_1	z_2	w_1	1
y_1	8	-1	1	3
z_1	13	-3	3	6
x_2	-2	1/3	-1/3	1
1	-6	2	-2	6

$T^{(3)}$:

	y_1	z_2	w_1	1
x_1	1/8	-1/8	1/8	3/8
z_1	-13/8	-11/8	11/8	9/8
x_2	1/4	1/12	-1/12	7/4
1	3/4	5/4	-5/4	33/4

$T^{(4)}$:

	y_1	z_2	z_1	1
x_1	3/11	0	-1/11	3/11
w_1	-13/11	-1	8/11	9/11
x_2	5/33	0	2/33	20/11
1	-8/11	0	10/11	102/11

$T^{(5)}$:

	x_1	z_2	z_1	1
y_1	11/3	0	-1/3	1
w_1	13/3	-1	1/3	2
x_2	-5/9	0	1/9	5/3
1	8/3	0	1/2	10

Recap of the calculations:

$T^{(1)}$: $x = (0;0)$; $\pi = 0$

$T^{(2)}$: $x = (0;1)$; $\pi = 6$

$T^{(3)}$: $x = (3/8;7/4)$; $\pi = 33/4$

$T^{(4)}$: $x = (3/11;20/11)$; $\pi = 102/11$

$T^{(5)}$: $\bar{x} = (0;5/3)$; $\bar{\pi} = 10$.

1.1.4 The Dual Simplex–Algorithm

Hypotheses

Given the following problem:

$$P: \left.\begin{array}{c}\min\\\max\end{array}\right\} \pi = c \cdot x$$

$$A \cdot x \leq b$$
$$x \geq 0$$
$$b \in \mathbb{R}^m$$

All systems of constraints may be written in this form through the appropriate transformations as follows:

$a \cdot x \geq b \quad \rightarrow \quad -a \cdot x \leq -b$

$a \cdot x = b \quad \rightarrow \quad a \cdot x \leq b \quad$ and $\quad -a \cdot x \leq -b$

Principle

The algorithm begins with a primal non-feasible basic solution $x=(0;\ldots;0)$. Maintaining the dual feasibility in each iteration (i.e. non-negative coefficients in the objective function row), the algorithm determines new solutions which do not decrease the value of the objective function. The first primal feasible solution discovered in this manner is necessarily an optimal solution.

Description

Step 1: Add the slack variables y_i to the inequalities and set up the initial tableau. (In a maximizing problem, the coefficients of the objective function are negated.)

Step 2: Is $c_j \geq 0 \ \forall j$?
If yes: Go to step 5.
If no : Go to step 3.

Step 3: Add one column and one row to the initial tableau; the additional constraint is

$\sum_{j=1}^{n} x_j + \tilde{y} = C$, where C is any very large positive number and $\tilde{y}$ is the additional slack variable.

Step 4: If $c_s := \max \{ c_j \mid c_j < 0 \}$ and the additional row is the $(m + 1)$-th row, then $a_{m+1,s}$ is the pivot-element.
Do one iteration as in step 6, Primal Simplex-Algorithm, without "go to"-statement.
Result: $c_j^* \geq 0 \; \forall j$.

Step 5: Is $b_i \geq 0 \; \forall i = 1,\ldots,m+1$?
If yes: Stop, the current solution is optimal.
If no : Go to step 6.

Step 6: (selection of the pivot-row r)
Select any row with $b_i < 0$ as the pivot-row.
Note: Preferably select row r with $b_r := \min_i \{b_i\}$.

Step 7: (selection of the pivot-column s)
From all negative elements of pivot-row r ,

$$\text{determine } \frac{c_s}{|a_{rs}|} := \min_{j: a_{rj} < 0} \left\{ \frac{c_j}{|a_{rj}|} \right\} .$$

$\exists \; a_{rj} < 0$?
If yes: s is the pivot-column and a_{rs} is the (always negative) pivot-element . Go to step 8.
If no :Stop, P has no feasible solution.

Step 8: Do one iteration as in step 6, Primal Simplex-Algorithm, without "go to"-statement. Go to step 5.

Example 1
Given the following problem P:

P: $\min \pi = 2 \cdot x_1 + 3 \cdot x_2$
(1) $2 \cdot x_1 + x_2 \geq 4$
(2) $-x_1 + x_2 \geq 2$
$x_1, x_2 \geq 0$

$\rightarrow$

P: $\min \pi = 2 \cdot x_1 + 3 \cdot x_2$
(1) $-2 \cdot x_1 - x_2 \leq -4$
(2) $x_1 - x_2 \leq -2$
$x_1, x_2 \geq 0$

36 Linear Programming

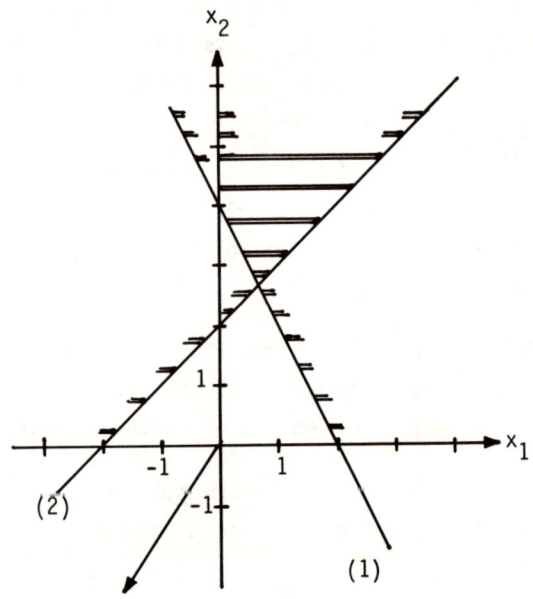

$T^{(1)}$:

x_1	x_2	y_1	y_2	1
-2	-1	1	0	-4
1	-1	0	1	-2
2	3	0	0	0

$T^{(2)}$:

x_1	x_2	y_1	y_2	1
1	1/2	-1/2	0	2
0	-3/2	1/2	1	-4
0	2	1	0	-4

$T^{(3)}$:

x_1	x_2	y_1	y_2	1
1	0	-1/3	1/3	2/3
0	1	-1/3	-2/3	8/3
0	0	5/3	4/3	-28/3

Recap of the calculations:
$T^{(1)}$: $x = (0;0)$; $\pi = 0$
$T^{(2)}$: $x = (2;0)$; $\pi = -4$
$T^{(3)}$: $\bar{x} = (2/3; 8/3)$; $\bar{\pi} = -28/3$

Example 2
Given the following problem P:

P: max $\pi = 2 \cdot x_1 + x_2$
  (1) $2 \cdot x_1 - x_2 \geq 5$
  (2) $x_1 + 3 \cdot x_2 \leq 6$
  $x_1, x_2 \geq 0$

P: max $\pi = 2 \cdot x_1 + x_2$
  (1) $-2 \cdot x_1 + x_2 \leq -5$
  (2) $x_1 + 3 \cdot x_2 \leq 6$
  $x_1, x_2 \geq 0$

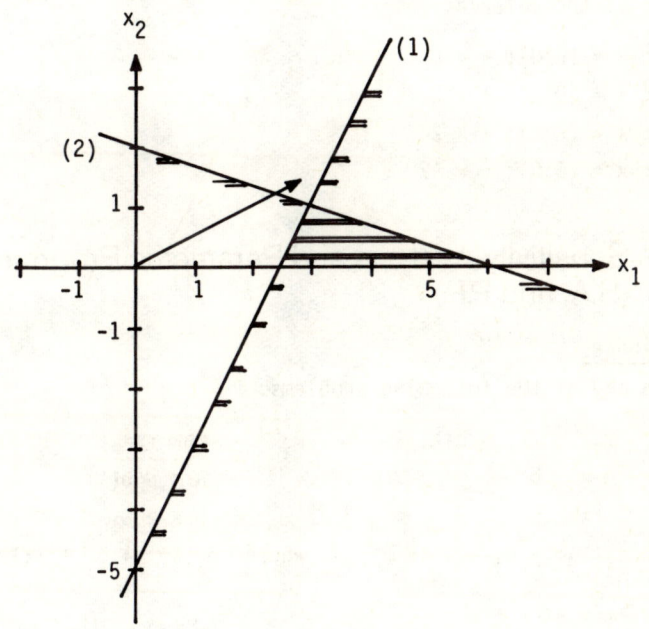

$T^{(1)}$:

x_1	x_2	y_1	y_2	1
-2	1	1	0	-5
1	3	0	1	6
-2	-1	0	0	0

$T^{(2)}$:

x_1	x_2	y_1	y_2	$\tilde{y}$	1
-2	1	1	0	0	-5
1	3	0	1	0	6
1	1	0	0	1	C
-2	-1	0	0	0	0

38 Linear Programming

$T^{(3)}$:

x_1	x_2	y_1	y_2	$\tilde{y}$	1
0	3	1	0	2	-5+2C
0	2	0	1	-1	6-C
1	1	0	0	1	C
0	1	0	0	2	2C

$T^{(4)}$:

x_1	x_2	y_1	y_2	$\tilde{y}$	1
0	7	1	2	0	7
0	-2	0	-1	1	C-6
1	3	0	1	0	6
0	5	0	2	0	12

Recap of the calculations:

$T^{(1)}$: $x = (0;0)$; $\pi = 0$
$T^{(2)}$: $x = (0;0)$; $\pi = 0$
$T^{(3)}$: $x = (C;0)$; $\pi = 2C$
$T^{(4)}$: $\bar{x} = (6;0)$; $\bar{\pi} = 12$

1.1.5 Sensitivity Analysis and Parametric Programming (S.A. and P.P.)

<u>Hypotheses</u>

Given any of the following problems:

$P_\lambda^{(1)}$: max $\pi(\lambda) = c(\lambda) \cdot x$
    $A \cdot x \leq b$
    $x \geq \Theta$,
where $c_j(\lambda) := c_j + \lambda; \lambda \in \mathbb{R}$

$P_\lambda^{(2)}$:= max $\pi = c \cdot x$
    $A \cdot x \leq b(\lambda)$
    $x \geq \Theta$,
where $b_i(\lambda) := b_i + \lambda; \lambda \in \mathbb{R}$

$P_\lambda^{(3)}$: max $\pi = c \cdot x$
    $A(\lambda) \cdot x \leq b$
    $x \geq \Theta$

In general we examine with the sensitivity analysis how far certain elements of an optimal tableau can be changed under the assumption that the tableau remains optimal. With parametric programming we determine a new optimal tableau, when one or more elements have been changed so that the original optimal tableau no longer holds for the new optimum.

Note: When changing more than one element, indexed parameters λ_i must be utilized, thereby increasing considerably the difficulty of determining the intervals.

1.1.5.1 S.A. and P.P. with Expected Alterations

Hypotheses

Given one of the problems P_λ in which a subsequent alteration of one element by λ^* is to be accomplished.

Note: The mark "-" over an element will indicate the value of that element in the optimal tableau.

Description

Step 1: Solve the problem by one of the simplex-methods, complete with the parameter λ, but without influence in selecting the pivot-column or -row.
Result: For $\lambda = 0$ the computed tableau is optimal (if a feasible solution does exist).

Step 2: (sensitivity analysis)
Determine the interval $[\underline{\lambda}; \bar{\lambda}]$, so that for all
$\tilde{\lambda} \in [\underline{\lambda}; \bar{\lambda}]\ \bar{c}_j(\tilde{\lambda}) \geq 0\ \forall\ j;\ \bar{b}_i(\tilde{\lambda}) \geq 0\ \forall\ i$.

Step 3: Is $\lambda^* \in [\underline{\lambda}; \bar{\lambda}]$?
If yes: Stop, the optimal tableau computed in step 1 remains optimal even after alteration of λ^*. The new parameters are calculated setting $\lambda := \lambda^*$.
If no : Go to step 4.

Step 4: Is $\lambda^* > \bar{\lambda}$?
If yes: Go to step 5.
If no : Go to step 8.

Step 5: Is $\bar{\lambda}$ determined by the element $\bar{c}_s(\lambda)$?
If yes: Select column s for pivot-column. Go to step 6.
If no : Select row r, that determined $\bar{\lambda}$, for pivot-row. Go to step 7.

Step 6: $\exists\ \bar{a}_{is} > 0$?

If yes: Determine the pivot-element in column s as in
the Primal Simplex-Algorithm and do one iteration.
Go to step 2.
If no : Stop, for $\lambda = \lambda^*$ there is no feasible solution
of P_λ .

Step 7: $\exists \ \bar{a}_{rj} < 0$?
If yes: Determine the pivot-element in row r as in the
Dual Simplex-Algorithm and do one iteration.
Go to step 2.
If no : Stop, for $\lambda = \lambda^*$ there is no feasible solution
of P_λ .

Step 8: Is $\underline{\lambda}$ determined by the element $\bar{c}_s(\lambda)$?
If yes: Select column s for pivot-column. Go to step 9.
If no : Select row r , that determined $\underline{\lambda}$, for pivot-row.
Go to step 10.

Step 9: $\exists \ \bar{a}_{is} > 0$?
If yes: Determine the pivot-element in column s as in
the Primal Simplex-Algorithm and do one iteration.
Go to step 2.
If no : Stop, for $\lambda = \lambda^*$ there is no feasible solution of P_λ .

Step 10: $\exists \ \bar{a}_{rj} < 0$?
If yes: Determine the pivot-element in row r as in the
Dual Simplex-Algorithm and do one iteration.
Go to step 2.
If no : Stop, for $\lambda = \lambda^*$ there is no feasible solution
of P_λ .

Example 1
Given the following problem $P_\lambda^{(1)}$:

$P_\lambda^{(1)}$: max $\pi(\lambda) = 3 \cdot x_1 + (2 + \lambda) \cdot x_2$

$\qquad 2 \cdot x_1 + x_2 \leq 4$

$\qquad x_1 - 2 \cdot x_2 \leq 6$

$\qquad x_1, x_2 \geq 0$

$T^{(1)}$:

x_1	x_2	y_1	y_2	1
2	1	1	0	4
1	-2	0	1	6
-3	-2-λ	0	0	0

$T^{(2)}$:

x_1	x_2	y_1	y_2	1
2	1	1	0	4
5	0	2	1	14
1+2λ	0	2+λ	0	8+4λ

$[\underline{\lambda}; \bar{\lambda}] = [-1/2; \infty)$

a) If $\lambda^* = 10$, then the tableau remains optimal and we have:
$\bar{\pi}(\lambda^*) = c(\lambda^*) \cdot \bar{x} = 48$

b) If $\lambda^* = -1$, then the column belonging to x_1 becomes pivot-column.

$T^{(3)}$:

x_1	x_2	y_1	y_2	1
1	1/2	1/2	0	2
0	-5/2	-1/2	1	4
0	-1/2-λ	3/2	0	6

$[\underline{\lambda}; \bar{\lambda}] = (-\infty; -1/2]$. Then: $\bar{x} = (2; 0)$; $\bar{\pi}(\lambda^*) = c(\lambda^*) \cdot \bar{x} = 6$

Example 2
Given the following problem $P_\lambda^{(2)}$:

$P_\lambda^{(2)}$: max $\pi = 3 \cdot x_1 + 2 \cdot x_2$

$\qquad 2 \cdot x_1 + x_2 \leq 4 + \lambda$

$\qquad x_1 - 2 \cdot x_2 \leq 6$

$\qquad x_1, x_2 \geq 0$

42 Linear Programming

$T^{(1)}$:

x_1	x_2	y_1	y_2	1
2	1	1	0	$4+\lambda$
1	-2	0	1	6
-3	-2	0	0	0

$T^{(2)}$:

x_1	x_2	y_1	y_2	1
2	1	1	0	$4+\lambda$
5	0	2	1	$14+2\lambda$
1	0	2	0	$8+2\lambda$

$[\underline{\lambda}; \bar{\lambda}] = [-4; \infty)$

a) If $\lambda^* = 5$, then the tableau remains optimal and we have:
$\bar{x}(\lambda^*) = (0;9)$; $\bar{\pi}(\lambda^*) = c \cdot \bar{x}(\lambda^*) = 18$.

b) If $\lambda^* = -5$, then the first row becomes pivot-row, but because $a_{1j} \geq 0 \; \forall j$ no pivot-element exists. For $\lambda^* < -4$ there is no feasible solution of $P_\lambda^{(2)}$.

Example 3

Given the following problem $P_\lambda^{(3)}$:

$P_\lambda^{(3)}$: max $\pi = 3 \cdot x_1 + 2 \cdot x_2$
$2 \cdot x_1 + \lambda \cdot x_2 \leq 4$
$x_1 - 2 \cdot x_2 \leq 6$
$x_1, x_2 \geq 0$

$T^{(1)}$:

x_1	x_2	y_1	y_2	1
2	λ	1	0	4
1	-2	0	1	6
-3	-2	0	0	0

$T^{(2)}$:

x_1	x_2	y_1	y_2	1
1	$\lambda/2$	1/2	0	2
0	$-2-\lambda/2$	-1/2	1	4
0	$-2+(3/2) \cdot \lambda$	3/2	0	6

$[\underline{\lambda}; \bar{\lambda}] = [4/3; \infty)$

a) If $\lambda^* = 3$, then the tableau remains optimal and we have:
$\bar{x}(\lambda^*) = (2;0)$; $\bar{\pi}(\lambda^*) = c \cdot \bar{x}(\lambda^*) = 6$.

b) If $\lambda^* = 1/2$, then the second column becomes pivot-column.

$T^{(3)}$:

x_1	x_2	y_1	y_2	1
$2/\lambda$	1	$1/\lambda$	0	$4/\lambda$
$1+4/\lambda$	0	$2/\lambda$	1	$6+8/\lambda$
$-3+4/\lambda$	0	$2/\lambda$	0	$8/\lambda$

$[\underline{\lambda} ; \bar{\lambda}] = (0; 4/3]$

For $\lambda^* = 1/2$ the optimal solution is $\bar{x}(\lambda^*) = (0;2)$; $\bar{\pi}(\lambda^*) = c \cdot \bar{x}(\lambda^*) = 16$

1.1.5.2 S.A. and P.P. with Unexpected Alterations

1.1.5.2.1 Subsequent Alterations of the Restriction Vector

<u>Hypotheses</u>

Given a problem P, for which the optimal tableau is known, determine what subsequent alterations of the element b_i can be accomplished without destroying the optimality of the tableau. If it is not possible for a given parameter λ^*, then a new restriction vector must be calculated. The slack variable y_i, belonging to the i-th restriction, can be found in the (n + i)-th column; an alteration λ^* has to be accomplished.
(The mark "-" over an element will indicate the value of that element in the optimal tableau.)

<u>Description</u>

Step 1: (sensitivity analysis)
    Determine the interval $[\underline{\lambda} ; \bar{\lambda}]$, so that

$$\underline{\lambda} := - \min_{k} \left\{ \frac{\bar{b}_k}{\bar{a}_{k,n+i}} \;\Big|\; \bar{a}_{k,n+i} > 0 \right\},$$

if $\not\exists\ \bar{a}_{k,n+i} > 0$, set $\underline{\lambda} := -\infty$;

$$\bar{\lambda} := \min_{k} \left\{ \frac{\bar{b}_k}{|\bar{a}_{k,n+i}|} \mid \bar{a}_{k,n+i} < 0 \right\},$$

if $\not\exists\ \bar{a}_{k,n+i} < 0$, set $\bar{\lambda} := \infty$.

Step 2: Is $\lambda^* \in [\underline{\lambda} ; \bar{\lambda}]$?

If yes: Stop, the current optimal tableau remains optimal, the new values for $\bar{b}_k$ are computed as follows:
$\bar{b}_k := \bar{b}_k + \lambda^* \cdot \bar{a}_{k,n+i}\ \forall\ k = 1,\ldots,m$.
If no : Go to step 3.

Step 3: Set $\bar{b}_i := \bar{b}_i + \lambda^*$; determine the pivot-element in row i; do dual simplex-iterations until a new optimal tableau is obtained, or it is shown that no feasible solutions exist.

Example 1
Given the following problem P:

P: max $\pi = 3 \cdot x_1 + 2 \cdot x_2$

$2 \cdot x_1 + x_2 \leq 4$

$x_1 - 2 \cdot x_2 \leq 6$

$x_1, x_2 \geq 0$

Determine an interval for the possible subsequent alterations of b_1. Afterwards alter b_1 with $\lambda^* = -2$.

$T^{(1)}$:

x_1	x_2	y_1	y_2	1
2	1	1	0	4
1	-2	0	1	6
-3	-2	0	0	0

$T^{(2)}$:

x_1	x_2	y_1	y_2	1
2	1	1	0	4
5	0	2	1	14
1	0	2	0	8

$\underline{\lambda} = -\min \{4/1; 14/2\} = -4$; $\bar{\lambda} = \infty$;
$[\underline{\lambda} ; \bar{\lambda}] = [-4; \infty)$

$$\bar{b} = \begin{pmatrix} 4 + (-2) \cdot 1 \\ 14 + (-2) \cdot 2 \end{pmatrix} = \begin{pmatrix} 2 \\ 10 \end{pmatrix}$$

Example 2

Given the problem as in example 1.
Determine the new restriction vector, where b_1 is to be altered.
Let $\lambda^* = -5$.
The optimal tableau is:

x_1	x_2	y_1	y_2	1
2	1	1	0	4
5	0	2	1	14
1	0	2	0	8

$\bar{b}_1 := \bar{b}_1 + \lambda^* = 4 - 5 = -1$. One dual simplex-iteration with the first row as pivot-row must be executed. Since $\not\exists \; \bar{a}_{1j} < 0$, there are no feasible solutions of P.

1.1.5.2.2 Subsequent Alterations of Coefficients of the Objective Function

Hypotheses

Given a (maximization-)problem P, for which an optimal tableau is known, determine what subsequent alterations of the element c_j can be accomplished without destroying the optimality of the tableau. An alteration by λ^* has to be accomplished. (The mark "-" over an element will indicate the value of that element in the optimal tableau.)

Description

Step 1: Is the variable x_j, belonging to c_j, a basis variable ?
   If yes: Go to step 2.
   If no : Go to step 5.

Step 2: Let x_j be the bv in column j and row r. Determine the interval $[\underline{\lambda} ; \bar{\lambda}]$, so that

$$\underline{\lambda} := \min_k \left\{ \frac{\bar{c}_k}{\bar{a}_{rk}} \mid \bar{a}_{rk} < 0 \right\},$$

if $\not\exists \ \bar{a}_{rk} < 0$, set $\underline{\lambda} := -\infty$;

$$\bar{\lambda} := \min_k \left\{ \frac{\bar{c}_k}{\bar{a}_{rk}} \mid \bar{a}_{rk} > 0, k \neq j \right\},$$

if $\not\exists \ \bar{a}_{rk} > 0$, set $\bar{\lambda} := \infty$.

Step 3: Is $\lambda^* \in [\underline{\lambda} ; \bar{\lambda}]$?
  If yes: Stop, the current optimal tableau remains optimal, the new coefficients of the objective function are calculated as follows: $\bar{c}_l := \bar{c}_l + \lambda^* \cdot \bar{a}_{rl}$.
  If no : Go to step 4.

Step 4: Set $\bar{c}_j := -\lambda^*$, select column j as pivot-column and do as many primal simplex-iterations as necessary. Then terminate.

Step 5: Determine the interval $[\underline{\lambda} ; \bar{\lambda}]$, so that $\underline{\lambda} := -\infty$; $\bar{\lambda} := \bar{c}_j$.

Step 6: Set $\bar{c}_j := \bar{c}_j - \lambda^*$ and do as many primal simplex-iterations as necessary. Then terminate.

<u>Example 1</u>
Given the following problem P:

P: max $\pi = 3 \cdot x_1 + 2 \cdot x_2$
  $2 \cdot x_1 + x_2 \leq 4$
  $x_1 - 2 \cdot x_2 \leq 6$
  $x_1, x_2 \geq 0$

Find the subsequent alteration of c_1 with $\lambda^* = 2$.

$T^{(1)}$:

x_1	x_2	y_1	y_2	1
2	1	1	0	4
1	-2	0	1	6
-3	-2	0	0	0

$T^{(2)}$:

x_1	x_2	y_1	y_2	1
2	1	1	0	4
5	0	2	1	14
1	0	2	0	8

$[\underline{\lambda}\,;\,\bar{\lambda}\,] = (-\infty;1]$; $\bar{c}_1: = \bar{c}_1 - \lambda^* = -1$

$T^{(3)}$:

x_1	x_2	y_1	y_2	1
2	1	1	0	4
5	0	2	1	14
-1	0	2	0	8

$T^{(4)}$:

x_1	x_2	y_1	y_2	1
1	1/2	1/2	0	2
0	-5/2	-1/2	1	4
0	1/2	5/2	0	10

Solution: $\bar{x} = (2;0)$; $\bar{\pi} = 10$

Example 2

Given the problem as in example 1, but find the subsequent alteration of c_2 with $\lambda^* = 1$.
The optimal tableau is:

$T^{(2)}$:

x_1	x_2	y_1	y_2	1
2	1	1	0	4
5	0	2	1	14
1	0	2	0	8

$T^{(3)}$:

x_1	x_2	y_1	y_2	1
2	1	1	0	4
5	0	2	1	14
1	-1	2	0	8

$[\underline{\lambda}\,;\,\bar{\lambda}\,] = (-\infty;\,1/2]$; $\lambda^* \notin [\underline{\lambda}\,;\,\bar{\lambda}\,]$; $\bar{c}_2: = -1$

$T^{(4)}$:

x_1	x_2	y_1	y_2	1
2	1	1	0	4
5	0	2	1	14
3	0	3	0	12

Solution: $\bar{x} = (0;4)$; $\bar{\pi} = 12$.

48 *Linear Programming*

1.2 Shortened Methods

1.2.1 The Transportation Problem

<u>Hypotheses</u>
Given m places A_i, $i = 1,\ldots,m$ and n places B_j, $j = 1,\ldots,n$. In each place A_i z_i units of a certain product are offered; in each place B_j w_j units of the same product are demanded. Let $c_{ij} \in \mathbb{N}_0$ be the transportation costs for shipping one unit of the product from A_i to B_j. These costs are given in the matrix C and all assume to increase linearly.

Furthermore $\sum_{i=1}^{m} z_i = \sum_{j=1}^{n} w_j$,

When necessary an additional place, where the surplus is demanded or vice versa, must be included in the problem to insure that the above equality holds. This fictions place is called "dummy". If a dummy is used, the costs to/from all other places may be arbitrarily selected. The problem is to formulate a transportation plan, which satisfies the demand in all places B_j and eliminates the supply in all places A_i. Define x_{ij} as the goods transported from A_i to B_j, then the problem is as follows:

$$P: \min \pi = C \cdot X = \sum_{i=1}^{m} \sum_{j=1}^{n} c_{ij} \cdot x_{ij}$$

$$\sum_{j=1}^{n} x_{ij} = z_i \quad \forall\ i$$

$$\sum_{i=1}^{m} x_{ij} = w_j \quad \forall\ j$$

$$x_{ij} \geq 0 \quad \forall\ i,j\ .$$

Note: From the structure of the problem P it is seen that the desired transportation plan could be formulated with one of the simplex-

Section 1.2.1 49

methods. Indeed, the steps in the following methods correspond to simplex-steps in a shortened simplex-tableau.

Principle
Using one of the methods 1.2.1.1 - 1.2.1.7 a feasible basic solution for the transportation problem is obtained. The Stepping Stone Method includes a test of the current transportation plan for optimality and gives an iterative method for generating new feasible transportation plans if the current plan is not optimal.

1.2.1.1 The Northwest-Corner Rule

Hypotheses
The index-sets I and J are defined as: $I:=\{i\}$; $J:=\{j\}$, for the initial transportation matrix $X : X_{[m \times n]}$ $x_{ij} = 0 \;\forall\; i,j$ holds.

Description
Step 1: Determine $r := \min \{i \mid i \in I\}$
$s := \min \{j \mid j \in J\}$.

Step 2: Determine $\varepsilon := \min \{z_r; w_s\}$
and set $x_{rs} := \varepsilon$
$z_r := z_r - \varepsilon$
$w_s := w_s - \varepsilon$.

Step 3: Define $I := \begin{cases} I - \{r\} \text{, if } z_r = 0 \\ I \text{, otherwise} \end{cases}$

$J := \begin{cases} J - \{s\} \text{, if } w_s = 0 \\ J \text{, otherwise} \end{cases}$

Step 4: Is $(I = \emptyset) \vee (J = \emptyset)$?
If yes: Stop, the matrix X yields a feasible transportation plan .
If no : Go to step 1.

Example

Given the following problem:

$$C = \begin{pmatrix} 2 & 5 & 7 \\ 3 & 6 & 1 \\ 9 & 6 & 4 \end{pmatrix} ; \quad z = (z_i) = (10;4;6); \quad w = (w_j) = (5;12;8).$$

Since the supply does not equal the demand, a dummy is introduced so that the new cost matrix reads:

$$C = \begin{pmatrix} 2 & 5 & 7 \\ 3 & 6 & 1 \\ 9 & 6 & 4 \\ 50 & 50 & 50 \end{pmatrix} \quad \text{The new vectors are:} \\ z = (z_i) = (10;4;6;5); \quad w = (w_j) = (5;12;8);$$

$I = \{1;2;3;4\}; \quad J = \{1;2;3\}$
$r = \min \{1;2;3;4\} = 1; \quad s = \min \{1;2;3\} = 1; \quad X = \begin{pmatrix} 0 & 0 & 0 \\ 0 & 0 & 0 \\ 0 & 0 & 0 \\ 0 & 0 & 0 \end{pmatrix}$

$\varepsilon = \min \{z_1;w_1\} = \min \{10;5\} = 5;$

$x_{11} = 5; \quad X = \begin{pmatrix} 5 & 0 & 0 \\ 0 & 0 & 0 \\ 0 & 0 & 0 \\ 0 & 0 & 0 \end{pmatrix}$

$z_1 = 10 - 5 = 5; \quad z = (5;4;6;5); \quad w_1 = 5 - 5 = 0; \quad w = (0;12;8);$

$I = \{1;2;3;4\} ; \quad J = \{2;3\}; \quad r = \min \{1;2;3;4\} = 1;$
$s = \min \{2;3\} = 2 ; \quad \varepsilon = \min \{z_1;w_2\} = \min \{5;12\} = 5 ;$

$x_{12} = 5; \quad X = \begin{pmatrix} 5 & 5 & 0 \\ 0 & 0 & 0 \\ 0 & 0 & 0 \\ 0 & 0 & 0 \end{pmatrix}$

$z_1 = 5 - 5 = 0; \quad z = (0;4;6;5); \quad w_2 = 12 - 5 = 7; \quad w = (0;7;8);$

$I = \{2;3;4\} ; \quad J = \{2;3\} ; \quad r = \min \{2;3;4\} = 2 ;$
$s = \min \{2;3\} = 2 ; \quad \varepsilon = \min \{z_2;w_2\} = \min \{4;7\} = 4 ;$

$x_{22} = 4; \quad X = \begin{pmatrix} 5 & 5 & 0 \\ 0 & 4 & 0 \\ 0 & 0 & 0 \\ 0 & 0 & 0 \end{pmatrix}$

$z_2 = 4 - 4 = 0$; $z = (0;0;6;5)$; $w_2 = 7 - 4 = 3$; $w = (0;3;8)$;

$I = \{3;4\}$; $J = \{2;3\}$; $r = \min \{3;4\} = 3$; $s = \min \{2;3\} = 2$;
$\varepsilon = \min \{z_3;w_2\} = \min \{6;3\} = 3$;

$x_{32} = 3$; $X = \begin{pmatrix} 5 & 5 & 0 \\ 0 & 4 & 0 \\ 0 & 3 & 0 \\ 0 & 0 & 0 \end{pmatrix}$

$z_3 = 6 - 3 = 3$; $z = (0;0;3;5)$; $w_2 = 3 - 3 = 0$; $w = (0;0;8)$;

$I = \{3;4\}$; $J = \{3\}$; $r = \min \{3;4\} = 3$; $s = \min \{3\} = 3$;
$\varepsilon = \min \{z_3;w_3\} = \min \{3;8\} = 3$;

$x_{33} = 3$; $X = \begin{pmatrix} 5 & 5 & 0 \\ 0 & 4 & 0 \\ 0 & 3 & 3 \\ 0 & 0 & 0 \end{pmatrix}$

$z_3 = 3 - 3 = 0$; $z = (0;0;0;5)$; $w_3 = 8 - 3 = 5$; $w = (0;0;5)$;

$I = \{4\}$; $J = \{3\}$; $r = \min \{4\} = 4$; $s = \min \{3\} = 3$;
$\varepsilon = \min \{z_4;w_3\} = \min \{5;5\} = 5$;

$x_{43} = 5$; $X = \begin{pmatrix} 5 & 5 & 0 \\ 0 & 4 & 0 \\ 0 & 3 & 3 \\ 0 & 0 & 5 \end{pmatrix}$

$z_4 = 5 - 5 = 0$; $z = (0;0;0;0)$; $w_3 = 5 - 5 = 0$; $w = (0;0;0)$;
$I = \emptyset$; $J = \emptyset$;

Stop, $X = \begin{pmatrix} 5 & 5 & 0 \\ 0 & 4 & 0 \\ 0 & 3 & 3 \\ 0 & 0 & 5 \end{pmatrix}$ is a feasible transportation plan.

1.2.1.2 The Row Minimum Method

<u>Hypotheses</u>

The index-sets I and T are defined as: $I := \{i\}$; $T := \emptyset$; for the initial transportation matrix $X : X_{[m \times n]}$ $x_{ij} = 0$ $\forall$ i,j holds.

Description

Step 1: Determine $r := \min \{i \mid i \in I\}$.

Step 2: Determine $c_{rs} := \min \{c_{rj} \mid (r,j) \notin T\}$

$\varepsilon := \min \{z_r; w_s\}$

Step 3: Set $x_{rs} := \varepsilon$

$z_r := z_r - \varepsilon$

$w_s := w_s - \varepsilon$.

Step 4: Define $I := \begin{cases} I - \{r\}, & \text{if } z_r = 0 \\ I, & \text{otherwise} \end{cases}$

$T := \begin{cases} T \cup \{(i,s)\} \; \forall \; i, & \text{if } w_s = 0 \\ T, & \text{otherwise} \end{cases}$

Step 5: Is $I = \emptyset$?

If yes: Stop, the current matrix X yields a feasible transportation plan.

If no : Go to step 1.

Example

Given the following problem:

$$C = \begin{pmatrix} 2 & 5 & 7 \\ 3 & 6 & 1 \\ 9 & 6 & 4 \\ 50 & 50 & 50 \end{pmatrix} \; ; \; z = (10;4;6;5); \; w = (5;12;8);$$

(See also the example "Northwest-Corner Rule")

$I = \{1;2;3;4\} \; ; \; T = \emptyset \; ;$
$$X = \begin{pmatrix} 0 & 0 & 0 \\ 0 & 0 & 0 \\ 0 & 0 & 0 \\ 0 & 0 & 0 \end{pmatrix}$$

$r = \min \{1;2;3;4\} = 1; \quad c_{1s} = \min \{c_{1j}\} = c_{11} \rightarrow s = 1 \; ;$
$\varepsilon = \min \{z_1; w_1\} = \min \{10;5\} = 5;$

$x_{11} = 5;$ $\quad X = \begin{pmatrix} 5 & 0 & 0 \\ 0 & 0 & 0 \\ 0 & 0 & 0 \\ 0 & 0 & 0 \end{pmatrix}$

$z_1 = 10 - 5 = 5;$ $\quad z = (5;4;6;5);$ $\quad w_1 = 5 - 5 = 0;$ $\quad w = (0;12;8);$

$I = \{1;2;3;4\}$;
$T = \emptyset \cup \{(1,1);(2,1);(3,1);(4,1)\} = \{(1,1);(2,1);(3,1);(4,1)\};$
$r = \min \{1;2;3;4\} = 1;$
$c_{1s} = \min \{c_{1j} \mid (1,j) \notin T\} = \min \{c_{12};c_{13}\} = \min \{5;7\} = 5 \to s = 2;$
$\varepsilon = \min \{z_1;w_2\} = \min \{5;12\} = 5;$

$x_{12} = 5;$ $\quad X = \begin{pmatrix} 5 & 5 & 0 \\ 0 & 0 & 0 \\ 0 & 0 & 0 \\ 0 & 0 & 0 \end{pmatrix}$

$z_1 = 5 - 5 = 0;$ $\quad z = (0;4;6;5);$ $\quad w_2 = 12 - 5 = 7;$ $\quad w = (0;7;8);$

$I = \{2;3;4\};$ $\quad T = \{(1,1);(2,1);(3,1);(4,1)\}$; $\quad r = \min \{2;3;4\} = 2$
$c_{2s} = \min \{c_{2j} \mid (2,j) \notin T\} = \min \{c_{22};c_{23}\} = \min \{6;1\} = 1 \to s = 3;$
$\varepsilon = \min \{z_2;w_3\} = \min \{4;7\} = 4;$

$x_{23} = 4;$ $\quad X = \begin{pmatrix} 5 & 5 & 0 \\ 0 & 0 & 4 \\ 0 & 0 & 0 \\ 0 & 0 & 0 \end{pmatrix}$

$z_2 = 4 - 4 = 0;$ $\quad z = (0;0;6;5);$ $\quad w_3 = 8 - 4 = 4;$ $\quad w = (0;7;4);$

$I = \{3;4\};$ $\quad T = \{(1,1); (2,1);(3,1);(4,1)\};$ $\quad r = \min \{3;4\} = 3;$
$c_{3s} = \min \{c_{3j} \mid (3,j) \notin T\} = \min \{c_{32};c_{33}\} = \min \{6;4\} = 4 \to s = 3;$
$\varepsilon = \min \{z_3;w_3\} = \min \{6;4\} = 4;$

$x_{33} = 4;$ $\quad X = \begin{pmatrix} 5 & 5 & 0 \\ 0 & 0 & 4 \\ 0 & 0 & 4 \\ 0 & 0 & 0 \end{pmatrix}$

$z_3 = 6 - 4 = 2;$ $\quad z = (0;0;2;5);$ $\quad w_3 = 4 - 4 = 0;$ $\quad w = (0;7;0);$

$I = \{4\}$; $T = \{(1,1);(1,3);(2,1);(2,3);(3,1);(3,3);(4,1);(4,3)\}$;
$r = \min \{c_{4j} \mid (4,j) \notin T\} = \min \{c_{42}\} = c_{42} = 50 \rightarrow s = 2$;
$\varepsilon = \min \{z_4; w_2\} = \min \{5; 5\} = 5$;

$x_{42} = 5$; $\quad X = \begin{pmatrix} 5 & 5 & 0 \\ 0 & 0 & 4 \\ 0 & 2 & 4 \\ 0 & 5 & 0 \end{pmatrix}$

$z_4 = 5 - 5 = 0$; $z = (0;0;0;0)$; $w_2 = 5 - 5 = 0$; $w = (0;0;0)$;

$I = \emptyset$; Stop, $X = \begin{pmatrix} 5 & 5 & 0 \\ 0 & 0 & 4 \\ 0 & 2 & 4 \\ 0 & 5 & 0 \end{pmatrix}$ is a feasible transportation plan.

1.2.1.3 The Column Minimum Method

Hypotheses

The index-sets J and T are defined as: $J := \{j\}$; $T := \emptyset$, for the initial tranportation plan $X : X_{[m \times n]}$ $x_{ij} = 0 \; \forall \; i,j$ holds.

Description

Step 1: Determine $s := \min \{j \mid j \in J\}$.

Step 2: Determine $c_{rs} := \min \{c_{is} \mid (i,s) \notin T\}$

$\varepsilon := \min \{z_r; w_s\}$.

Step 3: Set $x_{rs} := \varepsilon$
$z_r := z_r - \varepsilon$
$w_s := w_s - \varepsilon$.

Step 4: Define $J := \begin{cases} J - \{s\}, & \text{if } w_s = 0 \\ J, & \text{otherwise} \end{cases}$

$T := \begin{cases} T \cup \{(r,j)\} \; \forall \; j, & \text{if } z_r = 0 \\ T, & \text{otherwise} \end{cases}$.

Step 5: Is $J = \emptyset$?
If yes: Stop, the current matrix X yields a feasible

transportation plan.
If no : Go to step 1.

Example

Given the following problem:

$$C = \begin{pmatrix} 2 & 5 & 7 \\ 3 & 6 & 1 \\ 9 & 6 & 4 \\ 50 & 50 & 50 \end{pmatrix} ; \quad z = (10;4;6;5); \quad w = (5;12;8);$$

(See also the example "Northwest-Corner Rule").

$J = \{1;2;3\}$; $T = \emptyset$; $X = \begin{pmatrix} 0 & 0 & 0 \\ 0 & 0 & 0 \\ 0 & 0 & 0 \\ 0 & 0 & 0 \end{pmatrix}$

$s = \min \{1;2;3\} = 1$;

$c_{r1} = \min \{c_{i1} \mid (i,1) \notin T\} = \min \{c_{11};c_{21};c_{31};c_{41}\} = \min \{2;3;9;50\}$
$\qquad\qquad\qquad\qquad\qquad\qquad\qquad\qquad\qquad\qquad = 2 \rightarrow r = 1$;

$\varepsilon = \min \{z_1;w_1\} = \min \{10;5\} = 5$;

$x_{11} = 5$; $X = \begin{pmatrix} 5 & 0 & 0 \\ 0 & 0 & 0 \\ 0 & 0 & 0 \\ 0 & 0 & 0 \end{pmatrix}$

$z_1 = 10 - 5 = 5$; $z = (5;4;6;5)$; $w_1 = 5 - 5 = 0$; $w = (0;12;8)$;

$J = \{2;3\}$; $T = \emptyset$; $s = \min \{2;3\} = 2$;

$c_{r2} = \min \{c_{i2} \mid (i,2) \notin T\} = \min \{c_{12};c_{22};c_{32};c_{42}\} = \min \{5;6;6;50\}$
$\qquad\qquad\qquad\qquad\qquad\qquad\qquad\qquad\qquad\qquad = 5 \rightarrow r = 1$;

$\varepsilon = \min \{z_1;w_2\} = \min \{5;12\} = 5$;

$x_{12} = 5$; $X = \begin{pmatrix} 5 & 5 & 0 \\ 0 & 0 & 0 \\ 0 & 0 & 0 \\ 0 & 0 & 0 \end{pmatrix}$

$z_1 = 5 - 5 = 0$; $z = (0;4;6;5)$; $w_2 = 12 - 5 = 7$; $w = (0;7;8)$;

$J = \{2;3\}$; $T = \emptyset \cup \{(1,1);(1,2);(1,3)\} = \{(1,1);(1,2);(1,3)\}$;
$s = \min \{2;3\} = 2$;

56 Linear Programming

$c_{r2} = \min \{c_{i2} \mid (i,2) \ T\} = \min \{c_{12}; c_{22}; c_{32}; c_{42}\} = \min \{5;6;6;50\}$
$\qquad\qquad\qquad\qquad\qquad\qquad\qquad\qquad\qquad\qquad = 5 \to r = 1 \ ;$
$\varepsilon = \min \{z_1; w_2\} = \min \{5;12\} = 5;$

$x_{12} = 5; \qquad X = \begin{pmatrix} 5 & 5 & 0 \\ 0 & 0 & 0 \\ 0 & 0 & 0 \\ 0 & 0 & 0 \end{pmatrix}$

$z_1 = 5 - 5 = 0; \quad z = (0;4;6;5); \quad w_2 = 12 - 5 = 7; \quad w = (0;7;8);$

$J = \{2;3\}; \ T = \emptyset \cup \{(1,1);(1,2);(1,3)\} = \{(1,1);(1,2);(1,3)\} \ ;$
$s = \min \{2;3\} = 2;$
$c_{r2} = \min \{c_{i2} \mid (i,2) \notin T\} = \min \{c_{22}; c_{32}; c_{42}\} = \min \{6;6;50\}$
$\qquad\qquad\qquad\qquad\qquad\qquad\qquad\qquad\qquad\qquad = 6 \to r = 3 \ ;$
$\varepsilon = \min \{z_3; w_2\} = \min \{6;7\} = 6;$

$x_{32} = 6; \qquad X = \begin{pmatrix} 5 & 5 & 0 \\ 0 & 0 & 0 \\ 0 & 6 & 0 \\ 0 & 0 & 0 \end{pmatrix}$

$z_3 = 6 - 6 = 0; \quad z = (0;4;0;5); \quad w_2 = 7 - 6 = 1; \quad w = (0;1;8);$

$J = \{2;3\}; \ T = \{(1,1);(1,2);(1,3)\} \cup \{(3,1);(3,2);(3,3)\} =$
$\qquad\qquad = \{(1,1);(1,2);(1,3); (3,1);(3,2);(3,3)\};$
$s = \min \{2;3\} = 2;$
$c_{r2} = \min \{c_{i2} \mid (i,2) \notin T\} = \min \{c_{22}; c_{42}\} = \min \{6;50\} = 6 \to r = 2;$
$\varepsilon = \min \{z_2; w_2\} = \min \{4;1\} = 1;$

$x_{22} = 1; \qquad X = \begin{pmatrix} 5 & 5 & 0 \\ 0 & 1 & 0 \\ 0 & 6 & 0 \\ 0 & 0 & 0 \end{pmatrix}$

$z_2 = 4 - 1 = 3; \quad z = (0;3;0;5); \quad w_2 = 1 - 1 = 0; \quad w = (0;0;8);$

$J = \{3\}; \ T = \{(1,1);(1,2);(1,3);(3,1);(3,2);(3,3)\} \ ; \ s = \min \{3\} = 3;$
$c_{r3} = \min \{c_{i3} \mid (i,3) \notin T\} = \min \{c_{23}; c_{43}\} = \min \{1;50\} = 1 \to r = 2;$
$\varepsilon = \min \{z_2; w_3\} = \min \{3;8\} = 3;$

$x_{23} = 3;$ $\quad X = \begin{pmatrix} 5 & 5 & 0 \\ 0 & 1 & 3 \\ 0 & 6 & 0 \\ 0 & 0 & 0 \end{pmatrix}$

$z_2 = 3 - 3 = 0;$ $\quad z = (0;0;0;5);$ $\quad w_3 = 8 - 3 = 5;$ $\quad w = (0;0;5);$

$J = \{3\};$ $\quad T = \{(1,1);(1,2);(1,3);(3,1);(3,2);(3,3)\} \cup \{(2,1);(2,2);$
$\hfill (2,3)\}=$
$\hfill = \{(1,1);(1,2);(1,3);(2,1);(2,2);(2,3);(3,1);(3,2);$
$\hfill (3,3)\};$
$s = \min \{3\} = 3;$

$c_{r3} = \min \{c_{i3} \mid (i,3) \notin T\} = \min \{c_{43}\} = c_{43} = 50 \to r = 4;$

$\varepsilon = \min \{z_4; w_3\} = 5;$

$x_{43} = 5;$ $\quad X = \begin{pmatrix} 5 & 5 & 0 \\ 0 & 1 & 3 \\ 0 & 6 & 0 \\ 0 & 0 & 5 \end{pmatrix}$

$z_4 = 5 - 5 = 0;$ $\quad z = (0;0;0;0);$ $\quad w_3 = 5 - 5 = 0;$ $\quad w = (0;0;0);$

$J = \emptyset;$ Stop, $X = \begin{pmatrix} 5 & 5 & 0 \\ 0 & 1 & 3 \\ 0 & 6 & 0 \\ 0 & 0 & 5 \end{pmatrix}$ is a feasible transportation plan.

1.2.1.4 The Matrix Minimum Method

<u>Hypotheses</u>

The index-sets I, J and T are defined as: $I := \{i\};$ $J := \{j\};$ $T := \emptyset,$ for the initial transportation matrix $X : X_{[m \times n]}$ $x_{ij} = 0 \; \forall \; i,j$ holds.

<u>Description</u>

Step 1: Determine $c_{rs} := \min \{c_{ij} \mid (i,j) \notin T\}$

$\qquad \varepsilon := \min \{z_r; w_s\}$.

Step 2: Set $x_{rs} := \varepsilon$

$\qquad z_r := z_r - \varepsilon$

$\qquad w_s := w_s - \varepsilon$.

58 Linear Programming

Step 3: Define $I := \begin{cases} I - \{r\}, & \text{if } z_r = 0 \\ I, & \text{otherwise} \end{cases}$

$J := \begin{cases} J - \{s\}, & \text{if } w_s = 0 \\ J, & \text{otherwise} \end{cases}$

$T := \begin{cases} T \cup \{(r,j)\} \; \forall \; j, & \text{if } z_r = 0 \\ T \cup \{(i,s)\} \; \forall \; i, & \text{if } w_s = 0 \end{cases}$

Step 4: Is $(I = \emptyset) \vee (J = \emptyset)$?

If yes: Stop, the current matrix X yields a feasible transportation plan.

If no : Go to step 1 .

Example

Given the following problem:

$$C = \begin{pmatrix} 2 & 5 & 7 \\ 3 & 6 & 1 \\ 9 & 6 & 4 \\ 50 & 50 & 50 \end{pmatrix} \; ; \; z = (10;4;6;5); \; w = (5;12;8);$$

(See also example "Northwest-Corner Rule")

$I = \{1;2;3;4\}$; $J = \{1;2;3\}$; $T = \emptyset$; $\quad X = \begin{pmatrix} 0 & 0 & 0 \\ 0 & 0 & 0 \\ 0 & 0 & 0 \\ 0 & 0 & 0 \end{pmatrix}$

$c_{rs} = \min \{c_{11};c_{12};c_{13};c_{21};c_{22};c_{23};c_{31};c_{32};c_{33};c_{41};c_{42};c_{43}\} =$
$= \min \{2;5;7;3;6;1;9;6;4;50;50;50\} = 1 \rightarrow r = 2; \; s = 3;$

$\varepsilon = \min \{z_2;w_3\} = \min \{4;8\} = 4;$

$x_{23} = 4; \quad X = \begin{pmatrix} 0 & 0 & 0 \\ 0 & 0 & 4 \\ 0 & 0 & 0 \\ 0 & 0 & 0 \end{pmatrix}$

$z_2 = 4 - 4 = 0; \; z = (10;0;6;5); \; w_3 = 8 - 4 = 4; \; w = (5;12;4);$

$I = \{1;3;4\}$; $J = \{1;2;3\}$;
$T = \emptyset \cup \{(2,1);(2,2);(2,3)\} = \{(2,1);(2,2);(2,3)\}$;
$c_{rs} = \min \{c_{11};c_{12};c_{13};c_{31};c_{32};c_{33};c_{41};c_{42};c_{43}\} =$
$\quad = \min \{2;5;7;9;4;6;50;50;50\} = 2 \to r = 1; s = 1;$
$\varepsilon = \min \{z_1;w_1\} = \min \{10;5\} = 5;$

$x_{11} = 5;$ $\quad X = \begin{pmatrix} 5 & 0 & 0 \\ 0 & 0 & 4 \\ 0 & 0 & 0 \\ 0 & 0 & 0 \end{pmatrix}$

$z_1 = 10 - 5 = 5;$ $z = (5;0;6;5);$ $w_1 = 5 - 5 = 0;$ $w = (0;12;4);$

$I = \{1;3;4\};$ $J = \{2;3\}$;
$T = \{(2,1);(2,2);(2,3)\} \cup \{(1,1);(2,1);(3,1);(4,1)\} =$
$\quad = \{(1,1);(2,1);(2,2);(2,3);(3,1);(4,1)\}$;
$c_{rs} = \min \{c_{12};c_{13};c_{32};c_{33};c_{42};c_{43}\} = \min \{5;7;6;4;50;50\} = 4;$
$\quad\quad\quad\quad\quad\quad\quad\quad\quad\quad\quad\quad\quad\quad\quad \to r = 3; s = 3;$
$\varepsilon = \min \{z_3;w_3\} = \min \{6;4\} = 4;$

$x_{33} = 4;$ $\quad X = \begin{pmatrix} 5 & 0 & 0 \\ 0 & 0 & 4 \\ 0 & 0 & 4 \\ 0 & 0 & 0 \end{pmatrix}$

$z_3 = 6 - 4 = 2;$ $z = (5;0;2;5);$ $w_3 = 4 - 4 = 0;$ $w = (0;12;0);$

$I = \{1;3;4\}$; $J = \{2\};$
$T = \{(1,1);(2,1);(2,2);(2,3);(3,1);(4,1)\} \cup \{(1,3);(2,3);(3,3);(4,3)\}$
$\quad = \{(1,1);(1,3);(2,1);(2,2);(2,3);(3,1);(3,3);(4,1);(4,3)\}$;
$c_{rs} = \min \{c_{12};c_{32};c_{42}\} = \min \{5;6;50\} = 5 \to r = 1; s = 2$;
$\varepsilon = \min \{z_1;w_2\} = \min \{5;12\} = 5;$

$x_{12} = 5;$ $\quad X = \begin{pmatrix} 5 & 5 & 0 \\ 0 & 0 & 4 \\ 0 & 0 & 4 \\ 0 & 0 & 0 \end{pmatrix}$

$z_1 = 5 - 5 = 0;$ $z = (0;0;2;5);$ $w_2 = 12 - 5 = 7;$ $w = (0;7;0);$

$I = \{3;4\}$; $J = \{2\}$;
$T = \{(1,1);(1,3);(2,1);(2,2);(2,3);(3,1);(3,3);(4,1);(4,3)\} \cup$
$\hspace{5cm} \cup \{(1,1);(1,2);(1,3)\}$
$\hspace{0.5cm} = \{(1,1);(1,2);(1,3);(2,1);(2,2);(2,3);(3,1);(3,3);(4,1);(4,3)\}$;
$c_{rs} = \min \{c_{32};c_{42}\} = \min \{6;50\} = 6 \to r = 3; s = 2;$
$\varepsilon = \min \{z_3;w_2\} = \min \{2;7\} = 2;$

$x_{32} = 2;$ $\quad X = \begin{pmatrix} 5 & 5 & 0 \\ 0 & 0 & 4 \\ 0 & 2 & 4 \\ 0 & 0 & 0 \end{pmatrix}$

$z_3 = 2 - 2 = 0;$ $\quad z = (0;0;0;5);$ $\quad w_2 = 7 - 2 = 5;$ $\quad w = (0;5;0);$

$I = \{4\}$; $J = \{2\}$;
$T = \{(1,1);(1,2);(1,3);(2,1);(2,2);(2,3);(3,1);(3,3);(4,1);(4,3)\} \cup$
$\hspace{5cm} \cup \{(3,1);(3,2);(3,3)\}$
$\hspace{0.5cm} = \{(1,1);(1,2);(1,3);(2,1);(2,2);(2,3);(3,1);(3,2);(3,3);(4,1);$
$\hspace{10cm} (4,3)\}$;
$c_{rs} = \min \{c_{42}\} = c_{42} = 50 \to r = 4; s = 2;$
$\varepsilon = \min \{z_4;w_2\} = \min \{5;5\} = 5;$

$x_{42} = 5;$ $\quad X = \begin{pmatrix} 5 & 5 & 0 \\ 0 & 0 & 4 \\ 0 & 2 & 4 \\ 0 & 5 & 0 \end{pmatrix}$

$z_4 = 5 - 5 = 0;$ $\quad z = (0;0;0;0);$ $\quad w_2 = 5 - 5 = 0;$ $\quad w = (0;0;0);$

$I = \emptyset;$ $\quad J = \emptyset;$
Stop, $\quad X = \begin{pmatrix} 5 & 5 & 0 \\ 0 & 0 & 4 \\ 0 & 2 & 4 \\ 0 & 5 & 0 \end{pmatrix}$ is a feasible transportation plan.

1.2.1.5 The Double Preference Method

<u>Hypotheses</u>

The index-set T is defined as: $T := \emptyset$, for the initial transportation matrix $X : X_{[m \times n]}$ $\quad x_{ij} = 0 \; \forall \; i,j$ holds.

Description

Step 1: (determination of the row-minima)

Determine $c_{is} := \min_{j} \{c_{ij}\} \; \forall \; i = 1,\ldots,m$

and set $Q := \{c_{is}\}$.

Step 2: (determination of the column-minima)

Determine $c_{rj} := \min_{i} \{c_{ij}\} \; \forall j = 1,\ldots,n$

and set $S := \{c_{rj}\}$.

Step 3: Define

$$M^{(2)} := \{c_{ij} \mid (c_{ij} \in Q) \wedge (c_{ij} \in S)\}$$

$$M^{(1)} := \{c_{ij} \mid [(c_{ij} \in Q) \wedge (c_{ij} \notin S)] \vee [(c_{ij} \notin Q) \wedge (c_{ij} \in S)]\}$$

$$M^{(0)} := \{c_{ij} \mid (c_{ij} \notin Q) \wedge (c_{ij} \notin S)\}$$

and set the running indices $k := 1; \; l := 2$.

Step 4: (assignment of reference numbers)

Is $M^{(1)} = \emptyset$?

If yes: Go to step 6.

If no : Go to step 5.

Step 5: Determine $c_{rs} := \min \{c_{ij} \mid c_{ij} \in M^{(1)}\}$,

define $c_{rs}^{(k)} := c_{rs}$

$$M^{(1)} := M^{(1)} - \{c_{rs}\}$$

and set $k := k + 1$. Go to step 4

Step 6: Is $l = 0$?

If yes: Set the running index $p := 1$ and go to step 7.

If no : Set $l := l - 1$ and go to step 4.

Step 7: Determine the element $c_{rs}^{(p)}$.

62 *Linear Programming*

Step 8: Is $(r,s) \notin T$?
    If yes: Set $p := p + 1$ and go to step 10.
    If no : Go to step 9.

Step 9: Determine $\varepsilon := \min\{z_r; w_s\}$,
    define $x_{rs} := \varepsilon$
              $z_r := z_r - \varepsilon$
              $w_s := w_s - \varepsilon$

$$T := \begin{cases} T \cup \{(r,j)\} \; \forall \; j, & \text{if } z_r = 0 \\ T \cup \{(i,s)\} \; \forall \; i, & \text{if } w_s = 0 \end{cases}$$

and set $p := p + 1$. Go to step 10.

Step 10: Is $(z = \Theta) \wedge (w = \Theta)$?
    If yes: Stop, the current Matrix X yields a feasible transportation plan.
    If no : Go to step 7.

Example
Given the following problem:

$$C = \begin{pmatrix} 2 & 5 & 7 \\ 3 & 6 & 1 \\ 9 & 6 & 4 \\ 50 & 50 & 50 \end{pmatrix} ; \; z = (10;4;6;5); \; w = (5;12;8);$$

(See also example "Northwest-Corner Rule")

$c_{1s} = \min\{c_{11}; c_{12}; c_{13}\} = \min\{2;5;7\} = 2$; $c_{1s} = c_{11}$;
$c_{2s} = \min\{c_{21}; c_{22}; c_{23}\} = \min\{3;6;1\} = 1$; $c_{2s} = c_{23}$;
$c_{3s} = \min\{c_{31}; c_{32}; c_{33}\} = \min\{9;6;4\} = 4$; $c_{3s} = c_{33}$;
$c_{4s} = \min\{c_{41}; c_{42}; c_{42}\} = \min\{50;50;50\} = 50$; $c_{4s} = \{c_{41}; c_{42}; c_{43}\}$;
$Q = \{c_{11}; c_{23}; c_{41}; c_{42}; c_{43}\}$;

$c_{r1} = \min \{c_{11}; c_{21}; c_{31}; c_{41}\} = \min \{2;3;9;50\} = 2;$ $\quad c_{r1} = c_{11}$;

$c_{r2} = \min \{c_{12}; c_{22}; c_{32}; c_{42}\} = \min \{5;6;6;50\} = 5;$ $\quad c_{r2} = c_{12}$;

$c_{r3} = \min \{c_{13}; c_{23}; c_{33}; c_{43}\} = \min \{7;1;4;50\} = 1;$ $\quad c_{r3} = c_{23}$;

$S = \{c_{11}; c_{12}; c_{23}\}$;

$M^{(2)} = \{c_{11}; c_{23}\};$ $\quad M^{(1)} = \{c_{12}; c_{33}; c_{41}; c_{43}\}$;

$M^{(0)} = \{c_{13}; c_{21}; c_{22}; c_{31}; c_{32}\};$ $\quad l = 2;$ $\quad k = 1;$

$M^{(2)} \neq \emptyset \rightarrow c_{rs} = \min \{c_{ij} \mid c_{ij} \in M^{(2)}\} = \min \{c_{11}; c_{23}\} = \min\{2;1\} = 1;$

$c_{rs} = c_{23}; c_{rs}^{(1)} = c_{23};$

$M^{(2)} = \{c_{11}; c_{23}\} - \{c_{23}\} = \{c_{11}\};$ $\quad k = 2;$

$c_{rs} = \min \{c_{11}\} = 2;$ $\quad c_{rs} = c_{11};$ $\quad c_{rs}^{(2)} = c_{11};$ $\quad M^{(2)} = \emptyset;$

$k = 3;$ $\quad l = 1;$ $\quad M^{(1)} \neq \emptyset \rightarrow$

$c_{rs} = \min \{c_{ij} \mid c_{ij} \in M^{(1)}\} = \min \{c_{12}; c_{33}; c_{41}; c_{42}; c_{43}\} =$
$= \min \{5;4;50;50;50\} = 4;$ $\quad c_{rs} = c_{33};$ $\quad c_{rs}^{(3)} = c_{33};$

$M^{(1)} = \{c_{12}; c_{33}; c_{41}; c_{42}; c_{43}\} - \{c_{33}\} = \{c_{12}; c_{41}; c_{42}; c_{43}\}$;

$k = 4;$ $\quad c_{rs} = \min \{c_{12}; c_{41}; c_{42}; c_{43}\} = \min \{5;50;50;50\} = 5;$

$c_{rs} = c_{12};$ $\quad c_{rs}^{(4)} = c_{12};$

$M^{(1)} = \{c_{41}; c_{42}; c_{43}\}$; $\quad k = 5;$

$c_{rs} = \min \{c_{41}; c_{42}; c_{43}\} = \min \{50;50;50\} = 50;$ $\quad c_{rs} = c_{41};$ $\quad c_{rs}^{(5)} = c_{41};$

$M^{(1)} = \{c_{42}; c_{43}\};$ $\quad k = 6;$ $\quad c_{rs} = \min \{c_{42}; c_{43}\} = \min \{50;50\} = 50;$

$c_{rs} = c_{42}; c_{rs}^{(6)} = c_{42};$

$M^{(1)} = \{c_{43}\};$ $k = 7;$ $c_{rs} = \min \{c_{43}\} = c_{43};$ $c_{rs}^{(7)} = c_{43};$

$M^{(1)} = \emptyset;$ $\quad k = 8;$ $\quad l = 0;$

$M^{(0)} \neq \emptyset \rightarrow c_{rs} = \min \{c_{ij} \mid c_{ij} \in M^{(0)}\} = \min\{c_{13}; c_{21}; c_{22}; c_{31}; c_{32}\} =$

$= \min \{7;3;6;9;4\} = 3;\ c_{rs} = c_{21};\ c_{rs}^{(8)} = c_{21};$

$M^{(0)} = \{c_{13}; c_{22}; c_{31}; c_{32}\};\ k = 9;$

$c_{rs} = \min \{c_{13}; c_{22}; c_{31}; c_{32}\} = \min \{7;6;9;4\} = 4;$

$c_{rs} = c_{32};\ c_{rs}^{(9)} = c_{32};\ M^{(0)} = \{c_{13}; c_{22}; c_{31}\};\ k = 10;$

$c_{rs} = \min \{c_{13}; c_{22}; c_{31}\} = \min \{7;6;9\} = 6;$

$c_{rs} = c_{22};\ c_{rs}^{(10)} = c_{22};\ M^{(0)} = \{c_{13}; c_{31}\};\ k = 11;$

$c_{rs} = \min \{c_{13}; c_{31}\} = \min \{7;9\} = 7;$

$c_{rs} = c_{13};\ c_{rs}^{(11)} = c_{13};\ M^{(0)} = \{c_{31}\};\ k = 12;$

$c_{rs} = \min \{c_{31}\} = c_{31};\ c_{rs}^{(12)} = c_{31};\ M^{(0)} = \emptyset\ ;$

$p = 1;\ T = \emptyset\ ;$

$c_{rs}^{(1)} = c_{23};\ (2,3) \notin T;\ \varepsilon = \min \{z_2; w_3\} = \min \{4;8\} = 4;$

$x_{23} = 4;\quad X = \begin{pmatrix} 0 & 0 & 0 \\ 0 & 0 & 4 \\ 0 & 0 & 0 \\ 0 & 0 & 0 \end{pmatrix}$

$z_2 = 4 - 4 = 0;\ z = (10;0;6;5);\ w_3 = 8 - 4 = 4;\ w = (5;12;4);$

$T = \emptyset\ \cup\ \{(2,1);(2,2);(2,3)\} = \{(2,1);(2,2);(2,3)\}\ ;\ p = 2;$

$c_{rs}^{(2)} = c_{11};\ (1,1) \notin T;\ \varepsilon = \min \{z_1; w_1\} = \min \{10;5\} = 5;$

$x_{11} = 5;\quad X = \begin{pmatrix} 5 & 0 & 0 \\ 0 & 0 & 4 \\ 0 & 0 & 0 \\ 0 & 0 & 0 \end{pmatrix}$

$z_1 = 10 - 5 = 5;\ z = (5;0;6;5);\quad w_1 = 5 - 5 = 0;\ w = (0;12;4);$

$T = \{(1,1);(2,1);(2,2);(2,3);(3,1);(4,1)\};\ p = 3;$

$c_{rs}^{(3)} = c_{33};\ (3,3) \notin T\ ;\ \varepsilon = \min \{z_3; w_3\} = \min \{6;4\} = 4;$

$x_{33} = 4$; $\quad X = \begin{pmatrix} 5 & 0 & 0 \\ 0 & 0 & 4 \\ 0 & 0 & 4 \\ 0 & 0 & 0 \end{pmatrix}$

$z_3 = 6 - 4 = 2$; $\quad z = (5;0;2;5)$; $\quad w_3 = 4 - 4 = 0$; $\quad w = (0;12;0)$;

$T = \{(1,1);(1,3);(2,1);(2,2);(2,3);(3,1);(3,3);(4,1);(4,3)\}$; $p = 4$;

$c_{rs}^{(4)} = c_{12}$; $\quad (1,2) \notin T$; $\quad \varepsilon = \min\{z_1;w_2\} = \min\{5;12\} = 5$;

$x_{12} = 5$; $\quad X = \begin{pmatrix} 5 & 5 & 0 \\ 0 & 0 & 4 \\ 0 & 0 & 4 \\ 0 & 0 & 0 \end{pmatrix}$

$z_1 = 5 - 5 = 0$; $\quad z = (0;0;2;5)$; $\quad w_2 = 12 - 5 = 7$; $\quad w = (0;7;0)$;

$T = \{(1,1);(1,2);(1,3);(2,1);(2,2);(2,3);(3,1);(3,3);(4,1);(4,3)\}$; $p = 5$;

$c_{rs}^{(5)} = c_{41}$; $\quad (4,1) \in T$; $\quad p = 6$;

$c_{rs}^{(6)} = c_{42}$; $\quad (4,2) \notin T$; $\quad \varepsilon = \min\{z_4;w_2\} = \min\{5;7\} = 5$;

$x_{42} = 5$; $\quad X = \begin{pmatrix} 5 & 5 & 0 \\ 0 & 0 & 4 \\ 0 & 0 & 4 \\ 0 & 5 & 0 \end{pmatrix}$

$z_4 = 5 - 5 = 0$; $\quad z = (0;0;2;0)$; $\quad w_2 = 7 - 5 = 2$; $\quad w = (0;2;0)$;

$T = \{(1,1);(1,2);(1,3);(2,1);(2,2);(2,3);(3,1);(3,3);(4,1);(4,2);(4,3)\}$; $p = 7$;

$c_{rs}^{(7)} = c_{43}$; $\quad (4,3) \in T$; $\quad p = 8$;

$c_{rs}^{(8)} = c_{21}$; $\quad (2,1) \in T$; $\quad p = 9$;

$c_{rs}^{(9)} = c_{32}$; $\quad (3,2) \notin T$; $\quad \varepsilon = \min\{z_3;w_2\} = \min\{2;2\} = 2$;

$x_{32} = 2$; $\quad X = \begin{pmatrix} 5 & 5 & 0 \\ 0 & 0 & 4 \\ 0 & 2 & 4 \\ 0 & 5 & 0 \end{pmatrix}$

66 *Linear Programming*

$z_3 = 2 - 2 = 0$; $z = (0;0;0;0)$; $w_2 = 2 - 2 = 0$; $w = (0;0;0)$;

Stop, $X = \begin{pmatrix} 5 & 5 & 0 \\ 0 & 0 & 4 \\ 0 & 2 & 4 \\ 0 & 5 & 0 \end{pmatrix}$ is a feasible transportation plan.

1.2.1.6 VOGEL's Approximation Method (VAM)

Hypotheses

Let the set D be the set of all elements of the cost matrix C, i.e. D: = $\{c_{ij}\}$, for the initial transportation matrix $X : X_{[m \times n]}$ $x_{ij} = 0 \; \forall \; i,j$ holds.

Description

Step 1: Determine c_{is}: = min $\{c_{ij} \mid c_{ij} \in D\}$
c_{ik}: = min $\{c_{ij} \mid c_{ij} \in D; j \neq s\}$ $\Big\} \; \forall \; i = 1,\ldots,m$.

(If $c_{ij}(c_{ik})$ does not exist, set $c_{ij} = 0$ ($c_{ik} = 0$).)

Step 2: Calculate δ'_i: = $\mid c_{ik} - c_{is} \mid$ $\forall \; i = 1,\ldots,m$.

Step 3: Determine c_{sj}: = min $\{c_{ij} \mid c_{ij} \in D\}$
c_{kj}: = min $\{c_{ij} \mid c_{ij} \in D; i \neq s\}$ $\Big\} \; \forall \; j = 1,\ldots,n$.

(If $c_{sj}(c_{kj})$ does not exist, set $c_{sj} = 0$ ($c_{kj} = 0$).)

Step 4: Calculate δ''_j: = $\mid c_{kj} - c_{sj} \mid$ $\forall \; j = 1,\ldots,n$.

Step 5: Determine δ: = max $\{\delta'_i; \delta''_j\}$.

Step 6: Is $\delta \in \{\delta'_i\}$?
If yes: Go to step 7.
If no : Go to step 8.

Step 7: Let $\delta = \delta'_r$; determine c_{rs}: = min $\{c_{rj} \mid c_{rj} \in D\}$.
Go to step 9.

Step 8: Let $\delta = \delta''_s$; determine c_{rs}: = min $\{c_{is} \mid c_{is} \in D\}$.
Go to step 9.

Step 9: Determine $\varepsilon := \min\{z_r; w_s\}$
and define $x_{rs} := \varepsilon$

$$z_r := z_r - \varepsilon$$
$$w_s := w_s - \varepsilon$$
$$D := \begin{cases} D - \{c_{rj}\} \;\forall j, \text{ if } z_r = 0 \\ D - \{c_{is}\} \;\forall i, \text{ if } w_s = 0 \end{cases}$$

Step 10: Is $D = \emptyset$?
If yes: Stop, the current matrix X yields a feasible transportation plan.
If no : Go to step 1.

Example

Given the following problem:

$$C = \begin{pmatrix} 2 & 5 & 7 \\ 3 & 6 & 1 \\ 9 & 6 & 4 \\ 50 & 50 & 50 \end{pmatrix}; \; z = (10;4;6;5); \; w = (5;12;8);$$

(See also example "Northwest-Corner Rule")

$D = \{c_{11}; c_{12}; c_{13}; c_{21}; c_{22}; c_{23}; c_{31}; c_{32}; c_{33}; c_{41}; c_{42}; c_{43}\};$

$c_{1s} = \min\{c_{11}; c_{12}; c_{13}\} = \min\{2;5;7\} = 2; \quad c_{1s} = c_{11};$

$c_{1k} = \min\{c_{12}; c_{13}\} = \min\{5;7\} = 5; \quad c_{1k} = c_{12};$

$c_{2s} = c_{23}; \quad c_{2k} = c_{21};$

$c_{3s} = c_{33}; \quad c_{3k} = c_{32}; \quad c_{4s} = c_{41}; \quad c_{4k} = c_{42};$

$\delta'_1 = c_{12} - c_{11} = 5 - 2 = 3; \quad \delta'_2 = 3 - 1 = 2;$

$\delta'_3 = 6 - 4 = 2; \quad \delta'_4 = 50 - 50 = 0;$

$c_{s1} = \min\{c_{11}; c_{21}; c_{31}; c_{41}\} = \min\{2;3;9;50\} = 2; \quad c_{s1} = c_{11};$

$c_{k1} = \min\{c_{21}; c_{31}; c_{41}\} = \min\{3;9;50\} = 3; \quad c_{k1} = c_{21};$

$c_{s2} = c_{12}; \quad c_{k2} = c_{22}; \quad c_{s3} = c_{23}; \quad c_{k3} = c_{33};$

$\delta_1'' = c_{21} - c_{11} = 3 - 2 = 1;$ $\quad \delta_2'' = 6 - 5 = 1;$ $\quad \delta_3'' = 4 - 1 = 3;$

$\delta = \max \{(3;2;2;0);(1;1;3)\} = 3;$ $\quad$ let $\delta = \delta_1';$

$c_{1s} = \min \{c_{11}; c_{12}; c_{13}\} = \min \{2;5;7\} = 2;$ $\quad c_{1s} = c_{11};$

$\varepsilon = \min \{z_1; w_1\} = \min \{10;5\} = 5;$

$x_{11} = 5;$ $\quad X = \begin{pmatrix} 5 & 0 & 0 \\ 0 & 0 & 0 \\ 0 & 0 & 0 \\ 0 & 0 & 0 \end{pmatrix}$

$z_1 = 10 - 5 = 5;$ $\quad z = (5;4;6;5);$ $\quad w_1 = 5 - 5 = 0;$ $\quad w = (0;12;8);$

$D = \{c_{12}; c_{13}; c_{22}; c_{23}; c_{32}; c_{33}; c_{42}; c_{43}\};$

$c_{1s} = \min \{c_{12}; c_{13}\} = \min \{5;7\} = 5;$ $\quad c_{1s} = c_{12};$

$c_{1k} = \min \{c_{13}\} = c_{13};$ $\quad c_{2s} = c_{23};$ $\quad c_{2k} = c_{22};$

$c_{3s} = c_{33};$ $\quad c_{3k} = c_{32};$ $\quad c_{4s} = c_{42};$ $\quad c_{4k} = c_{43};$

$\delta_1' = c_{13} - c_{12} = 7 - 5 = 2;$ $\quad \delta_2' = 6 - 1 = 5;$

$\delta_3' = 6 - 4 = 2;$ $\quad\quad\quad\quad\quad\quad \delta_4' = 50 - 50 = 0;$

$c_{s1} = 0;$ $\quad c_{k1} = 0;$

$c_{s2} = \min \{c_{12}; c_{22}; c_{32}; c_{42}\} = \min \{5;6;6;50\} = 5;$ $\quad c_{s2} = c_{12};$

$c_{k2} = \min \{c_{22}; c_{32}; c_{42}\} = \min \{6;6;50\} = 6;$ $\quad c_{k2} = c_{22};$

$c_{s3} = c_{23};$ $\quad c_{k3} = c_{33};$ $\quad \delta_1'' = 0 - 0 = 0;$

$\delta_2'' = 6 - 5 = 1;$ $\quad\quad\quad\quad \delta_3'' = 4 - 1 = 3;$

$\delta = \max \{(2;5;2;0);(0;1;3)\} = 5;$ $\quad \delta = \delta_2';$

$c_{2s} = \min \{c_{22}; c_{23}\} = \min \{6;1\} = 1;$ $\quad c_{2s} = c_{23};$

$\varepsilon = \min \{z_2; w_3\} = \min \{4;8\} = 4;$

$x_{23} = 4;$ $\quad X = \begin{pmatrix} 5 & 0 & 0 \\ 0 & 0 & 4 \\ 0 & 0 & 0 \\ 0 & 0 & 0 \end{pmatrix}$ $\quad \begin{array}{l} z_2 = 4 - 4 = 0; \quad z = (5;0;6;5); \\ w_3 = 8 - 4 = 4; \quad w = (0;12;4); \end{array}$

$D = \{c_{12}; c_{13}; c_{32}; c_{33}; c_{42}; c_{43}\}$;

$c_{1s} = \min \{c_{12}; c_{13}\} = \min \{5;7\} = 5; \quad c_{1s} = c_{12};$

$c_{1k} = \min \{c_{13}\} = c_{13}; \quad c_{2s} = 0; \quad c_{2k} = 0;$

$c_{3s} = c_{33}; \quad c_{3k} = c_{32}; \quad c_{4s} = c_{42}; \quad c_{4k} = c_{43};$

$\delta'_1 = c_{13} - c_{12} = 7 - 5 = 2; \quad \delta'_2 = 0 - 0 = 0;$

$\delta'_3 = 6 - 4 = 2; \quad\quad \delta'_4 = 50 - 50 = 0;$

$c_{s1} = 0; \quad c_{k1} = 0;$

$c_{s2} = \min \{c_{12}; c_{32}; c_{42}\} = \min \{5;6;50\} = 5; \quad c_{s2} = c_{12};$

$c_{k2} = \min \{c_{32}; c_{42}\} = \min \{6;50\} = 6; \quad c_{k2} = c_{32};$

$c_{s3} = c_{33}; \quad c_{k3} = c_{13}; \quad \delta''_1 = 0 - 0 = 0;$

$\delta''_2 = 6 - 5 = 1; \quad\quad \delta''_3 = 7 - 4 = 3;$

$\delta = \max \{(2;0;2;0);(0;1;3)\} = 3; \quad \delta = \delta''_3 ;$

$c_{r3} = \min \{c_{13}; c_{33}; c_{43}\} = \min \{7;4;50\} = 4; \quad c_{r3} = c_{33}; \quad \varepsilon = \min \{z_3; w_3\}$
$\quad\quad\quad\quad\quad\quad\quad\quad\quad\quad\quad\quad\quad\quad\quad\quad\quad\quad\quad\quad = \min \{6;4\} = 4 ;$

$x_{33} = 4; \quad X = \begin{pmatrix} 5 & 0 & 0 \\ 0 & 0 & 4 \\ 0 & 0 & 4 \\ 0 & 0 & 0 \end{pmatrix}$

$z_3 = 6 - 4 = 2; \quad z = (5;0;2;5); \quad w_3 = 4 - 4 = 0; \quad w = (0;12;0);$

$D = \{c_{12}; c_{32}; c_{42}\}; \quad c_{1s} = \min \{c_{12}\} = c_{12}; \quad c_{1k} = 0;$

$c_{2s} = 0; \quad c_{2k} = 0; \quad c_{3s} = c_{32}; \quad c_{3k} = 0; \quad c_{4s} = c_{42}; \quad c_{4k} = 0;$

$\delta'_1 = |0 - 5| = 5; \quad \delta'_2 = 0 - 0 = 0; \quad \delta'_3 = |0 - 6| = 6; \quad \delta'_4 = |0 - 50| = 50;$

$c_{s1} = 0; \quad c_{k1} = 0;$

$c_{s2} = \min \{c_{12}; c_{32}; c_{42}\} = \min \{5;6;50\} = 5; \quad c_{s2} = c_{12};$

$c_{k2} = \min \{c_{32}; c_{42}\} = \min \{6;50\} = 6; \quad c_{k2} = c_{32};$

$c_{s3} = 0; \quad c_{k3} = 0;$

$\delta''_1 = 0 - 0 = 0; \quad \delta''_2 = 6 - 5 = 1; \quad \delta''_3 = 0 - 0 = 0;$
$\delta = \max \{(5;0;6;50); (0;1;0)\} = 50; \quad \delta = \delta'_4 ;$

70   Linear Programming

$c_{4s} = \min \{c_{42}\} = c_{42}$; $\varepsilon = \min \{z_4;w_2\} = \min \{5;12\} = 5$;

$x_{42} = 5$; $X = \begin{pmatrix} 5 & 0 & 0 \\ 0 & 0 & 4 \\ 0 & 0 & 4 \\ 0 & 5 & 0 \end{pmatrix}$

$z_4 = 5 - 5 = 0$; $z = (5;0;2;0)$; $w_2 = 12 - 5 = 7$; $w = (0;7;0)$;

$D = \{c_{12};c_{32}\}$; $c_{1s} = \min \{c_{12}\} = c_{12}$; $c_{1k} = 0$;

$c_{2s} = 0$; $c_{2k} = 0$; $c_{3s} = c_{32}$; $c_{3k} = 0$; $c_{4s} = 0$; $c_{4k} = 0$;

$\delta_1' = |0 - 5| = 5$; $\delta_2' = 0 - 0 = 0$; $\delta_3' = |0 - 6| = 6$; $\delta_4' = 0 - 0 = 0$;

$c_{s1} = 0$; $c_{k1} = 0$;

$c_{s2} = \min \{c_{12};c_{32}\} = \min \{5;6\} = 5$; $c_{s2} = c_{12}$;

$c_{k2} = \min \{c_{32}\} = c_{32}$; $c_{s3} = 0$; $c_{k3} = 0$;

$\delta_1'' = 0 - 0 = 0$; $\delta_2'' = 6 - 5 = 1$; $\delta_3'' = 0 - 0 = 0$;

$\delta = \max \{(5;0;6;0);(0;1;0)\} = 6$; $\delta = \delta_3'$;

$c_{3s} = \min \{c_{32}\} = c_{32}$; $\varepsilon = \min \{z_3;w_2\} = \min \{2;7\} = 2$;

$x_{32} = 2$; $X = \begin{pmatrix} 5 & 0 & 0 \\ 0 & 0 & 4 \\ 0 & 2 & 4 \\ 0 & 5 & 0 \end{pmatrix}$

$z_3 = 2 - 2 = 0$; $z = (5;0;0;0)$; $w_2 = 7 - 2 = 5$; $w = (0;5;0)$;

$D = \{c_{12}\}$; x_{12} is the remaining free element, so :

$\varepsilon = \min \{z_1;w_2\} = \min \{5;5\} = 5$;

$x_{12} = 5$; $X = \begin{pmatrix} 5 & 5 & 0 \\ 0 & 0 & 4 \\ 0 & 2 & 4 \\ 0 & 5 & 0 \end{pmatrix}$

$z_1 = 5 - 5 = 0$; $z = (0;0;0;0)$; $w_2 = 5 - 5 = 0$; $w = (0;0;0)$;

$D = \emptyset$;Stop, $X = \begin{pmatrix} 5 & 5 & 0 \\ 0 & 0 & 4 \\ 0 & 2 & 4 \\ 0 & 5 & 0 \end{pmatrix}$ is a feasible transportation plan .

1.2.1.7 The Frequency Method

Hypotheses

The Frequency Method generates from the cost matrix C a more realistic pseudo-cost matrix $\tilde{C}$. In order to formulate a feasible transportation plan, any of the above described methods may be utilized on $\tilde{C}$.

Description

Step 1: Calculate $r_i := \frac{1}{n} \cdot \sum_{j=1}^{n} c_{ij}$ $\forall\, i = 1,\ldots,m$

$s_j := \frac{1}{m} \cdot \sum_{i=1}^{m} c_{ij}$ $\forall\, j = 1,\ldots,n$

$\tilde{c}_{ij} := (r_i + s_j) - c_{ij}$ $\forall\, i,j$.

Step 2: Is $\tilde{c}_{ij} \geq 0$ $\forall\, i,j$?
  If yes: Stop, the matrix $\tilde{C}$ is on hand. Work any other of the above methods using $\tilde{C}$.
  If no: Go to step 3.

Step 3: Let $\tilde{c}_{rs}$ be the minimal element of the matrix $\tilde{C}$. Construct the constant matrix $K: K_{[m \times n]}$, so that $K := (|\tilde{c}_{rs}|)$ and compute $\tilde{C} := \tilde{C} + K$.
  Stop, $\tilde{C}$ is on hand.

Example

Given the following matrix C:

$$C = \begin{pmatrix} 2 & 5 & 7 \\ 3 & 6 & 1 \\ 9 & 6 & 4 \\ 50 & 50 & 50 \end{pmatrix}$$

We find: $r_1 = 14/3$; $r_2 = 10/3$; $r_3 = 19/3$; $r_4 = 50$;

$s_1 = 16$; $s_2 = 67/4$; $s_3 = 31/2$;

72 Linear Programming

$$\tilde{C} = \begin{pmatrix} 56/3 & 197/12 & 79/6 \\ 49/3 & 169/12 & 107/6 \\ 40/3 & 205/12 & 107/6 \\ 16 & 67/4 & 31/2 \end{pmatrix}$$

Since $\tilde{c}_{ij} \geq 0 \; \forall \; i,j$ holds $\rightarrow$ Stop !

1.2.1.8 The Stepping-Stone Method

Hypotheses

Given a feasible solution X to the transportation problem, find an optimal (minimal cost) transportation plan.

Description

Step 1: (test for optimality of the current solution)
Compute the matrix C^* including the values u_i and v_j. C^* looks as follows:

	$v_1 \cdots$	v_j	$\cdots \; v_n$	
C^*:				u_1
				$\vdots$
		c^*_{ij}		u_i
				$\vdots$
				u_m

Consider the current transportation matrix X .
If $x_{ij} > 0$, then $c^*_{ij} := 0$.

Step 2: Set $v_1 := 0$ and compute all other u_i and v_j, so that $u_i + v_j := c_{ij} \; \forall \; c^*_{ij} = 0$. If it is not possible to determine all u_i and v_j in this manner, then set the element(s) $c^*_{ij} := 0$, so that all u_i and v_j can be computed. At the end of this calculation are at least (m + n - 1) elements $c^*_{ij} := 0$.

Step 3: Compute the remaining elements of the matrix C^* as follows:
$c^*_{ij} := u_i + v_j - c_{ij}$.
Result: A matrix C^*, which is called opportunity-cost matrix .

Step 4: In the transportation matrix X mark all elements x_{ij} with "∿" for which $x_{ij} = c^*_{ij} = 0$.
(If this occurs, the problem is called degenerate.)

Step 5: Determine $c^*_{rs} := \max_{i,j} \{c^*_{ij}\}$.

Step 6: Is $c^*_{rs} \leq 0$?
If yes: Stop, the current transportation plan X is optimal. The total costs C_{tot} are defined as

$$C_{tot} := \sum_{i=1}^{m} \sum_{j=1}^{n} x_{ij} \cdot c_{ij} \quad m.u.$$

Here the costs from/to the dummies are omitted.
If no : Go to step 7.

Step 7: (determination of a new solution)
a) Label the element x_{rs} with "+" .
b) Find an element greater than zero in row r , in whose corresponding column is at least one more element marked with "∿" or "greater than zero" . This element is labelled with "-" .
c) In a so determined column find an element "∿" or "greater than zero", in whose corresponding row is at least one more element "greater than zero" . This element is labelled with "+" . Continue this labelling procedure, until one element in column s is labelled with "-" .

74 Linear Programming

Step 8: Determine the element with minimal absolute value among those elements labelled with "-". Let this element be x_{kl}, then the new tranportation matrix reads as follows:

$$x_{ij} := \begin{cases} x_{ij} + x_{kl}, & \text{if } x_{ij} \text{ is labelled with "+"} \\ x_{ij} - x_{kl}, & \text{if } x_{ij} \text{ is labelled with "-"} \\ x_{ij}, & \text{otherwise} \end{cases}$$

Delete all labels and go to step 1.

Example

Given three stores with the inventory of (10;4;6) units, and three places with the demand of (5;12;8) units. The transportation costs per unit matrix C is:

$$C : \begin{array}{|ccc|} \hline 2 & 5 & 7 \\ 3 & 6 & 1 \\ 9 & 6 & 4 \\ \hline \end{array}$$

The initial feasible solution is the Northwest-Corner Solution:

$$X^{(1)} = \begin{array}{|ccc|} \hline 5 & 5 & 0 \\ 0 & 4^- & 0^+ \\ 0 & 3^+ & 3^- \\ 0 & 0 & 5 \\ \hline \end{array} \qquad C^* = \begin{array}{|ccc|c} \hline 0 & 3 & 1 & \\ 0 & 0 & -4 & 2 \\ 0 & 0 & 3 & 3 \\ -6 & 0 & 0 & 3 \\ \hline -1 & -1 & 0 & 49 \\ \end{array}$$

$$c^*_{rs} = c^*_{23} ;$$

$$X^{(2)} = \begin{array}{|ccc|} \hline 5 & 5 & 0 \\ 0 & 1^- & 3^+ \\ 0 & 6 & 0 \\ 0 & 0^+ & 5^- \\ \hline \end{array} \qquad C^* = \begin{array}{|ccc|c} \hline 0 & 3 & -2 & \\ 0 & 0 & -7 & 2 \\ 0 & 0 & 0 & 3 \\ -6 & 0 & -5 & 3 \\ \hline 2 & 5 & 0 & 52 \\ \end{array}$$

$$c^*_{rs} = c^*_{42} ;$$

$$X^{(3)} = \begin{bmatrix} 5 & 5 & 0 \\ 0 & 0 & 4 \\ 0 & 6^- & 0^+ \\ 0 & 1^+ & 4^- \end{bmatrix} \qquad C^* = \begin{bmatrix} 0 & 3 & 3 & \\ 0 & 0 & -2 & 2 \\ -5 & -5 & 0 & -2 \\ -6 & 0 & 2 & 3 \\ -3 & 0 & 0 & 47 \end{bmatrix}$$

$$c^*_{rs} = c^*_{33} ;$$

$$X^{(4)} = \begin{bmatrix} 5 & 5 & 0 \\ 0 & 0 & 4 \\ 0 & 2 & 4 \\ 0 & 5 & 0 \end{bmatrix} \qquad C^* = \begin{bmatrix} 0 & 3 & 1 & \\ 0 & 0 & -4 & 2 \\ -3 & -3 & 0 & 0 \\ -6 & 0 & 0 & 3 \\ -3 & 0 & -2 & 47 \end{bmatrix}$$

$c^*_{ij} \leq 0 \; \forall \; i,j \to$ in $X^{(4)}$ the optimal assignment of the goods to the transportation routes under the given conditions is obtained.

1.2.2 The Hungarian Method (Kuhn)

<u>Hypotheses</u>
Given n elements z_i, i = 1,...,n (for example workers) and n elements w_j, j = 1,...,n (for example machines). Exactly one w_j has to be assigned to each z_i and exactly one z_i has to be assigned to each w_j. The cost for each assignment is given in the matrix $A : A_{[n \times n]}$, with

a_{ij}: cost of assigning z_i to w_j
$a_{ij} \geq 0 \quad \forall \; i,j$

(If necessary a "constant" matrix, one consisting of the same value in all elements, must be added to A.) The problem is to find a feasible assignment with minimal assignment costs.

Linear Programming

Define x_{ij}, so that

$$x_{ij} := \begin{cases} 1, & \text{if } z_i \text{ is assigned to } w_j \\ 0, & \text{if } z_i \text{ is not assigned to } w_j \end{cases}.$$

Now the problem formally reads:

$$\min \pi = \sum_{i=1}^{n} \sum_{j=1}^{n} a_{ij} \cdot x_{ij}$$

$$\sum_{i=1}^{n} x_{ij} = \sum_{j=1}^{n} x_{ij} = 1$$

$$x_{ij} = 0 \vee 1 .$$

Principle

Beginning with a lower bound of assignment costs which cannot be undercut, attempt to determine a complete feasible assignment. If it is impossible, a new lower bound is established by attacking the minimal additional assignment costs to the old lower bound and trying again. Repeat the procedure as long as necessary.

Description

Step 1: Apply FLOOD's Technique to the matrix A.
Result: A matrix $\tilde{A}$.

Step 2: Define $b_{ij} := (\tilde{a}_{ij} | \tilde{a}_{ij} = 0) \; \forall \; i,j$

$$M := \{b_{ij}\}$$

$$P := Q := S := \emptyset .$$

Step 3: (assignment)
Determine row r, for which
$$|\{b_{rj}\}| := \min_i |\{b_{ij} \in M\}|$$
and select one element b_{rs}.

Section 1.2.2   77

Step 4: Set $Q := Q \cup \{b_{rs}\}$
$S := \{b_{is} \in M \mid i \neq r\} \cup \{b_{rj} \in M \mid j \neq s\}$
$P := P \cup S$
$M := M - S - \{b_{rs}\}$.

Step 5: Is $M = \emptyset$?
If yes: Go to step 6.
If no : Go to step 3.

Step 6: Is $|Q| = n$?
If yes: Go to step 12.
If no : Go to step 2.

Step 7: Define the index sets J and L , so that
$J := \{i \mid b_{ij} \in P\};$ $L := \{l \mid b_{lj} \notin Q\}$.

Step 8: (labelling of rows and columns)
Define the index sets H and L^* , so that
$H := \{h \mid \exists\, b_{lh}$ with $l \in L$ and $h \in J\}$
$L^* := L \cup \{l \mid \exists b_{lh}$ with $h \in H\}$.

Step 9: Is $|L^*| = |L|$?
If yes: Go to step 10.
If no : Set $L := L^*$ and go to step 8.

Step 10: Consider the matrix $\tilde{A} = (\tilde{a}_{ij})$ and define the sets $K^{(1)}$ and $K^{(2)}$, so that
$K^{(1)} := \{\tilde{a}_{ij} \mid i \notin L;\ j \notin H\} \cup \{\tilde{a}_{ij} \mid i \in L;\ j \in H\}$
$K^{(2)} := \{\tilde{a}_{ij} \mid i \notin L;\ j \in H\}$.

Step 11: (matrix alteration)
Determine $\tilde{a}_{rs} := \min \{\tilde{a}_{ij} \mid \tilde{a}_{ij} \notin (K^{(1)} \cup K^{(2)})\}$
and calculate the matrix $\tilde{A}^*$, so that

78 Linear Programming

$$\tilde{a}_{ij}^{*} := \begin{cases} \tilde{a}_{ij}, & \text{if } \tilde{a}_{ij} \in K^{(1)} \\ \tilde{a}_{ij} + \tilde{a}_{rs}, & \text{if } \tilde{a}_{ij} \in K^{(2)} \\ \tilde{a}_{ij} - \tilde{a}_{rs}, & \text{otherwise} \end{cases}$$

Set $\tilde{A} := \tilde{A}^{*}$ and go to step 2.

Step 12: The optimal (minimal cost) assignment is given by the set Q with :

$$x_{ij} := \begin{cases} 1, & \text{if } b_{ij} \in Q \\ 0, & \text{otherwise} \end{cases}$$

The assignment costs are $\pi := \sum_{i=1}^{n} \sum_{j=1}^{n} a_{ij} \cdot x_{ij}$.

Note: If each assignment yields a profit, then a feasible assignment must be found that maximizes profit. In this case transform the matrix A into $\hat{A}$.
Let $a_{kl} := \max_{i,j} \{a_{ij}\}$ and $A_{(kl)} := (a_{kl})$
the appropriate "constant" matrix, then $\hat{A} := A_{(kl)} - A$.
Start the above method as explained but using $\hat{A}$.

Example
Given the matrix A :

A :

	w_1	w_2	w_3	w_4
z_1	2	3	5	8
z_2	1	7	9	6
z_3	9	6	8	6
z_4	2	4	7	4

; $\tilde{A}$: $\begin{pmatrix} 0 & 1 & 1 & 6 \\ 0 & 6 & 6 & 5 \\ 3 & 0 & 0 & 0 \\ 0 & 2 & 3 & 2 \end{pmatrix}$

$M = \{b_{11}; b_{21}; b_{32}; b_{33}; b_{34}; b_{41}\}$; $P = \emptyset$; $Q = \emptyset$;
$1 = \min \{1;1;3;1\} \rightarrow \text{row } r := \text{row } 1$; $b_{rs} = b_{11}$;
$Q = \{b_{11}\}$; $S = \{b_{21}; b_{41}\}$; $P = \{b_{21}; b_{41}\}$;

$M = \{b_{32}; b_{33}; b_{34}\}$; $\to$ $3 = \min \{3\} \to$ row $r := $ row 3; $b_{rs} = b_{32}$;
$Q = \{b_{11}; b_{32}\}$; $S = \{b_{33}; b_{34}\}$; $P = \{b_{21}; b_{33}; b_{34}; b_{41}\}$;
$M \stackrel{!}{=} \emptyset$; $|Q| < n \to$ $J = \{1; 3; 4\}$; $L = \{2; 4\}$;
$H = \{1\}$; $L^* = \{1; 2; 4\}$; $|L^*| \neq |L|$;
$L := L^*$; $H = \{1\}$; $L^* = \{1; 2; 4\}$;
$|L^*| \stackrel{!}{=} |L|$; $K^{(1)} = \{\tilde{a}_{11}; \tilde{a}_{21}; \tilde{a}_{32}; \tilde{a}_{33}; \tilde{a}_{34}; \tilde{a}_{41}\}$;
$\qquad K^{(2)} = \{\tilde{a}_{31}\}$; $\tilde{a}_{rs} = \tilde{a}_{12} = \tilde{a}_{13} = 1$;

$$\tilde{A}^* = A : \begin{pmatrix} 0 & 0 & 0 & 5 \\ 0 & 5 & 5 & 4 \\ 4 & 0 & 0 & 0 \\ 0 & 1 & 2 & 1 \end{pmatrix}$$

$M = \{b_{11}; b_{12}; b_{13}; b_{21}; b_{32}; b_{33}; b_{34}; b_{41}\}$; $Q = \emptyset$; $P = \emptyset$; $S = \emptyset$;
$1 = \min \{3; 1; 3; 1\} \to$ row $r := $ row 2;
$b_{rs} = b_{21}$; $Q = \{b_{21}\}$; $S = \{b_{11}; b_{41}\}$; $P = \{b_{11}; b_{41}\}$;
$M = \{b_{12}; b_{13}; b_{32}; b_{33}; b_{34}\}$; $2 = \min \{2; 3\} \to$ row $r := $ row 1;
$b_{rs} = b_{12}$; $Q = \{b_{12}; b_{21}\}$; $S = \{b_{13}; b_{32}\}$; $P = \{b_{11}; b_{13}; b_{32}; b_{41}\}$;
$M = \{b_{33}; b_{34}\}$; $2 = \min \{2\} \to$ row $r := $ row 3;
$b_{rs} = b_{33}$; $Q = \{b_{12}; b_{21}; b_{33}\}$; $S = \{b_{34}\}$; $P = \{b_{11}; b_{13}; b_{32}; b_{34};$
$M \stackrel{!}{=} \emptyset$; $|Q| < n \to$ $\hfill b_{41}\}$;
$J = \{1; 2; 3; 4\}$; $L = \{4\}$; $H = \{1\}$;
$L^* = \{2; 4\}$; $|L^*| \neq |L| \to L := L^*$;
$H = \{1\}$; $L^* = \{2; 4\}$; $|L^*| \stackrel{!}{=} |L|$;
$K^{(1)} = \{\tilde{a}_{12}; \tilde{a}_{13}; \tilde{a}_{14}; \tilde{a}_{21}; \tilde{a}_{32}; \tilde{a}_{33}; \tilde{a}_{34}; \tilde{a}_{41}\}$; $K^{(2)} = \{\tilde{a}_{11}; \tilde{a}_{31}\}$;
$\tilde{a}_{rs} = \tilde{a}_{42} = \tilde{a}_{44} = 1$;

80 *Linear Programming*

$$\tilde{A}^* = \tilde{A} : \begin{pmatrix} 1 & 0 & 0 & 5 \\ 0 & 4 & 4 & 3 \\ 5 & 0 & 0 & 0 \\ 0 & 0 & 1 & 0 \end{pmatrix}$$

$M = \{b_{12}; b_{13}; b_{21}; b_{32}; b_{33}; b_{34}; b_{41}; b_{42}; b_{44}\}$; $Q = \emptyset$; $S = \emptyset$; $P = \emptyset$;
$1 = \min \{2; 1; 3; 3\} \to$ row $r: =$ row 2;
$b_{rs} = b_{21}$; $Q = \{b_{21}\}$; $S = \{b_{41}\}$; $P = \{b_{41}\}$;
$M = \{b_{12}; b_{13}; b_{32}; b_{33}; b_{34}; b_{42}; b_{44}\}$;
$2 = \min \{2; 3; 2\} \to$ row $r: =$ row 1;
$b_{rs} = b_{13}$; $Q = \{b_{13}; b_{21}\}$; $S = \{b_{12}; b_{33}\}$; $P = \{b_{12}; b_{33}; b_{41}\}$;
$M = \{b_{32}; b_{34}; b_{42}; b_{44}\}$; $2 = \min \{2; 2\} \to$ row $r: =$ row 3;
$b_{rs} = b_{34}$; $Q = \{b_{13}; b_{21}; b_{34}\}$; $S = \{b_{32}; b_{44}\}$;
$P = \{b_{12}; b_{32}; b_{33}; b_{41}; b_{44}\}$; $M = \{b_{42}\}$;
$1 = \min \{1\} \to$ row $r: =$ row 4;
$b_{rs} = b_{42}$; $Q = \{b_{13}; b_{21}; b_{34}; b_{42}\}$; $S = \emptyset$;
$P = \{b_{12}; b_{32}; b_{33}; b_{41}; b_{44}\}$; $M \stackrel{!}{=} \emptyset$; $|Q| \stackrel{!}{=} n \to$

Stop, an optimal solution has been found, the assignment is:

$$x_{13} = 1 \;\hat{=}\; z_1 \leftrightarrow w_3 ;$$
$$x_{21} = 1 \;\hat{=}\; z_2 \leftrightarrow w_1 ;$$
$$x_{34} = 1 \;\hat{=}\; z_3 \leftrightarrow w_4 ;$$
$$x_{42} = 1 \;\hat{=}\; z_4 \leftrightarrow w_2 ;$$

the assignment costs are 16 m.u.

1.2.3 The Decomposition Principle (Dantzig; Wolfe)

<u>Hypotheses</u>
Given a linear programming problem of the following form:

$$P_1: \min \pi = \sum_{k=1}^{r} c^{(k)} \cdot x^{(k)}$$

$$\sum_{k=1}^{r} V^{(k)} \cdot x^{(k)} = b^{(0)}$$

$$A^{(k)} \cdot x^{(k)} = b^{(k)} \quad \forall k$$

$$x^{(k)} \geq \Theta \quad \forall k$$

where $c^{(k)} : c^{(k)}_{[1 \times n_k]}$; $V^{(k)} : V^{(k)}_{[m_0 \times n_k]}$;

$x^{(k)} : x^{(k)}_{[n_k \times 1]}$; $A^{(k)} : A^{(k)}_{[m_k \times n_k]}$;

$b^{(k)} : b^{(k)}_{[m_k \times 1]}$.

The corresponding tableau looks as follows:

$V^{(1)}$	$V^{(2)}$	. . .	$V^{(r)}$	$b^{(0)}$
$A^{(1)}$				$b^{(1)}$
	$A^{(2)}$	Θ		$b^{(2)}$
Θ		.		.
		.	$A^{(r)}$	$b^{(r)}$
$c^{(1)}$	$c^{(2)}$	. . .	$c^{(r)}$	π

<u>Principle</u>
The method starts by solving all subproblems. If there exist feasible solutions for all subproblems, then an iterative process is

82 *Linear Programming*

utilized to examine combinations of basic solutions of the subproblems which fulfill certain "connecting equations".

Description

Step 1: Solve all subproblems

$$c^{(k)} \cdot x^{(k)} \Rightarrow \min$$
$$A^{(k)} \cdot x^{(k)} = b^{(k)}$$
$$x^{(k)} \geq 0 \quad \forall \quad k=1,\ldots,r \;.$$

Step 2: Does an optimal solution exist for each subproblem ?

If yes: Lt $\bar{x}^{(k)}$ be the optimal solution of the k-th subproblem. Go to step 3.

If no : Stop, P_1 has no feasible solution.

Step 3: Select $\nu_k > 1$ out of the set of basic solutions of the k-th subproblem, so that $\sum_{k=1}^{r} \nu_k := r + m_0$.

Let Q be the set of these basic solutions, so that $Q := \{x_{ik}\}$, where x_{ik} is the i-th selected basic solution of the k-th subproblem, for all $i = 1,\ldots,\nu_k$; $k = 1,\ldots,r$. It is possible to determine a set Q, which has not yet been considered ?

If yes: Go to step 4.

If no : Stop, P_1 has no feasible solution.

Step 4: Select m_0 subsets $s'_t := \{x_{ik}\} \subset Q$, $k = 1,\ldots,r$; $t = 1,\ldots,m_0$; with $s'_t \neq s'_{t+1}$, so that

$$\sum_{k=1}^{r} v^{(k)} \cdot x_{ik}^{(t)} \leq b^{(0)} \quad \forall \quad t = 1,\ldots,m_0 \;.$$

Is $\sum_{k=1}^{r} v^{(k)} \cdot x_{ik}^{(t)} \leq b^{(0)} \quad \forall \quad t = 1,\ldots,m_0$?

If yes: Go to step 5.

If no : Go to step 3 and determine another, not yet considered set Q .

Step 5: Select m_0 subsets $s_t'' := \{x_{ik}\} \subset Q$, $k = 1,\ldots,r$; $t = 1,\ldots,m_0$, with $s_t'' \neq s_{t+1}''$, so that

$$\sum_{k=1}^{r} v^{(k)} \cdot x_{ik}^{(t)} \geq b_0 \quad \forall\, t = 1,\ldots,m_0.$$

Is $\sum_{k=1}^{r} v^{(k)} \cdot x_{ik}^{(t)} \geq b_0 \quad \forall\, t = 1,\ldots,m_0$?

If yes: Go to step 6.

If no : Go to step 3 and determine another, not yet considered set Q.

Step 6: Determine $\quad v_{ik} := v^{(k)} \cdot x_{ik} \quad \forall\, k = 1,\ldots,r;$
$\qquad\qquad\qquad\quad c_{ik} := c^{(k)} \cdot x_{ik} \quad\;\, i = 1,\ldots,\nu_k.$

Step 7: Set up the problem P_2 considering the following sequence of constraints:

$$P_2: \quad \sum_{k=1}^{r} \sum_{i=1}^{\nu_k} c_{ik} \cdot w_{ik} \Rightarrow \min$$

$$\sum_{k=1}^{r} \sum_{i=1}^{\nu_k} v_{ik} \cdot w_{ik} = b^{(0)}$$

$$\sum_{i=1}^{\nu_k} w_{ik} = 1 \quad \forall\, k = 1,\ldots,r$$

$$w_{ik} \geq 0 \quad \forall\, i,k$$

and solve this problem by the Two-Phase Method.

Step 8: Let the current optimal solution of P_2 be $\bar{w}$, the dual solution $\bar{u} = (u';\tilde{u})$, where $u' : u'_{[1 \times m_0]}$ and $\tilde{u} : \tilde{u}_{[1 \times r]}$.

Compute $[c^{(k)} + u' \cdot v^{(k)}] \cdot x^{(k)} \quad \forall\, k = 1,\ldots,r$.

Step 9: Set the index $p := 1$.

84 Linear Programming

Step 10: Set up the problems $P_3^{(k)}$:

$$P_3^{(k)}: \quad [c^{(k)} + u' \cdot V^{(k)}] \cdot x^{(k)} \Rightarrow \min$$
$$A^{(k)} \cdot x^{(k)} = b^{(k)}$$
$$x^{(k)} \geq \Theta$$
$$\forall \ k = 1,\ldots,r \ .$$

Step 11: Solve the problem $P_3^{(p)}$.
Result: An optimal solution $\tilde{x}^{(p)}$.

Step 12: Is $d^{(p)} := ([c^{(p)} + u' \cdot V^{(p)}] \cdot \tilde{x}^{(p)} + \tilde{u}_p) < 0$?
If yes: Go to step 15.
If no : Go to step 13.

Step 13: Is $p < r$?
If yes: Set $p := p + 1$ and go to step 11.
If no : Go to step 14.

Step 14: Compute the optimal solution of P_1 :
$\bar{x} := (\bar{x}^{(1)};\ldots;\bar{x}^{(r)})$, where
$\bar{x}^{(k)} := \sum_i x_{ik} \cdot \bar{w}_{ik}$; then terminate.

Step 15: Until now assume that q basic solutions have been used for the p-th subproblem. Define $x_{q+1,p} := \tilde{x}^{(p)}$ and determine the vector $l_{q+1,p} := V^{(p)} \cdot x_{q+1,p}$,
where $l_{q+1,p}: l_{q+1,p\,[m_o \times 1]}$.

Let $e^{(p)}$ be the unit vector of dimension $[r \times 1]$ with "1" on p-th position and B^{-1} the coefficient matrix of the artificial variables of the current optimal tableau of the problem P_2 .
Determine the column vector $B^{-1} \cdot (l_{q+1,p}; e^{(p)})$ and add this vector with the corresponding variable $w_{q+1,p}$ and

with the corresponding coefficient of the objective function $d^{(p)}$ to problem P_2. Determine the pivot-element in this new column and do one iteration according to the Primal Simplex-Algorithm. Eliminate from the problem that variable x_j which leaves the basis, along with the corresponding column vector. Go to step 8 .

Example
Given the following problem P_1:

$P_1:$ min $\pi = 4 \cdot x_1 + 3 \cdot x_2 + 12 \cdot x_3 + 6 \cdot x_4 + 10 \cdot x_5$

$2 \cdot x_1 + x_2 + 4 \cdot x_3 + x_4 + 2 \cdot x_5 = 13$
$2 \cdot x_1 + 3 \cdot x_2 + 6 \cdot x_3 = 12$
$4 \cdot x_1 + 10 \cdot x_2 + 14 \cdot x_3 = 28$
$2 \cdot x_4 + x_5 = 4$
$x_j \geq 0 \ \forall \ j = 1,\ldots,5 \ .$

We have:

$c^{(1)} = (4;3;12); \quad c^{(2)} = (6;10); \quad v^{(1)} = (2;1;4); \quad v^{(2)} = (1;2);$

$A^{(1)} = \begin{pmatrix} 2 & 3 & 6 \\ 4 & 10 & 14 \end{pmatrix}; \quad A^{(2)} = (2;1);$

$b^{(0)} = 13; \quad b^{(1)} = \begin{pmatrix} 12 \\ 28 \end{pmatrix}; \quad b^{(2)} = 4; \quad m_0 = 1;$

Solution of the first subproblem:

x_1	x_2	x_3	1
2	3	6	12
4	10	14	28
4	3	12	0

x_1	x_2	x_3	1
1/3	1/2	1	2
-2/3	3	0	0
0	-3	0	-24

86 *Linear Programming*

x_1	x_2	x_3	1
0	2	1	2
1	-9/2	0	0
0	-3	0	-24

x_1	x_2	x_3	1
0	1	1/2	1
1	0	9/4	9/2
0	0	3/2	-21

Optimal! Feasible basic solutions are:

$x_{11} = (0;0;2)$; $x_{21} = (9/2;1;0)$;

Solution of the second subproblem:

x_4	x_5	1
2	1	4
6	10	0

x_4	x_5	1
2	1	4
-11	0	40

x_4	x_5	1
1	1/2	2
0	7	-12

Optimal! Feasible basic solutions are:

$x_{12} = (0;4)$; $x_{22} = (2;0)$;

Let $r + m_0 = 3$ be the necessary basic solutions:

$Q = \{x_{11}; x_{12}; x_{22}\} = \{(0;0;2);(0;4);(2;0)\}$.

Now we select: $S_1' = \{x_{11}; x_{22}\}$; we have:

$\sum_{k=1}^{2} v^{(k)} \cdot x_{ik}^{(1)} = (2;1;4) \cdot \begin{pmatrix} 0 \\ 0 \\ 2 \end{pmatrix} + (1;2) \cdot \begin{pmatrix} 2 \\ 0 \end{pmatrix} = 10 \leq 13$.

Now we select: $S_1'' = \{x_{11}; x_{12}\}$; we have:

$\sum_{k=1}^{2} v^{(k)} \cdot x_{ik}^{(1)} = (2;1;4) \cdot \begin{pmatrix} 0 \\ 0 \\ 2 \end{pmatrix} + (1;2) \cdot \begin{pmatrix} 0 \\ 4 \end{pmatrix} = 16 \geq 13$;

$v_{11} = (2;1;4) \cdot \begin{pmatrix} 0 \\ 0 \\ 2 \end{pmatrix} = 8$; $c_{11} = (4;3;12) \cdot \begin{pmatrix} 0 \\ 0 \\ 2 \end{pmatrix} = 24$;

$v_{12} = (1;2) \cdot \begin{pmatrix} 0 \\ 4 \end{pmatrix} = 8$; $c_{12} = (6;10) \cdot \begin{pmatrix} 0 \\ 4 \end{pmatrix} = 40$;

$v_{22} = (1;2) \cdot \begin{pmatrix} 2 \\ 0 \end{pmatrix} = 2$; $c_{22} = (6;10) \cdot \begin{pmatrix} 2 \\ 0 \end{pmatrix} = 12$;

Problem P_2 states as:

P_2:

w_{11}	w_{12}	w_{22}	z_1	z_2	z_3	1
8	8	2	1	0	0	13
1	0	0	0	1	0	1
0	1	1	0	0	1	1
-9	-9	-3	0	0	0	-15
24	40	12	0	0	0	0

After three iterations we have the follwoing optimal tableau:

w_{11}	w_{12}	w_{22}	z_1	z_2	z_3	1
0	1	0	1/6	-4/3	-1/3	1/2
1	0	0	0	1	0	1
0	0	1	-1/6	4/3	4/3	1/2
0	0	0	1	1	1	0
0	0	0	-14/3	40/3	-8/3	-50

$\bar{w} = (\bar{w}_{11}; \bar{w}_{12}; \bar{w}_{22}) =$
$= (1; 1/2; 1/2) ;$
$\bar{u} = (u'; \tilde{u}) =$
$= (-14/3; (40/3; -8/3));$

$((4; 3; 12) + (-14/3) \cdot (2; 1; 4)) \cdot x^{(1)} =$
$= ((4; 3; 12) - (28/3; 14/3; 56/3)) \cdot x^{(1)} = (-16/3; -5/3; -20/3) \cdot \begin{pmatrix} x_1 \\ x_2 \\ x_3 \end{pmatrix}$

$P_3^{(1)}$:

x_1	x_2	x_3	1
2	3	6	12
4	10	14	28
-16/3	-5/3	-20/3	0

After two iterations we have the following optimal tableau:

x_1	x_2	x_3	1
1	0	9/4	9/2
0	1	1/2	1
0	0	37/6	77/3

$\tilde{x}^{(1)} = (9/2; 1; 0) ;$

88 Linear Programming

$d^{(1)} = (-16/3 \cdot 9/2 - 5/3 \cdot 1 - 20/3 \cdot 0) + 40/3 = (-37/3) < 0$.

Be x_{21}: = $\tilde{x}^{(1)}$ = (9/2;1;0); l_{21}: = $v^{(1)} \cdot x_{21}$ = $(2;1;4) \cdot \begin{pmatrix} 9/2 \\ 1 \\ 0 \end{pmatrix}$ = 10;

$e_1 = \begin{pmatrix} 1 \\ 0 \end{pmatrix}$;

$B^{-1} \cdot \begin{Bmatrix} l_{21} \\ e_1 \end{Bmatrix} = \begin{pmatrix} 1/6 & -4/3 & -1/3 \\ 0 & 1 & 0 \\ -1/6 & 4/3 & 4/3 \end{pmatrix} \cdot \begin{pmatrix} 10 \\ 1 \\ 0 \end{pmatrix} = \begin{pmatrix} 1/3 \\ 1 \\ -1/3 \end{pmatrix}$.

The expanded optimal tableau of P_2 is :

w_{11}	w_{21}	w_{12}	w_{22}	z_1	z_2	z_3	1
0	1/3	1	0	1/6	-4/3	-1/3	1/2
1	1	0	0	0	1	0	1
0	-1/3	0	1	-1/6	4/3	4/3	1/2
0	-37/3	0	0	-14/3	40/3	-8/3	-50

w_{11}	w_{21}	w_{12}	w_{22}	z_1	z_2	z_3	1
-1/3	0	1	0	1/6	-5/3	-1/3	1/6
1	1	0	0	0	1	0	1
1/3	0	0	1	-1/6	5/3	4/3	5/6
37/3	0	0	0	-14/3	77/3	-8/3	-113/3

$\bar{w} = (\bar{w}_{21}; \bar{w}_{12}; \bar{w}_{22}) =$
$= (1; 1/6; 5/6)$;
$\bar{u} = (-14/3; (77/3; -8/3))$;

$((4;3;12) + (-14/3) \cdot (2;1;4)) \cdot x^{(1)} = (-16/3; -5/3; -20/3) \cdot \begin{pmatrix} x_1 \\ x_2 \\ x_3 \end{pmatrix}$

In relation to the last iteration $P_3^{(1)}$ is unchanged, therefore:
$\tilde{x}^{(1)} = (9/2; 1; 0)$;

$d^{(1)} = (-16/3 \cdot 9/2 - 5/3 \cdot 1 - 20/3 \cdot 0) + 77/3 = 0$.

$((6;10) + (-14/3) \cdot (1;1)) \cdot x^{(2)} = ((6;10) - (14/3;14/3)) \cdot x^{(2)} =$

$= (4/3;16/3) \cdot \begin{pmatrix} x_4 \\ x_5 \end{pmatrix}$.

$P_3^{(2)}$:

	x_4	x_5	1
	2	1	4
	4/3	16/3	0

	x_4	x_5	1
	1	1/2	2
	0	14/3	8/3

$\tilde{x}^{(2)} = (2;0)$;

$d^{(2)} = (4/3 \cdot 2 + 16/3 \cdot 0) + (-8/3) = 0$.

Computation of the optimal solution of P_1 :

$\bar{x}^{(1)} := \bar{w}_{21} \cdot x_{21} = 1 \cdot (9/2;1;0) = (9/2;1;0)$;

$\bar{x}^{(2)} := \bar{w}_{12} \cdot x_{12} + \bar{w}_{22} \cdot x_{22} = 1/6 \cdot (0;4) + 5/6 \cdot (2;0) = (5/3;2/3)$;

$\bar{x} = (\bar{x}^{(1)};\bar{x}^{(2)}) = (9/2;1;0;5/3;2/3)$;

control:

$(2 \cdot 9/2 + 1 \cdot 1 + 4 \cdot 0 + 1 \cdot 5/3 + 2 \cdot 2/3) = 13$;

$(2 \cdot 9/2 + 3 \cdot 1 + 6 \cdot 0) = 12$;

$(4 \cdot 9/2 + 10 \cdot 1 + 14 \cdot 0) = 28$;

$(2 \cdot 5/3 + 1 \cdot 2/3) = 4$;

$\pi = (4 \cdot 9/2 + 3 \cdot 1 + 12 \cdot 0 + 6 \cdot 5/3 + 10 \cdot 2/3) = 113/3$.

1.2.4 FLOOD's Technique

<u>Hypotheses</u>
Given a matrix $A : A_{[m \times n]}$, find a matrix $\hat{A} : \hat{A}_{[m \times n]}$, which contains at least one element equal to zero in each column and in each row.

<u>Description</u>
Step 1: Determine the vector of the row-minima $a^{(r)}$, so that $a^{(r)T} := (a_1^{(r)};...;a_m^{(r)})$, where $a_i^{(r)} := \min_j \{a_{ij}\}$.

90 Linear Programming

Step 2: Determine the matrix $A^{(r)} : A^{(r)}_{[m \times n]}$, so that

$$A^{(r)} := (a^{(r)}_{ij}), \text{ where } a^{(r)}_{ij} := a^{(r)}_i \;\; \forall j = 1,\ldots,n; \;\; \forall i = 1,\ldots,m,$$

and compute the matrix $\hat{A}$, so that $\hat{A} := A - A^{(r)}$.

Step 3: Determine the vector of the column-minima $a^{(c)}$, so that

$$a^{(c)} := (a^{(c)}_1;\ldots;a^{(c)}_n), \text{ where } a^{(c)}_j := \min_i \{\hat{a}_{ij}\} .$$

Step 4: Determine the matrix $A^{(c)} : A^{(c)}_{[m \times n]}$, so that

$$A^{(c)} := (a^{(c)}_{ij}), \text{ where } a^{(c)}_{ij} := a^{(c)}_j \;\; \forall i = 1,\ldots,m; \;\; \forall j = 1,\ldots,n,$$

and compute the matrix $\tilde{A}$, so that $\tilde{A}: \hat{A} - A^{(c)}$.
The reductionsconstant r_0 has the following value:

$$r_0 := \sum_{i=1}^{m} a^{(r)}_i + \sum_{j=1}^{n} a^{(c)}_j .$$

Example

$$A : \begin{pmatrix} 5 & 2 & 4 \\ 2 & 3 & 1 \\ 4 & 2 & 2 \\ 2 & 1 & 3 \end{pmatrix} ; \;\; a^{(r)} : \begin{pmatrix} 2 \\ 1 \\ 2 \\ 1 \end{pmatrix} ; \;\; A^{(r)} : \begin{pmatrix} 2 & 2 & 2 \\ 1 & 1 & 1 \\ 2 & 2 & 2 \\ 1 & 1 & 1 \end{pmatrix} ;$$

$$\hat{A} := A - A^{(r)} : \begin{pmatrix} 3 & 0 & 2 \\ 1 & 2 & 0 \\ 2 & 0 & 0 \\ 1 & 0 & 2 \end{pmatrix} ; \;\; a^{(c)} : (1;0;0) ;$$

$$A^{(c)} : \begin{pmatrix} 1 & 0 & 0 \\ 1 & 0 & 0 \\ 1 & 0 & 0 \\ 1 & 0 & 0 \end{pmatrix} ; \;\; \tilde{A} := \hat{A} - A^{(c)} : \begin{pmatrix} 2 & 0 & 2 \\ 0 & 2 & 0 \\ 1 & 0 & 0 \\ 0 & 0 & 2 \end{pmatrix} ;$$

$r_0 = ((2 + 1 + 2 + 1) + (1 + 0 + 0)) = 7$.

1.3 Theorems and Rules

1.3.1 The Dual Problem

In some cases it is convenient not to solve the given primal problem, but the corresponding dual problem. The dual problem can be determined from the primal problem as follows:

Description

Step 1: Transform the primal problem into the following form:

$$P : \max \pi = \sum_{j=1}^{n} c_j \cdot x_j$$

$$\sum_{j=1}^{n} a_{ij} \cdot x_j \leq b_i \quad \forall\, i = 1,\ldots,k$$

$$\sum_{j=1}^{n} a_{ij} \cdot x_j = b_i \quad \forall\, i = k+1,\ldots,m, \text{ where } k \leq m$$

$$x_j \geq 0 \quad \forall\, j = 1,\ldots,l$$

$$x_j \text{ unbounded } \forall\, j = l+1,\ldots,n, \text{ where } l \leq n$$

or in matrix-notation:

$$P : \max \pi = c \cdot x$$
$$A' \cdot x \leq b'$$
$$A'' \cdot x = b''$$
$$x \gtrless \theta \;.$$

Step 2: Construct the dual problem P_D, so that

$$P_D: \min \pi = \sum_{i=1}^{m} u_i \cdot b_i$$

$$\sum_{i=1}^{m} a_{ij} \cdot u_i \geq c_j \quad \forall\, j = 1,\ldots,l$$

$$\sum_{i=1}^{m} a_{ij} \cdot u_i = c_j \quad \forall j = 1+1,\ldots,n, \text{ where } 1 \leq n$$

$u_i \geq 0 \quad \forall i = 1,\ldots,k$

u_i unbounded $\quad \forall i = k+1,\ldots,m,$ where $k \leq m$

or in matrix-notation:

P_D: min $\pi = u \cdot b$
$\quad\quad u \cdot A' \geq c'$
$\quad\quad u \cdot A'' = c''$
$\quad\quad u \gtreqless \Theta$, i.e.

a) A dual constrained variable $u_i \geq 0$ is assigned to each type I inequality in the primal problem.

b) A dual unbounded variable u_i is assigned to each equality in the primal problem.

c) A dual type II inequality is assigned to each constrained primal variable $x_j \geq 0$.

d) A dual equality is assigned to each unbounded primal variable x_j.

e) The primal objective function max $\pi = c \cdot x$ is replaced by the dual objective function min $\pi = u \cdot b$.

Note: A type I inequality is of type "≤", a type II inequality is of type "≥".

Example

Given the following problem $\tilde{P}$:

$\tilde{P}$: min $\pi = 3 \cdot x_1 - 4 \cdot x_2$
$\quad$ (1) $\quad 2 \cdot x_1 + x_2 \geq 6$
$\quad$ (2) $\quad x_1 - 2 \cdot x_2 \leq 10 \quad \Rightarrow$
$\quad$ (3) $\quad 5 \cdot x_1 + 7 \cdot x_2 = 13$
$\quad\quad\quad\quad x_1 \geq 0$
$\quad\quad x_2$ unbounded

P: max $\pi = (-3 \cdot x_1 + 4 \cdot x_2)$
$\quad$ (1) $\quad -2 \cdot x_1 - x_2 \leq -6$
$\quad$ (2) $\quad x_1 - 2 \cdot x_2 \leq 10$
$\quad$ (3) $\quad 5 \cdot x_1 + 7 \cdot x_2 = 13$
$\quad\quad\quad\quad x_1 \geq 0$
$\quad\quad x_2$ unbounded

P_D : min $\pi = (-6 \cdot u_1 + 10 \cdot u_2 + 13 \cdot u_3)$
  (1) $- 2 \cdot u_1 + u_2 + 5 \cdot u_3 \geq -3$
  (2) $- u_1 - 2 \cdot u_2 + 7 \cdot u_3 = 4$
  $u_1, u_2 \geq 0$
  u_3 unbounded

$\tilde{P}_D$: max $\pi = 6 \cdot u_1 - 10 \cdot u_2 - 13 \cdot u_3$
  (1) $2 \cdot u_1 - u_2 - 5 \cdot u_3 \leq 3$
  (2) $-u_1 - 2 \cdot u_2 + 7 \cdot u_3 = 4$
  $u_1, u_2 \geq 0$
  u_3 unbounded

1.3.2 Theorems of Duality

The following relationships exist between the primal problem and the dual problem that was developed in 1.3.1 .

a) For each linear programming problem P there exists a corresponding dual problem P_D .
b) $(P_D)_D = P$
c) $\exists$ optimal solution of P $\Leftrightarrow$ $\exists$ optimal solution of P_D .
d) P has an unbounded solution $\Leftrightarrow$ $\nexists$ feasible solution of P_D .
e) P_D has an unbounded solution $\Leftrightarrow$ $\nexists$ feasible solution of P .
f) For the initial and the optimal tableau the following conditions hold:
  1. primal values of P $\equiv$ dual values of P_D
  2. dual values of P $\equiv$ primal values of P_D .

1.3.3 The Lexicographic Selection Rule

Hypotheses

Given a primal feasible simplex-tableau with column s selected as pivot-column, we have

$$\frac{b_{r_1}}{a_{r_1 s}} = \frac{b_{r_2}}{a_{r_2 s}} = \ldots = \frac{b_{r_p}}{a_{r_p s}} = \min_i \left\{ \frac{b_i}{a_{is}} \mid a_{is} > 0 \right\},$$

i.e. there is more than one row equally eligible for selection as the pivot-row.

Description

Step 1: Let a_{r_μ} be the r_μ-th row vector of the above simplex-tableau. Compute the vectors a'_{r_μ}, so that

$$a'_{r_\mu} := \frac{1}{a_{r_\mu s}} \cdot a_{r_\mu} \quad \forall \mu = 1, \ldots, p \;.$$

Step 2: Determine $a'_{r_\nu} := \text{lex min } \{a'_{r_\mu}\}$,

i.e. the lexicographic minimum of all alternative row vectors a'_{r_μ}. The pivot-element is given by $a_{r_\nu s}$.

Note:

a) When more than one column is equally eligible for pivot-column in the dual problem, proceed in the corresponding way as described above.

b) The pivot-row could also be chosen at random.
(Proof of convergence in [1.B.7])

Example

Given the following simplex-tableau:

x_1	x_2	x_3	y_1	y_2	y_3	y_4	1
2	-1	5	1	0	0	0	2
5	0	7/2	0	1	0	0	5
3	3	2	0	0	1	0	3
1	4	1/4	0	0	0	1	1/2
-5	-2	-3	0	0	0	0	0

column 1 is selected as pivot-column:

$$\frac{b_1}{a_{11}} = \frac{b_2}{a_{21}} = \frac{b_3}{a_{31}} = 1$$

$a_1 = (2;-1;5;1;0;0;0;2);$ $a_1' = (1;-1/2;5/2;1/2;0;0;0;1)$

$a_2 = (5;0;7/2;0;1;0;0;5);$ $a_2' = (1;0;7/10;0;1/5;0;0;1)$

$a_3 = (3;3;2;0;0;1;0;3);$ $a_3' = (1;1;2/3;0;0;1/3;0;1)$

a_1' is the lexicographic minimum, a_{11} is pivot-element.

2. Integer Programming

2.1 Cutting Plane Methods

2.1.1 The GOMORY-I-All Integer Method

Hypotheses
Given the following problem P :

$$P : \begin{matrix} \min \\ \max \end{matrix} \} \quad \pi = c \cdot x$$
$$A \cdot x \gtreqless b$$
$$x \in \mathbb{N}_0^n$$

Principle
Beginning with a continuous optimal solution, if it exists, new solutions are determined in each iteration by adding new restrictions (cuts) to the problem.

Description
Step 1: Solve the problem without the integer condition by one of the simplex-methods.
Result: A solution $\bar{x}$ with the value of the objective function $\bar{\pi} = c \cdot \bar{x}$ (if a feasible solution exists) .
Set the running index $p := 1$.

Step 2: Is $\bar{x}_j \in \mathbb{N}_0 \quad \forall j = 1,\ldots,n$?
If yes: Stop, $\bar{x}$ is an optimal solution of P .
If no : Go to step 3 .

Step 3: Determine $\tilde{r}_{q_0} := \min_i \{r_{i_0} \mid r_{i_0} := b_i - [b_i]; r_{i_0} > 0\}$,

row q is called the source row.

Step 4: Form a new restriction (a so-called cut) from the row q with the additional slack variable y_p^*:

$$y_p^* - \sum_{j:nbv} r_{qj} \cdot x_j = -\tilde{r}_{q_0}, \text{ where } r_{qj} := a_{qj} - [a_{qj}].$$

Note: The summation-index j holds for <u>all</u> nbv.

Add the cut to the current tableau; determine the pivot-element according to the Dual Simplex-Algorithm, and perform as many dual simplex-iterations as necessary.

Step 5: Does a feasible solution of the expanded problem exist?
If yes: Let $\bar{x}$ be the solution. Set p := p + 1 and go to step 2.
If no : Stop, there is no feasible integer solution of P.

Note: If a slack variable y_p^* is basis-variable for the second time, then the row and column corresponding to y_p^* can be eliminated. Therefore the maximal size of the tableau is (m + n + 1) rows and (m + 2·n + 1) columns.

Example

Given the following problem P:

P : max $\pi = 2 \cdot x_1 + x_2$
(1) $x_1 - x_2 \leq 5$
(2) $4 \cdot x_1 + 3 \cdot x_2 \leq 10$
           $x_1, x_2 \in \mathbb{N}_0$

98 Integer Programming

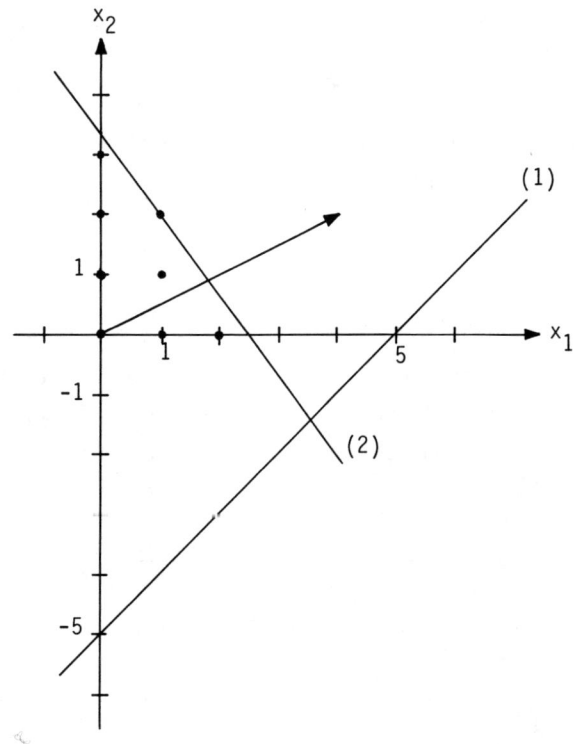

$T^{(1)}$:

x_1	x_2	y_1	y_2	1
1	-1	1	0	5
4	3	0	1	10
-2	-1	0	0	0

$T^{(2)}$:

x_1	x_2	y_1	y_2	1
0	-7/4	1	-1/4	5/2
1	3/4	0	1/4	5/2
0	1/2	0	1/2	5

$\bar{x} = (5/2; 0);$
$\bar{\pi} = 5;$

x_1 does not fulfill the integer condition, the second row is the source row. Gomory-I-cut:

$y_1^* - 3/4 \cdot x_2 - 1/4 \cdot y_2 = -1/2$

$T^{(3)}$:

x_1	x_2	y_1	y_2	y_1^*	1
0	-7/4	1	-1/4	0	5/2
1	3/4	0	1/4	0	5/2
0	-3/4	0	-1/4	1	-1/2
0	1/2	0	1/2	0	5

$T^{(4)}$:

x_1	x_2	y_1	y_2	y_1^*	1
0	0	1	1/3	-7/3	11/3
1	0	0	0	1	2
0	1	0	1/3	-4/3	2/3
0	0	0	1/3	2/3	14/3

Current solution: $x = (2; 2/3)$; the integer condition is not fulfilled by x_2, the third row is the source row.
Gomory-I-cut: $y_2^* - 1/3 \cdot y_2 - 2/3 \cdot y_1^* = -2/3$.

$T^{(5)}$:

x_1	x_2	y_1	y_2	y_1^*	y_2^*	1
0	0	1	1/3	-7/3	0	11/3
1	0	0	0	1	0	2
0	1	0	1/3	-4/3	0	2/3
0	0	0	-1/3	-2/3	1	-2/3
0	0	0	1/3	2/3	0	14/3

Here both elements, -1/3 as well as -2/3 may be selected for pivot-element. The sixth tableau corresponds to selecting -1/3 for pivot, the seventh tableau corresponds to selecting -2/3 for pivot.

$T^{(6)}$:

x_1	x_2	y_1	y_2	y_1^*	y_2^*	1
0	0	1	0	-3	1	3
1	0	0	0	1	0	2
0	1	0	0	-2	1	0
0	0	0	1	2	-3	2
0	0	0	0	0	1	4

First optimal integer solution:
$\bar{x} = (2; 0)$; $\bar{\pi} = c \cdot \bar{x} = 4$.

100   *Integer Programming*

$T^{(7)}$:

x_1	x_2	y_1	y_2	y_1	y_2	1
0	0	1	3/2	0	-7/2	6
1	0	0	-1/2	0	3/2	1
0	1	0	1	0	-2	2
0	0	0	1/2	1	-3/2	1
0	0	0	0	0	1	4

Second optimal integer solution:
$\bar{x} = (1;2)$; $\bar{\pi} = c \cdot \bar{x} = 4$.

2.1.2 The GOMORY-II-All Integer Method

Hypotheses
Given the following problem $\tilde{P}$:

$\tilde{P}$: min $\pi = c \cdot x$

$\tilde{A} \cdot x \leq \tilde{b}$

$x \in \mathbb{N}_0^n$

$\tilde{b} \in \mathbb{R}^m$.

Principle
The method consists of the repeated application of the Dual Simplex-Algorithm on an integer tableau. If it is not guaranteed that the tableau will remain integer, a new restriction (cut) is added.

Description
Step 1: Is $(\tilde{a}_{ij} \in \mathbb{Z}) \wedge (\tilde{b}_i \in \mathbb{Z}) \; \forall \; i,j$?
    If yes: Define $a_{ij} := \tilde{a}_{ij}$; $b_i := \tilde{b}_i$ $\forall \; i,j$ and go to step 3 .
    If no : Go to step 2 .

Step 2: Consider the coefficients of the i-th restriction, $\forall \; i = 1,\ldots,m$ and determine their common denominator, say q_i.

Compute $\left. \begin{array}{l} a_{ij} := \tilde{a}_{ij} \cdot q_i \\ b_i := \tilde{b}_i \cdot q_i \end{array} \right\} \forall\ i,j$.

Step 3: Set up a simplex-tableau for problem P :

$$P : \min\ \pi = c \cdot x + \Theta \cdot y$$
$$A \cdot x + y = b$$
$$x \in \mathbb{N}_0^n$$
$$y \in \mathbb{R}_+^m$$
$$b \in \mathbb{R}^m$$

and set the running index $p := 1$.

Step 4: Is $b_i \geq 0\ \forall\ i = 1,\ldots,m$?
If yes: Stop, the optimal integer solution for $\hat{P}$ has been found.
If no : Go to step 5 .

Step 5: Determine the provisional pivot-element $a_{r's}$ according to the Dual Simplex-Algorithm .

Step 6: $\exists\ a_{r's} < 0$?
If yes: Go to step 7 .
If no : Stop, there is no feasible solution of $\hat{P}$.

Step 7: Is $a_{r's} = (-1)$?
If yes: Set $a_{rs} := a_{r's}$ and go to step 9 .
If no : Go to step 8.

Step 8: Form a new restriction (a so-called cut) from the row r' with the additional slack variable y_p^* :

$$\left[\frac{a_{r'j}}{|\lambda|} \right] \cdot x_j + y_p^* = \left[\frac{b_{r'}}{|\lambda|} \right] ,\ \text{where}$$

$\lambda := \min_{j:nbv} \{a_{r'j}\}$ and add this cut to the current tableau. Let this be the r-th row. Disregard the provisional pivot-element $a_{r's}$ and determine the pivot-element a_{rs} according to the Dual Simplex-Algorithm. Set p: = p + 1.

Step 9: Perform one dual simplex-iteration and go to step 4.

Example
Given the following problem $\tilde{P}$:

$\tilde{P}$: min $\pi = 5 \cdot x_1 + 10 \cdot x_2$ min $\pi = 5 \cdot x_1 + 10 \cdot x_2$
(1) $x_1 + 4 \cdot x_2 \geq 2$ (1) $-x_1 - 4 \cdot x_2 \leq -2$
(2) $-x_1 + 3 \cdot x_2 \geq 1$ $\Rightarrow$ (2) $x_1 - 3 \cdot x_2 \leq -1$
       $x_1, x_2 \in \mathbb{N}_0$ $x_1, x_2 \in \mathbb{N}_0$

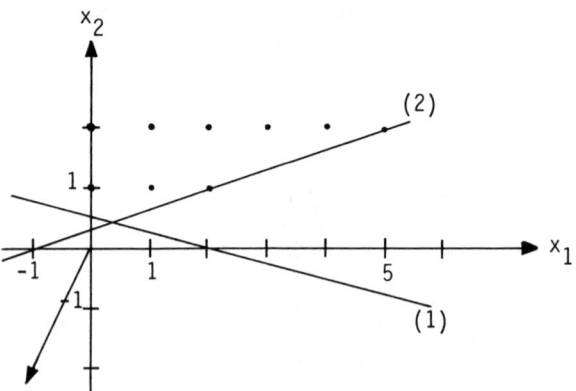

$T^{(1)}$:

x_1	x_2	y_1	y_2	1
-1	-4	1	0	-2
1	-3	0	1	-1
5	10	0	0	0

$\left[\dfrac{-1}{|\lambda|}\right] \cdot x_1 + \left[\dfrac{-4}{|\lambda|}\right] \cdot x_2 + y_1^* = \left[\dfrac{-2}{|\lambda|}\right]$;

$\lambda = -4$; Gomory-II-cut:
$-x_1 - x_2 + y_1^* = -1$;

$T^{(2)}$:

x_1	x_2	y_1	y_2	y_1^*	1
-1	-4	1	0	0	-2
1	-3	0	1	0	-1
-1	-1	0	0	1	-1
5	10	0	0	0	0

$T^{(3)}$:

x_1	x_2	y_1	y_2	y_1^*	1
0	-3	1	0	-1	-1
0	-4	0	1	1	-2
1	1	0	0	-1	1
0	5	0	0	5	-5

$\begin{bmatrix} -4 \\ |\lambda| \end{bmatrix} \cdot x_2 + \begin{bmatrix} 1 \\ |\lambda| \end{bmatrix} \cdot y_1^* + y_2^* = \begin{bmatrix} -2 \\ |\lambda| \end{bmatrix}$; $\lambda = -4$;

Gomory-II-cut: $-x_2 + y_2^* = -1$;

$T^{(4)}$:

x_1	x_2	y_1	y_2	y_1^*	y_2^*	1
0	-3	1	0	-1	0	-1
0	-4	0	1	1	0	-2
1	1	0	0	-1	0	1
0	-1	0	0	0	1	-1
0	5	0	0	5	0	-5

$T^{(5)}$:

x_1	x_2	y_1	y_2	y_1^*	y_2^*	1
0	0	1	0	-1	-3	2
0	0	0	1	1	-4	2
1	0	0	0	-1	1	0
0	1	0	0	0	-1	1
0	0	0	0	5	5	-10

The optimal integer solution is:
$\bar{x} = (0;1)$; $\bar{\pi} = c \cdot \bar{x} = 10$.

2.1.3 The GOMORY-III-Mixed Integer Method

<u>Hypotheses</u>

Given the following problem P :

P: max $\pi = c \cdot x$
$\quad\quad A \cdot x \leq b$

104 *Integer Programming*

$$x_j \in \mathbb{N}_0 \quad \forall j = 1,\ldots,k$$
$$x_j \geq 0 \quad \forall j = k+1,\ldots,n, \text{ where } k \leq n.$$

Note: If a system of restrictions $A \cdot x \gtreqless b$ is given, then the method starts with a feasible basic solution (phase I of the Two-Phase Method). The existence of a basic feasible solution does not guarantee the existence of a feasible integer solution of P.

Principle   See 2.1.1

Description

Step 1: Solve the problem without the integer condition by one of the simplex-methods.
Result: A solution $\bar{x}$ with the value of the objective function $\bar{\pi} = c \cdot \bar{x}$ (if a feasible solution does exist).
Set the running index $p := 1$.

Step 2: Is $\bar{x}_j \in \mathbb{N}_0 \quad \forall j = 1,\ldots,k$?
If yes: Stop, $\bar{x}$ is an optimal mixed-integer solution of P.
If no : Go to step 3.

Step 3: Let $r_{i_0} := b_i - [b_i]$.
Determine $r_{q_0} := \underset{i: b_i = x_j | x_j : bv; j=1,\ldots,k}{\text{maximum}} \{r_{i_0}\}$,
row q is called the source row.

Step 4: Form a new restriction (a so-called cut) from the row q with the additional slack variable y_p^*:

$$-\sum_{j:\text{nbv};a_{qj}\geq 0} a_{qj} \cdot x_j - \sum_{j:\text{nbv};a_{qj}<0} \left(\frac{r_{q_0}}{r_{q_0}-1} \cdot a_{qj} \right) \cdot x_j + y_p^* = -r_{q_0}$$

Note: The summation-index j holds for <u>all</u> nbv.

Section 2.1.3   105

Add the cut to the current tableau. Let this be the r-th row. Determine the pivot-element a_{rs} according to the Dual Simplex-Algorithm.

Step 5: $\exists\ a_{rs} < 0$?
If yes: Go to step 6 .
If no : Stop, there is no feasible mixed-integer solution of P .

Step 6: Perform one dual simplex-iteration.
Result: A solution $\bar{x}$. Set p: = p + 1 and go to step 2 .

Example
Given the following problem P:

P: max $\pi = 2 \cdot x_1 + x_2$

(1) $x_1 + x_2 \leq 5$

(2) $4 \cdot x_1 - 2 \cdot x_2 \leq 10$

$x_1 \geq 0$

$x_2 \in \mathbb{N}_0$

$T^{(1)}$:

x_1	x_2	y_1	y_2	1
1	1	1	0	5
4	-2	0	1	10
-2	-1	0	0	0

$T^{(2)}$:

x_1	x_2	y_1	y_2	1
0	3/2	1	-1/4	5/2
1	-1/2	0	1/4	5/2
0	-2	0	1/2	5

$T^{(3)}$:

x_1	x_2	y_1	y_2	1
0	1	2/3	-1/6	5/3
1	0	1/3	1/6	10/3
0	0	4/3	1/6	25/3

Optimal continuous solution:
$\bar{x} = (10/3; 5/3)$; $\bar{\pi} = c \cdot \bar{x} = 25/3$

The integer condition is not fulfilled by x_2. The first row is the source row with $r_{q_0} = 2/3$.

Gomory-III-cut: $-2/3 \cdot y_1 - 1/3 \cdot y_2 + y_1^* = -2/3$

$T^{(4)}$:

x_1	x_2	y_1	y_2	y_1^*	1
0	1	2/3	-1/6	0	5/3
1	0	1/3	1/6	0	10/3
0	0	-2/3	-1/3	1	-2/3
0	0	4/3	1/6	0	25/3

$T^{(5)}$:

x_1	x_2	y_1	y_2	y_1^*	1
0	1	1	0	-1/2	2
1	0	0	0	1/2	3
0	0	2	1	-3	2
0	0	1	0	1/2	8

The optimal mixed-integer solution is: $\bar{x} = (3;2)$; $\bar{\pi} = c \cdot \bar{x} = 8$.

2.1.4 The GOMORY-III-Mixed Integer Method with Intensified Cuts

Generally a solution of the problem P is obtained faster by modifying step 4 of the Gomory-III-Mixed Integer Method as follows:

Step 4': Let $r_{qj} := a_{qj} - [a_{qj}]$.

Form a new restriction (a so-called cut) from the row q with the additional slack variable y_p^*:

$$-\sum_{j:nbv} \lambda_j \cdot x_j + y_p^* = -r_{q_0}, \text{ where}$$

$$\lambda_j := \begin{cases} a_{qj}, & \text{if } (x_j \notin \mathbb{N}_0) \wedge (a_{qj} \geq 0) \\ \left(\dfrac{r_{q_0}}{r_{q_0}-1}\right), & \text{if } (x_j \notin \mathbb{N}_0) \wedge (a_{qj} < 0) \\ r_{qj}, & \text{if } (x_j \in \mathbb{N}_0) \wedge (r_{qj} \leq r_{q_0}) \\ \left[(r_{qj} - 1) \cdot \dfrac{r_{q_0}}{r_{q_0}-1}\right], & \text{if } (x_j \in \mathbb{N}_0) \wedge (r_{qj} > r_{q_0}) \end{cases}$$

Note: The summation-index j holds for <u>all</u> nbv.

Add the cut to the current tableau. Let this be the r-th row. Determine the pivot-element a_{rs} according to the Dual Simplex-Algorithm.

Example

Given the following problem P :

P: max $\pi = 2 \cdot x_1 + 3 \cdot x_2 + 4 \cdot x_3$

(1) $x_1 + 4 \cdot x_2 + 5 \cdot x_3 \leq 9$
(2) $4 \cdot x_1 \quad\quad\quad + 3 \cdot x_3 \leq 10$

$\quad\quad\quad x_1, x_3 \in \mathbb{N}_0$
$\quad\quad\quad x_2 \geq 0$

$T^{(1)}$:

x_1	x_2	x_3	y_1	y_2	1
1	4	5	1	0	9
4	0	3	0	1	10
-2	-3	-4	0	0	0

$T^{(2)}$:

x_1	x_2	x_3	y_1	y_2	1
1/4	1	5/4	1/4	0	9/4
4	0	3	0	1	10
-5/4	0	-1/4	3/4	0	27/4

$T^{(3)}$:

x_1	x_2	x_3	y_1	y_2	1
0	1	1	1/4	-1/16	13/8
1	0	3/4	0	1/4	5/2
0	0	11/16	3/4	5/16	79/8

Optimal continuous solution:
$\bar{x} = (5/2; 13/8; 0)$;
$\bar{\pi} = c \cdot \bar{x} = 79/8$;

The variable x_1 does not fulfill the integer condition, the second row is the source row with $r_{q_0} = 1/2$.

Intensified Gomory-III-cut:

$- (3/4-1) \cdot \dfrac{1/2}{1/2-1} \cdot x_3 - 1/4 \cdot y_2 + y_1^* = - 1/2$

$\rightarrow - 1/4 \cdot x_3 - 1/4 \cdot y_2 + y_1^* = - 1/2$

$T^{(4)}$:

x_1	x_2	x_3	y_1	y_2	y_1^*	1
0	1	1	1/4	-1/16	0	13/8
1	0	3/4	0	1/4	0	5/2
0	0	-1/4	0	-1/4	1	-1/2
0	0	11/16	3/4	5/16	0	79/8

$T^{(5)}$:

x_1	x_2	x_3	y_1	y_2	y_1^*	1
0	1	15/16	1/4	0	-1/4	7/4
1	0	1/2	0	0	1	2
0	0	1	0	1	-4	2
0	0	3/8	3/4	0	5/4	37/4

The optimal mixed-integer solution is:
$\bar{x} = (2; 7/4; 0)$;
$\bar{\pi} = c \cdot \bar{x} = 37/4$.

2.1.5 The Primal Cutting Plane Method (Young; Glover; Ben-Israel; Charnes)

<u>Hypotheses</u>

Given the following problem P :

$$P : \max \quad \pi = c \cdot x$$
$$A \cdot x \leq b$$
$$x \in \mathbb{N}_0^n$$
$$(a_{ij} \in \mathbb{Z}) \wedge (b_i \in \mathbb{N}_0) \quad \forall \; i,j$$

Note: When necessary the last conditions may be fulfilled by multiplying the restriction with an appropriate factor (common denominator).

Section 2.1.5   109

<u>Principle</u>   See 2.1.2 , however this is a primal method.

<u>Description</u>

Step 1: Set the running index p: = 1 .

Step 2: Is $c_j \geq 0 \; \forall \; j$?
If yes: Stop, an optimal solution of P has been found.
If no : Go to step 3 .

Step 3: Select the pivot-column s as well as the provisional pivot-row r' according to the Primal Simplex-Algorithm. Then the provisional pivot-element is given by $a_{r's}$.

Step 4: $\exists \; a_{r's} > 0$?
If yes: Go to step 5 .
If no : Stop, P has no feasible integer solution .

Step 5: Is $a_{r's} = 1$?
If yes: Define $a_{rs} := a_{r's}$. Go to step 7.
If no : Go to step 6 .

Step 6: Form a new restriction (a so-called cut) from the row r' with the additional slack variable y_p^* :

$$\sum_{j:nbv} \left[\frac{a_{r'j}}{\lambda} \right] \cdot x_j + y_p^* = \left[\frac{b_{r'}}{\lambda} \right], \text{ where}$$

$$\lambda := \max_{j:nbv} \{a_{r'j} \mid a_{r'j} > 0\} .$$

Note: The summation-index j holds for <u>all</u> nbv .

Add the cut to the current tableau. Let this be the r-th row, then the pivot-element $a_{rs} = 1$ is uniquely defined. Set p: = p + 1 .

Step 7: Perform one primal simplex-iteration. Go to step 2.

Note: The note pertaining to the hypotheses of the GOMORY-III-Mixed Integer Method holds in this case also.

Example

Given the following problem P :

P: max $\pi = x_1 + 3 \cdot x_2$

(1) $2 \cdot x_1 + 5 \cdot x_2 \leq 8$

(2) $8 \cdot x_1 - x_2 \leq 12$

$x_1, x_2 \in \mathbb{N}_0$

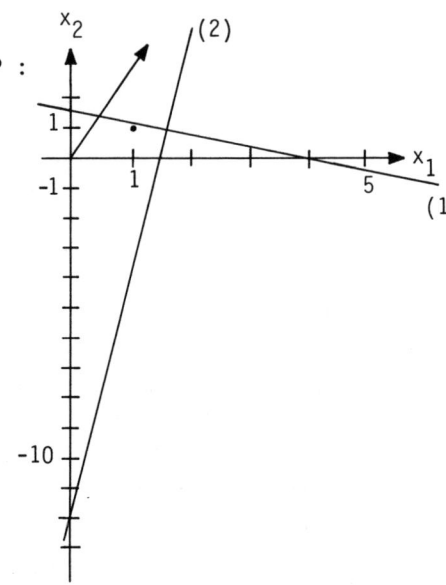

$T^{(1)}$:

x_1	x_2	y_1	y_2	1
2	5	1	0	8
8	-1	0	1	12
-1	-3	0	0	0

$\left[\dfrac{2}{\lambda}\right] \cdot x_1 + \left[\dfrac{5}{\lambda}\right] \cdot x_2 + y_1^* = \left[\dfrac{8}{\lambda}\right]$; $\lambda = 5 \rightarrow x_2 + y_1^* = 1$

$T^{(2)}$:

x_1	x_2	y_1	y_2	y_1^*	1
2	5	1	0	0	8
8	-1	0	1	0	12
0	1	0	0	1	1
-1	-3	0	0	0	0

$T^{(3)}$:

x_1	x_2	y_1	y_2	y_1^*	1
2	0	1	0	-5	3
8	0	0	1	1	13
0	1	0	0	1	1
-1	0	0	0	3	3

$\left[\dfrac{2}{\lambda}\right] \cdot x_1 + \left[\dfrac{-5}{\lambda}\right] \cdot y_1^* + y_2^* = \left[\dfrac{3}{\lambda}\right]$; $\lambda = 2 \rightarrow x_1 - 3 \cdot y_1^* + y_2^* = 1$

$T^{(4)}$:

x_1	x_2	y_1	y_2	y_1^*	y_2^*	1
2	0	1	0	-5	0	3
8	0	0	1	1	0	13
0	1	0	0	1	0	1
1	0	0	0	-3	1	1
-1	0	0	0	3	0	3

$T^{(5)}$:

x_1	x_2	y_1	y_2	y_1^*	y_2^*	1
0	0	1	0	1	-2	1
0	0	0	1	25	-8	5
0	1	0	0	1	0	1
1	0	0	0	-3	1	1
0	0	0	0	0	1	4

The optimal integer solution is: $\bar{x} = (1;1)$; $\bar{\pi} = c \cdot \bar{x} = 4$.

2.2 Branch and Bound Methods

2.2.1 The Method of LAND and DOIG

<u>Hypotheses</u>

Given the following problem P :

P: $\max \quad \pi = c \cdot x$
$\quad A \cdot x \gtreqless b$
$\quad x_j \in \mathbb{N}_0 \quad \forall j = 1,\ldots,k$
$\quad x_j \geq 0 \ \forall \ j = k+1,\ldots,n, \ \text{where} \ k \leq n$.

112 Integer Programming

The sets M and $I(0)$ are defined as follows: $M := I(0) := \emptyset$; furthermore let $t := 1$; $\pi^* := (-\infty)$. The problem $P(0)$ is given by the problem P without the integer condition.

Principle
Beginning with a continuous optimal solution, if it exists, a sequence of new problems is generated by setting one variable, not yet fulfilling the appropriate integer condition equal to some fixed integer. Ramify the specially chosen solutions of these problems until an integer solution with a maximal value of the objective function is reached or the non-existence of an integer solution is shown.

Description

Step 1: Solve the problem $P(0)$ by one of the simplex-methods.
Result: A solution $\bar{x}(0)$ with the value of the objective function $\bar{\pi}(0) = c \cdot \bar{x}(0)$ (if a feasible solution exists).

Step 2: Is $\bar{x}_j(0) \in \mathbb{N}_0 \ \forall \ j = 1,\ldots,k$?
If yes: Stop, $\bar{x}(0)$ is an optimal mixed-integer solution of P.
If no : Go to step 3.

Step 3: Let $\Delta := \{t; t+1\}$; $\bar{x}(\tau) := \bar{x}(0)$.
Select the variable $x_s(\tau) \notin \mathbb{N}_0$, where $s \in [1;k]$.
Formulate and solve the following problems:
$P(t)$: max $\pi = c \cdot x$; $A \cdot x \gtreqless b$; $x_s = [\bar{x}_s(\tau)]$; $x \geq \theta$;
$P(t+1)$: max $\pi = c \cdot x$; $A \cdot x \gtreqless b$; $x_s = \langle \bar{x}_s(\tau) \rangle$; $x \geq \theta$.
Result: The solutions $\bar{x}(\delta)$, where $\delta \in \Delta$.

Step 4: $\exists$ optimal solution $\bar{x}(\delta) \ \forall \ \delta \in \Delta$?
If yes: Go to step 5.
If no : Set $\bar{\pi}(\delta) := (-\infty)$. Go to step 11.

Step 5: $\exists \ (\bar{x}(\delta) | \ \bar{x}_j(\delta) \in \mathbb{N}_0 \ \ \forall \ j = 1,\ldots,k) \ \forall \ \delta \in \Delta$?
  If yes: Go to step 6.
  If no : Go to step 9.

Step 6: Determine $\bar{\pi}(\tau) := \max_{\delta} \{\bar{\pi}(\delta) \ | \ \bar{x}_j(\delta) \in \mathbb{N}_0 \ \ \forall \ j = 1,\ldots,k\}$.

Step 7: Is $\bar{\pi}(\tau) > \pi^*$?
  If yes: Go to step 8.
  If no : Go to step 9.

Step 8: Set $x^* := \bar{x}(\tau)$; $\pi^* := \bar{\pi}(\tau)$.

Step 9: Define
  $M := M \cup \{\delta | [\bar{x}_j(\delta) \notin \mathbb{N}_0 \ \forall \ j=1,\ldots,k] \vee [\bar{\pi}(\delta) \geq \pi^*]\} \ \forall \ \delta \in \Delta$
  $I(t) := I(t+1) := I(\tau) \cup \{x_s\}$
  $I(t+2) := \begin{cases} I(\tau), & \text{if } t > 1 \\ \emptyset, & \text{if } t = 1 \end{cases}$
  $J(t) := J(t+1) := x_s$
  $J(t+2) := \begin{cases} J(\tau), & \text{if } t > 1 \\ \text{not defined, if } t = 1 \end{cases}$
  $\varepsilon(t) := (-1); \quad \varepsilon(t+1) := 1$
  $\varepsilon(t+2) := \begin{cases} \varepsilon(\tau), & \text{if } t > 1 \\ \text{not defined, if } t = 1 \end{cases}$

Step 10: Is $t = 1$?
  If yes: Set $t := 3$. Go to step 11.
  If no : Set $t := t + 3$. Go to step 11.

Step 11: Is $M = \emptyset$?
  If yes: Go to step 16.
  If no : Go to step 12.

Step 12: Determine $\bar{\pi}(\tau) := \max \{\bar{\pi}(\delta) \ | \ \delta \in M\}$;
  set $M := M - \{\tau\}$.

Step 13: Is $\bar{\pi}(\tau) < \pi^*$?
        If yes: Go to step 11.
        If no : Go to step 14.

Step 14: Let $\Delta := \{t;\ t+1;\ t+2\}$.
        Select the variable $x_s(\tau) \notin \mathbb{N}_0$, where $s \in [1;k]$.

Step 15: Formulate and solve the following problems:

$P(t)$:   max $\pi = c \cdot x$; $A \cdot x \gtreqless b$; $x \geq \Theta$;
           $x_s = [\bar{x}_s(\tau)]$; $x_i =$ const. $\forall\ i \in I(\tau)$

$P(t+1)$: max $\pi = c \cdot x$; $A \cdot x \gtreqless b$; $x \geq \Theta$;
           $x_s = \langle \bar{x}_s(\tau) \rangle$; $x_i =$ const. $\forall\ i \in I(\tau)$

$P(t+2)$: max $\pi = c \cdot x$; $A \cdot x \gtreqless b$; $x \geq \Theta$;
           $x_k = x_k + \varepsilon(\tau)$; $x_i =$ const. $\forall\ i \in [I(\tau) - \{x_k\}]$,
         where $x_k = J(\tau)$.

Result: The solutions $\bar{x}(\delta)$, where $\delta \in \Delta$. Go to step 4.
Note: If x_k in problem $P(t+2)$ becomes negative because of the restriction $x_k = x_k + \varepsilon(\tau)$, then set $x_k := 0$.

Step 16: Is $\pi^* = (-\infty)$?
        If yes: Stop, P has no feasible mixed-integer solution.
        If no : Stop, x^* is an optimal solution of P , the
             value of the objective function is $\pi^* = c \cdot x^*$.

Example

Given the following problem P :

P: max $\pi = 2 \cdot x_1 + 6 \cdot x_2$
   (1) $3/2 \cdot x_1 + 4 \cdot x_2 \leq 10$
   (2) $-x_1 + 3 \cdot x_2 \leq 7$
         $x_1,\ x_2 \in \mathbb{N}_0$

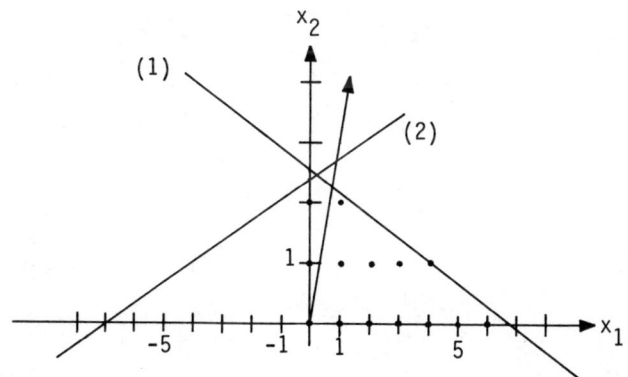

$T^{(1)}$:

x_1	x_2	y_1	y_2	1
3/2	4	1	0	10
-1	3	0	1	7
-2	-6	0	0	0

$T^{(2)}$:

x_1	x_2	y_1	y_2	1
1	8/3	2/3	0	20/3
0	17/3	2/3	1	41/3
0	-2/3	4/3	0	40/3

$T^{(3)}$:

x_1	x_2	y_1	y_2	1
1	0	6/17	-8/17	4/17
0	1	2/17	3/17	41/17
0	0	24/17	2/17	254/17

$\bar{x}(0) = (4/17; 41/17);\ \bar{\pi}(0) = 254/17$
$\bar{x}_j(0) \notin \mathbb{N}_0\ \forall\ j = 1,2\ \rightarrow\ M = \emptyset;\ I(0) = \emptyset;\ t = 1;$
$\pi^* = (-\infty);\ \Delta = \{1;2\}\ ;\ \bar{x}(\tau) = \bar{x}(0);\ x_s(0) = x_2(0);$

P(1): max $\pi = 2 \cdot x_1 + 6 \cdot x_2$ }   P(1): max $\pi = 2 \cdot x_1 + 12$
 (1) $3/2 \cdot x_1 + 4 \cdot x_2 \leq 10$ →   (1) $3/2 \cdot x_1 \leq 2$
 (2) $-x_1 + 3 \cdot x_2 \leq 7$ (2) $-x_1 \leq 1$
 (3) $x_2 = 2$ $x_1 \geq 0$
     $x_1, x_2 \geq 0$

P(2): max $\pi = 2 \cdot x_1 + 6 \cdot x_2$ P(2): max $\pi = 2 \cdot x_1 + 18$
 (1) $3/2 \cdot x_1 + 4 \cdot x_2 \leq 10$ (1) $3/2 \cdot x_1 \leq -2$
 (2) $-x_1 + 3 \cdot x_2 \leq 7$ → (2) $-x_1 \leq -2$
 (3) $x_2 = 3$ $x_1 \geq 0$
     $x_1, x_2 \geq 0$

$\bar{x}(1) = (4/3; 2);\ \bar{\pi}(1) = 44/3;$
$\not\exists$ optimal solution $\bar{x}(2);\ \bar{\pi}(2) = (-\infty);$
$\bar{x}_j(1) \notin \mathbb{N}_0\ \forall\ j = 1,\ldots,k \rightarrow M := \emptyset \cup \{1\} = \{1\}\ ;$
$I(1) = \{x_2\}\ ;\ I(2) = \{x_2\}\ ;\ I(3) = \emptyset\ ;$
$J(1) = x_2\ ;\ J(2) = x_2\ ;\ \varepsilon(1) = (-1);\ \varepsilon(2) = 1\ ;$
$t = 3;\ M \neq \emptyset;\ \bar{\pi}(\tau) = \bar{\pi}(1);\ M = M - \{1\} = \emptyset\ ;$
$\Delta = \{3; 4; 5\}\ ;\ x_s(1) = x_1(0)\ ;$

P(3): max $\pi = 2 \cdot x_1 + 6 \cdot x_2$ P(3): max $\pi = 14$
 (1) $3/2 \cdot x_1 + 4 \cdot x_2 \leq 10$ (1) $19/2 \leq 10$
 (2) $-x_1 + 3 \cdot x_2 \leq 7$ → (2) $5 \leq 7$
 (3) $x_2 = 2$
 (4) $x_1 \quad\ = 1$
     $x_1, x_2 \geq 0$

P(4): max $\pi = 2 \cdot x_1 + 6 \cdot x_2$
  (1) $3/2 \cdot x_1 + 4 \cdot x_2 \leq 10$
  (2) $-x_1 + 3 \cdot x_2 \leq 7$
  (3) $x_2 = 2$
  (4) $x_1 = 2$
  $x_1, x_2 \geq 0$

$\rightarrow$

P(4): max $\pi = 16$
  (1) $11 \not\leq 10$
  (2) $4 \leq 7$

P(5): max $\pi = 2 \cdot x_1 + 6 \cdot x_2$
  (1) $3/2 \cdot x_1 + 4 \cdot x_2 \leq 10$
  (2) $-x_1 + 3 \cdot x_2 \leq 7$
  (3) $x_2 = 1$
  $x_1, x_2 \geq 0$

$\rightarrow$

P(5): max $\pi = 2 \cdot x_1 + 6$
  (1) $3/2 \cdot x_1 \leq 6$
  (2) $-x_1 \leq 4$
  $x_1 \geq 0$

$\bar{x}(3) = (1;2)$; $\bar{\pi}(3) = 14$; $\not\exists$ optimal solution $\bar{x}(4)$; $\bar{\pi}(4) = (-\infty)$;
$\bar{x}(5) = (4;1)$; $\bar{\pi}(5) = 14$;
$\bar{x}_j(3) \in \mathbb{N}_0 \ \forall \ j = 1,2$; $\bar{x}_j(5) \in \mathbb{N}_0 \ \forall \ j = 1,2$;
$\bar{\pi}(\tau) = \bar{\pi}(3) = \bar{\pi}(5)$; $x_1^* = \bar{x}(3)$; $x_2^* = \bar{x}(5)$
$\pi^* = \bar{\pi}(3) = \bar{\pi}(5)$; $M = M \cup \emptyset = \emptyset$
$I(3) = \{x_1; x_2\}$; $I(4) = \{x_1; x_2\}$; $I(5) = \{x_2\}$; $J(3) = x_1$;
$J(4) = x_1$; $J(5) = x_2$; $\varepsilon(3) = (-1)$; $\varepsilon(4) = 1$; $\varepsilon(5) = \varepsilon(1) = (-1)$;
$t = 6$; $M \stackrel{!}{=} \emptyset \rightarrow$ Stop, $x_1^* = \bar{x}(3) = (1;2)$ and $x_2^* = \bar{x}(5) = (4;1)$
are optimal solutions of P; the value of the objective function is
$\pi^* = 14$.

Solution tree:

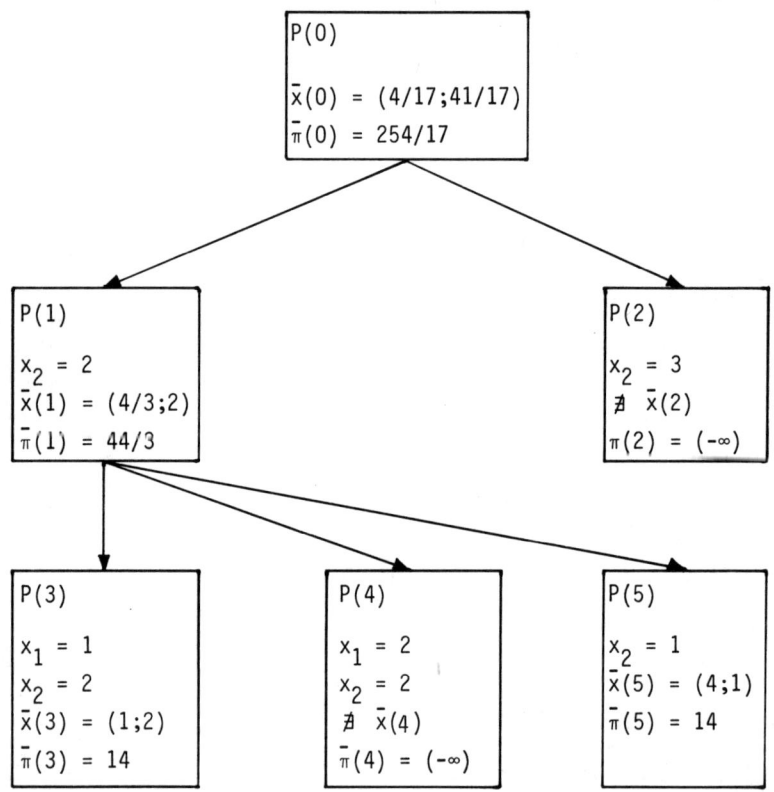

2.2.2 The Method of DAKIN

Hypotheses
Given the following problem P :

P: $\max \pi = c \cdot x$
$A \cdot x \leq b$
$x_j \in \mathbb{N}_0 \quad \forall \, j = 1,\ldots,k$
$x_j \geq 0 \quad \forall \, j = k+1,\ldots,n$, where $k \leq n$.

M and π^* are defined as: $M := \emptyset$; $\pi^* := (-\infty)$, the running indices are $t := 1$; $p := 1$. The problem $P(0)$ is given by the problem P without the integer condition.

Principle See 2.2.1

Description

Step 1: Solve the problem $P(0)$ by one of the simplex-methods.
Result: A solution $\bar{x}(0)$ with the value of the objective function $\bar{\pi}(0) = c \cdot \bar{x}(0)$ (if a feasible solution exists).

Step 2: Is $\bar{x}_j(0) \in \mathbb{N}_0 \; \forall j = 1,\ldots,k$?
If yes: Stop, $\bar{x}(0)$ is an optimal solution of P.
If no : Go to step 3.

Step 3: Define $\bar{x}(\tau) := \bar{x}(0)$; select the variable $x_s(\tau) \notin \mathbb{N}_0$, where $s \in [1;k]$.

Step 4: Set up the problem $P(t)$ as follows:

$P(t) := P(\tau) \cup \{x_s \leq [\bar{x}_s(\tau)]\}$.

This means: Add the restriction

$$-\sum_{j:nbv} a_{ij} \cdot x_j - \sum_{j:nbv} a_{ij} \cdot y_j + y_p^* = [\bar{x}_s(\tau)] - \bar{b}_i$$

to the optimal tableau of the problem $P(\tau)$ and do as many dual simplex-iterations as necessary.

Step 5: Set up the problem $P(t + 1)$ as follows:
$P(t + 1) := P(\tau) \cup \{x_s \geq <\bar{x}_s(\tau)>\}$.
This means: Add the restriction

$$-\sum_{j:nbv} a_{ij} \cdot x_j + \sum_{j:nbv} a_{ij} \cdot y_j + y_p^* = \bar{b}_i - <\bar{x}_s(\tau)>$$

to the optimal tableau of the problem $P(\tau)$ and do as many dual simplex-iterations as necessary.

Step 6: $\exists$ feasible solutions $\bar{x}(l) \; \forall \; l = t, t + 1$?
If yes: Go to step 7.

If no : Go to step 11.

Step 7: Is $\bar{\pi}(l) > \pi^*$ $\forall$ $l = t, t+1$?
If yes: Go to step 8.
If no : Go to step 11.

Step 8: Is $(\bar{x}_j(l) \in \mathbb{N}_0$ $\forall$ $j = 1,\ldots,k)$ $\forall$ $l = t, t+1$?
If yes: Go to step 9.
If no : Go to step 10.

Step 9: Determine $\bar{\pi}(r) := \max_{l} \{\bar{\pi}(l) | \bar{x}_j(l) \in \mathbb{N}_0 \; \forall \; j = 1,\ldots,k\}$
and set $x^* := \bar{x}(r)$; $\pi^* := \bar{\pi}(r)$.

Step 10: Define $M := M \cup \{l \; | \; (\bar{x}_j(l) \notin \mathbb{N}_0 \; \forall \; j = 1,\ldots,k) \forall \; l = t, t+1\}$.

Step 11: Is $M = \emptyset$?
If yes: Go to step 15.
If no : Go to step 12.

Step 12: Determine $\bar{\pi}(\tau) := \max \{\bar{\pi}(i) | i \in M\}$.

Step 13: Is $\bar{\pi}(\tau) < \pi^*$?
If yes: Set $M := M - \{\tau\}$. Go to step 11.
If no : Go to step 14.

Step 14: Select the variable $x_s(\tau) \notin \mathbb{N}_0$, where $s \in [1;k]$;
set $t := t + 2$; $p := p + 1$. Go to step 4.

Step 15: Is $\pi^* = (-\infty)$?
If yes: Stop, P has no feasible solution.
If no : Stop, x^* is an optimal solution of P , the value
of the objective function is $\pi^* = c \cdot x^*$.

Example

Given the following problem P :

P: max $\pi = x_1 + 3 \cdot x_2$

(1) $4 \cdot x_1 - x_2 \leq 6$

(2) $2 \cdot x_1 + 3 \cdot x_2 \leq 10$

$x_1 \geq 0; x_2 \in \mathbb{N}_0$

M: = $\emptyset$; π^*: = $(-\infty)$;
t: = 1 ; p : = 1 ;

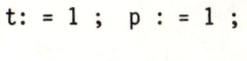

$T^{(0)}$:

x_1	x_2	y_1	y_2	1
4	-1	1	0	6
2	3	0	1	10
-1	-3	0	0	0

$T^{(0)}_{opt}$:

x_1	x_2	y_1	y_2	1
14/3	0	1	1/3	28/3
2/3	1	0	1/3	10/3
1	0	0	1	10

Solution of P(0): $\bar{x}(0) = (0;10/3)$; $\bar{\pi}(0) = 10$; $\bar{x}_2(0) \notin \mathbb{N}_0$;
→ $\bar{x}(\tau) := \bar{x}(0); x_s = x_2$.

P(1): = P(0) ∪ {$x_s \leq 3$}, i.e., the additional restriction is:
$-2/3 \cdot x_1 - 1/3 \cdot y_2 + y_1^* = -1/3$

P(2): = P(0) ∪ {$x_s \geq 4$}, i.e., the additional restriction is:
$2/3 \cdot x_1 + 1/3 \cdot x_2 + y_1^* = -2/3$

122 *Integer Programming*

$T^{(1)}$:

x_1	x_2	y_1	y_2	y_1^*	1
14/3	0	1	1/3	0	28/3
2/3	1	0	1/3	0	10/3
-2/3	0	0	-1/3	1	-1/3
1	0	0	1	0	10

$T_{opt}^{(1)}$:

x_1	x_2	y_1	y_2	y_1^*	1
0	0	1	-2	7	7
0	1	0	0	1	3
1	0	0	1/2	-3/2	1/2
0	0	0	1/2	3/2	19/2

$\bar{x}(1) = (1/2;3)$; $\bar{\pi}(1) = 19/2$; $\bar{\pi}(1) > \pi^*$; $\bar{x}_2(1) \in \mathbb{N}_0$

→ $\pi^* := \bar{\pi}(1)$; $x^* := \bar{x}(1)$; $M \stackrel{!}{=} \emptyset$; $\pi^* \neq (-\infty)$

→ Stop, $x^* = (1/2;3)$ is an optimal solution of P, the value of the objective function is $\pi^* = 19/2$.

$T^{(2)}$:

x_1	x_2	y_1	y_2	y_1^*	1
14/3	0	1	1/3	0	28/3
2/3	1	0	1/3	0	10/3
2/3	0	0	1/3	1	-2/3
1	0	0	1	0	10

There is no negative pivot-element → P(2) has no feasible soltuion.

2.2.3 The Method of DRIEBEEK

<u>Hypotheses</u>
Given the following problem P :

P: max $\pi = c \cdot x$
    $A \cdot x \leq b$
    $x_j \in \mathbb{N}_0 \quad \forall \; j = 1,\ldots,k$
    $x_j \geq 0 \quad \forall \; j = k+1,\ldots,n$, where $k < n$.

Note: This method is especially efficient for problems having only a few variables which are required to be integers. Furthermore each of these variables should be bounded above by κ and below by λ so that the interval is small. I.e. $d_j := \kappa_j - \lambda_j \leq 5$ j where d_j is the interval length for the variable x_j which is required to be an integer.

Principle

In this method, the variables, which are required to be integers, are partitioned into new variables in the unit interval and a continuous solution for this new problem is determined. If one exists, the variables with the imposed integer condition are fixed according to the cost penalty table. The continuous optimal solution is tested with a new vector $\tilde{b}$. The method terminates in an optimal solution.

Description

Step 1: Replace each integer variable x_j, where $x_j \in [\lambda_j ; \kappa_j]$ by:

$$\sum_{p=0}^{d_j} x_{jp} \; ;$$ to these new variables correspond the coefficients a_{ij} or c_j, respectively.

Step 2: For each integer variable x_j, where $x_j \in [\lambda_j ; \kappa_j]$ add the following equations to the problem P :

$$x_{jo} = \lambda_j$$

$$x_{j1} + y_{jo} = 1$$

$$x_{j,p+1} - x_{jp} + y_{jp} = 0 \;\; \forall \; p = 1,\ldots,(d_j-1)$$

$$-x_{jd_j} + y_{jd_j} = 0$$

Complete the simplex-tableau of the newly defined problem $\bar{P}$ by addition of slack variables to the original restrictions.

Step 3: Solve the problem $\bar{P}$ without the integer condition by the Primal Simplex-Algorithm.

124 *Integer Programming*

Does an optimal solution of $\bar{P}$ exist ?
If yes: Go to step 4.
If no : Stop, P has no feasible mixed-integer solution.

Step 4: Is $(\bar{x}_j := \sum_{p=0}^{d_j} \bar{x}_{jp}) \in \mathbb{N}_o \quad \forall \; j = 1,\ldots,k$?

If yes: Stop, an optimal mixed-integer solution of P has been found.

If no : Let $\bar{c}$ be the value of the objective function, then $\pi^* := (-\infty)$; i: = 1. Go to step 5

Step 5: Set up the cost penalty table

	variable	x_1	x_2	...	x_k
Real cost penalty	value: λ_j	$\bar{s}_{10}$	$\bar{s}_{20}$	.	$\bar{s}_{k0}$
	λ_j+1	$\bar{s}_{11}$	$\bar{s}_{21}$	.	$\bar{s}_{k1}$
	.	.	.	.	.
	κ_j	$\bar{s}_{1d_1}$	$\bar{s}_{2d_2}$	.	$\bar{s}_{kd_k}$
Pseudo cost penalty	value: λ_j	$\tilde{s}_{10}$	$\tilde{s}_{20}$	.	$\tilde{s}_{k0}$
	λ_j+1	$\tilde{s}_{11}$	$\tilde{s}_{21}$	.	$\tilde{s}_{k1}$
	.	.	.	.	.
	κ_j	$\tilde{s}_{1d_1}$	$\tilde{s}_{2d_2}$	.	$\tilde{s}_{kd_k}$

as follows:

- If the slack variable y_{jq} is an nbv in the optimal tableau without the integer condition, then $\bar{s}_{jq} := \bar{c}_{jq}$, where $\bar{c}_{jq}$ is the coefficient of y_{jq} in the objective function, and $\bar{s}_{jq} := \infty$, otherwise .

-If the slack variable y_{jq} is a bv in the optimal tableau without the integer condition, with "1" in row r, then

$$\tilde{s}_{jq} := (\bar{b}_r - 1) \cdot \left(\max \left\{ \frac{\bar{c}_v}{\bar{a}_{rv}} \mid \bar{a}_{rv} < 0 \right\} \right);$$

$\tilde{s}_{jq} := \infty$, otherwise.
(The values marked with "‾" are values of the optimal tableau of $\bar{P}$). Set the running index $t := 1$.

Step 6: Determine an element $\bar{s}$ or $\tilde{s}$ for each integer variable x_j in the cost penalty table, so that
$\hat{s} := \{$"sum of the selected $\bar{s}_{jq}$ and/or $\tilde{s}_{jq}$"$\}$
is the t-th smallest element of all possible $\bar{s}/\tilde{s}$-combinations. Furthermore note the corresponding slack variables y_{jq}. Has this combination of $\bar{s}_{jq}/\tilde{s}_{jq}$-values already been considered?
If yes: Go to step 7.
If no : Go to step 8.

Step 7: Do other $\bar{s}/\tilde{s}$-combinations exist with the same sum $\hat{s}$?
If yes: Go to step 6 and select one of these combinations.
If no : Set $t := t + 1$. Go to step 6.

Step 8: Is $\hat{s} \geq \bar{c} - \pi^*$?
If yes: Stop, x^* is an optimal mixed-integer solution of P, the corresponding value of the objective function is π^*.
If no : Go to step 9.

Step 9: Construct the column vector Δb as follows:
$\Delta b^T := (y_1; \ldots; y_m; y_{10}; \ldots; y_{1d_1}; y_{20}; \ldots; y_{jq}; \ldots; y_{kd_k})$.
The sequence of the slack variables is identical to that of the initial tableau of $\bar{P}$. The elements of this vector are "1" for all y_{jq}, whose corresponding

$\tilde{\tilde{s}}_{jq}$ or $\tilde{s}_{jq}$ were selected as minimal, and "0" otherwise.

Step 10: Determine $\tilde{b}$: $= \bar{b} - B^{-1} \cdot \Delta b$. Replace $\bar{b}$ in the optimal tableau $T_{opt}(cont)$ belonging to $\bar{P}$ by $\tilde{b}$. (B^{-1} is the matrix of coefficients of the slack variables y in $T_{opt}(cont)$).

Step 11: Perform dual simplex-iterations until:

    a) the tableau is optimal with the solution $\bar{\bar{x}}(i)$, all variables $\bar{\bar{x}}_j$: $= \sum_{p=0}^{d_j} \bar{\bar{x}}_{jp}$ are integer, the value of the current solution is $\bar{\bar{c}}(i) > \pi^*$. Then go to step 12.

    or b) the tableau is dual unrestricted. Then go to step 7.

    or c) a dual simplex-iteration leads to a value of the objective function $\bar{\bar{c}}(i) \leq \pi^*$. Then go to step 7.

Step 12: Store the current mixed-integer solution x^*: $= \bar{\bar{x}}(i)$; π^*: $= \bar{\bar{c}}(i)$ and set i: $= i + 1$. Go to step 7.

<u>Example</u>

Given the following problem P :

P: max $\pi = x_1 + x_2$

$\quad x_1 + 2/3 \cdot x_2 \leq 4$

$-1/2 \cdot x_1 + x_2 \leq 5/2$

$\quad\quad\quad\quad\quad x_1 \leq 3$

$\quad\quad\quad\quad\quad x_1 \geq 1$

$x_1 \in \mathbb{N}$; $x_2 \geq 0$

Define $x_1 := \sum_{p=0}^{2} x_{1p} := x_{10} + x_{11} + x_{12}$ and formulate the problem $\bar{P}$ under the following additional restrictions:

$x_{10} = 0$; $x_{11} + y_{10} = 1$; $x_{12} - x_{11} + y_{11} = 0$; $-x_{12} + y_{12} = 0$.

$\bar{P}$: max $\pi = x_{10} + x_{11} + x_{12} + x_2$

$x_{10} + x_{11} + x_{12} + 2/3 \cdot x_2 + y_1 = 4$

$-1/2 \cdot x_{10} - 1/2 \cdot x_{11} - 1/2 \cdot x_{12} + x_2 + y_2 = 5/2$

$x_{10} = 1$

$x_{11} + y_{10} = 1$

$x_{12} - x_{11} + y_{11} = 0$

$-x_{12} + y_{12} = 0$

$x_{1p}, x_2, y_1, y_2, y_{1p} \geq 0 \; \forall \; p$

After two iterations we obtain the optimal tableau $T_{opt}(cont)$ of $\bar{P}$:

$T_{opt}(cont)$:

x_{11}	x_{12}	x_2	y_1	y_2	y_{10}	y_{11}	y_{12}	
1	1	0	3/4	-1/2	0	0	0	3/4
0	0	1	3/8	3/4	0	0	0	27/8
0	-1	0	-3/4	1/2	1	0	0	1/4
0	2	0	3/4	-1/2	0	1	0	3/4
0	-1	0	0	0	0	0	1	0
0	0	0	9/8	1/4	0	0	0	41/8

The part of $T_{opt}(cont)$, inclosed in the dotted lines, is the matrix B^{-1}.

The solution is: $\bar{x}_{10} = 1$; $\bar{x}_{11} = 3/4$; $\bar{x}_{12} = 0$; $\bar{x}_2 = 27/8$; $\bar{c} = 41/8$.

$\bar{x}_1 := \sum_{p=0}^{2} \bar{x}_{1p} = 7/4 \notin \mathbb{N}$; $\pi^* = (-\infty)$; $i = 1$;

128 *Integer Programming*

cost penalty table:

	value			value		
	1	2	3	1	2	3
x_1	∞	∞	∞	0	1/8	0

$\tilde{s}_{10} = (1/4 - 1) \cdot (\max\{ \frac{0}{-1} ; \frac{9/8}{-3/4} \}) = (-3/4) \cdot 0 = 0;$

$\tilde{s}_{11} = (3/4 - 1) \cdot (\max\{ \frac{1/4}{-1/2} \}) = (-1/4) \cdot (-1/2) = 1/8 ;$

$\tilde{s}_{12} = (0 - 1) \cdot (\max\{ \frac{0}{-1} \}) = (-1) \cdot 0 = 0 .$

Select: $\tilde{s}_{12}$, where $\hat{s} = 0$; $\Delta b^T = (0;0;0;0;1)$;

$$\hat{b} = \bar{b} - B^{-1} \cdot \Delta b = \begin{pmatrix} 3/4 \\ 27/8 \\ 1/4 \\ 3/4 \\ 0 \end{pmatrix} - \begin{pmatrix} 0 \\ 0 \\ 0 \\ 0 \\ 1 \end{pmatrix} = \begin{pmatrix} 3/4 \\ 27/8 \\ 1/4 \\ 3/4 \\ -1 \end{pmatrix}$$

$\hat{b}$ is the new right-hand-side in $T_{opt}(cont)$; after two dual simplex-iterations we obtain the following integer solution:

$\bar{\bar{x}}_{10} = 1; \bar{\bar{x}}_{11} = 1; \bar{\bar{x}}_{12} = 1; \bar{\bar{x}}_2 = 3/2; \bar{\bar{x}}_1 := \sum_{p=0}^{2} \bar{\bar{x}}_{1p} = 3; \bar{\bar{c}}(1) = 9/2;$

$x^* = (3;3/2); \pi^* = 9/2; i = 2 .$

Select: $\tilde{s}_{10}$, where $\hat{s} = 0$; $\Delta b^T = (0;0;1;0;0)$;

$$\hat{b} = \begin{pmatrix} 3/4 \\ 27/8 \\ 1/4 \\ 3/4 \\ 0 \end{pmatrix} - \begin{pmatrix} 0 \\ 0 \\ 1 \\ 0 \\ 0 \end{pmatrix} = \begin{pmatrix} 3/4 \\ 27/8 \\ -3/4 \\ 3/4 \\ 0 \end{pmatrix}$$

$\hat{b}$ is the new right-hand-side in $T_{opt}(cont)$; after two dual simplex-iterations we obtain the following integer solution:

$\bar{\bar{x}}_{10} = 1; \bar{\bar{x}}_{11} = 0; \bar{\bar{x}}_{12} = 0; \bar{\bar{x}}_2 = 3; \rightarrow \bar{\bar{x}}_1 = 1; \bar{\bar{c}}(2) = 4;$

this solution is not as good as the other, so it is rejected.

Select: $\tilde{s}_{11}$, where $\hat{s} = 1/8$; $\Delta b^T = (0;0;0;1;0)$;

$$\tilde{b} = \begin{pmatrix} 3/4 \\ 27/8 \\ 1/4 \\ 3/4 \\ 0 \end{pmatrix} - \begin{pmatrix} 0 \\ 0 \\ 0 \\ 1 \\ 0 \end{pmatrix} = \begin{pmatrix} 3/4 \\ 27/8 \\ 1/4 \\ -1/4 \\ 0 \end{pmatrix}$$

$\tilde{b}$ is the new right-hand-side in T_{opt}(cont); after one dual simplex-iteration we obtain the following integer solution:

$\bar{\bar{x}}_{10} = 1$; $\bar{\bar{x}}_{11} = 1$; $\bar{\bar{x}}_{12} = 0$; $\bar{\bar{x}}_2 = 3$; $\rightarrow$ $\bar{\bar{x}}_1 = 2$; $\bar{\bar{c}}(2) = 5$;

$x^* = (2;3)$; $\pi^* = 5$; $i = 3$.

Select: $\bar{s}_{12}$, where $\hat{s} = \infty$;
we have: $\hat{s} > \bar{c} - \pi^*$ $\rightarrow$ Stop, an optimal mixed-integer solution of P has been found.

2.2.4 The Additive Algorithm (Balas)

<u>Hypotheses</u>
Given the following problem $\tilde{P}$:

$\tilde{P}$: min $\tilde{\pi} = \tilde{c} \cdot \tilde{x}$
$\tilde{A} \cdot \tilde{x} \geq \tilde{b}$
$\tilde{x}_j = 0 \lor 1$ $\forall j = 1,\ldots,n$.

<u>Principle</u>
The algorithm enumerates (after an eventual sorting and indexing) solutions of the problem. As soon as a feasible solution is found, certain sets of solutions with special properties are rejected. The remaining set of solutions is systematically examined.

130 *Integer Programming*

Description

Step 1: Is $\tilde{c}_j \geq 0 \ \forall \ j = 1,\ldots,n$?
  If yes: Set $\hat{c}_j := \tilde{c}_j \ \forall \ j = 1,\ldots,n$. Go to step 3.
  If no : Go to step 2.

Step 2: Define $\hat{x}_j := \begin{cases} \tilde{x}_j, & \text{if } \tilde{c}_j \geq 0 \\ \tilde{x}_j - 1, & \text{if } \tilde{c}_j < 0 \end{cases}$

$\hat{c}_j := \begin{cases} \tilde{c}_j, & \text{if } \tilde{c}_j \geq 0 \\ -\tilde{c}_j, & \text{if } \tilde{c}_j < 0 \end{cases}$.

Step 3: Is $\hat{c}_i \leq \hat{c}_j \ \forall \ i < j$?
  If yes: Set $x_j := \hat{x}_j \ \forall \ j = 1,\ldots,n$. Go to step 5.
  If no : Go to step 4.

Step 4: Define $c_r := r\text{-min}\{\hat{c}_j\} \ \forall \ r = 1,\ldots,n;$
  $x_j := \hat{x}_r \ \forall \ j = 1,\ldots,n; \ r = 1,\ldots,n; \ j = r$.

Step 5: Now the following problem P is given :
  P: $\min \ \pi = c \cdot x - \varepsilon$
  $A \cdot x \geq b$
  $x_j = 0 \lor 1 \ \forall \ j = 1,\ldots,n$,
  where $\varepsilon = \text{const.}$

  Define $x_j := 0$, if $a_{ij} \leq 0 \ \forall \ i = 1,\ldots,m;$
  set the running indices $l := 0; \ k := 0;$ define the set
  Q, so that $Q := \emptyset$.

Step 6: (vertical rejecting)
  Define $M^{(1)} := \{x^{(1i)} \mid (x^{(1i)} \in S^n) \land (\sum_j x_j^{(1i)} = 1)\} - \{x^{(1q)} \mid J' := \{j \ x_j^{(1q)} = 1\} \supset J^{(p)} \ \forall \ p \mid x^{(p)} \in Q\}$,
  where $S^n := \{x^{(1i)} = (x_1^{(1i)}; \ldots; x_n^{(1i)}) \mid x_j^{(1i)} = 0 \lor 1 \lor j\}$
  defines the n-dimensional shift-space.

Step 7: Is $M^{(1)} = \emptyset$?
If yes: Go to step 13.
If no : Go to step 8.

Step 8: Determine $x^{(k)} := \text{lex max}_i \{x^{(1i)} | x^{(1i)} \in M^{(1)}\}$.

Step 9: Is $R^{(k)} := A \cdot x^{(k)} - b \geq \Theta$?
If yes: Go to step 11.
If no : Define $M^{(1)} := M^{(1)} - \{x^{(k)}\}$. Go to step 10.

Step 10: Is $M^{(1)} = \emptyset$?
If yes: Set $l := l + 1$; $k := k + 1$. Go to step 6.
If no : Set $k := k + 1$. Go to step 8.

Step 11: (horizontal rejecting)
Determine the index-vectors
$J^{(k)} := ((j_\alpha) | (x_{j_\alpha}^{(k)} = 1) \wedge (j_\alpha < j_\beta \vee \alpha < \beta))$;
$\bar{J}^{(1i)} := ((j_\alpha) | (x^{(1i)} \in M^{(1)} - \{x^{(k)}\}) \wedge (x_{j_\alpha}^{(1i)} = 1) \wedge (j_\alpha < j_\beta \vee \alpha < \beta))$,
define $Q := Q \cup \{x^{(k)}\}$;
$M^{(1)} := M^{(1)} - \{x^{(k)}\} - \{x^{(1i)} | J^{(k)} - \bar{J}^{(1i)} \leq \Theta \;\forall\; i\}$.

Step 12: Is $M^{(1)} = \emptyset$?
If yes: Set $l := l + 1$; $k := k + 1$. Go to step 6.
If no : Set $k := k + 1$. Go to step 8.

Step 13: Is $Q = \emptyset$?
If yes: Stop, $\tilde{P}$ has no feasible solution.
If no : Go to step 14.

Step 14: Determine the solution vector $x^{(r)}$, so that
$\pi^{(r)} := \min \{\pi^{(p)} | x^{(p)} \in Q\}$. Return $x^{(r)}$ to the form $\tilde{x}^{(r)}$ and determine the value of the objective function
$\tilde{\pi} = \tilde{c} \cdot \tilde{x}^{(r)}$.

Example Given the following problem $\tilde{P}$:
$\tilde{P}$: min $\tilde{\pi} = 2 \cdot \tilde{x}_1 - 3 \cdot \tilde{x}_2 + \tilde{x}_3 + 5 \cdot \tilde{x}_4$
(1) $5 \cdot \tilde{x}_1 - 3 \cdot \tilde{x}_2 + \tilde{x}_3 - \tilde{x}_4 \geq 1$

(2) $\quad -2\cdot\tilde{x}_1 + \quad \tilde{x}_2 + 6\cdot\tilde{x}_3 - 4\cdot\tilde{x}_4 \geq 1$

(3) $\quad \tilde{x}_1 + 3\cdot\tilde{x}_2 - 2\cdot\tilde{x}_3 - 2\cdot\tilde{x}_4 \geq 0$

$\quad \tilde{x}_1, \tilde{x}_2, \tilde{x}_3, \tilde{x}_4 = 0 \vee 1$

$\hat{x}_1 = \tilde{x}_1;\ \hat{x}_2 = (\tilde{x}_2-1);\ \hat{x}_3 = \tilde{x}_3;\ \hat{x}_4 = \tilde{x}_4;$

$\hat{c}_1 = \tilde{c}_1;\ \hat{c}_2 = (-\tilde{c}_2);\ \hat{c}_3 = \tilde{c}_3;\ \hat{c}_4 = \tilde{c}_4;$

after the transformation we have:

$\hat{P}$: $\min \hat{\pi} = 2\cdot\hat{x}_1 + 3\cdot\hat{x}_2 + \hat{x}_3 + 5\cdot\hat{x}_4 - 3$

(1) $\quad 5\cdot\hat{x}_1 + 3\cdot\hat{x}_2 + \quad \hat{x}_3 - \quad \hat{x}_4 \geq 4$

(2) $\quad -2\cdot\hat{x}_1 - \quad \hat{x}_2 + 6\cdot\hat{x}_3 - 4\cdot\hat{x}_4 \geq 0$

(3) $\quad \hat{x}_1 - 3\cdot\hat{x}_2 - 2\cdot\hat{x}_3 - 2\cdot\hat{x}_4 \geq -3$

$\quad \hat{x}_1, \hat{x}_2, \hat{x}_3, \hat{x}_4 = 0 \vee 1$

$x_1 = \hat{x}_3;\ x_2 = \hat{x}_1;\ x_3 = \hat{x}_2;\ x_4 = \hat{x}_4;$

$c_1 = \hat{c}_3;\ c_2 = \hat{c}_1;\ c_3 = \hat{c}_2;\ c_4 = \hat{c}_4;$

now problem P is :

P: $\min \pi = x_1 + 2\cdot x_2 + 3\cdot x_3 + 5\cdot x_4 - 3$

(1) $\quad x_1 + 5\cdot x_2 + 3\cdot x_3 - \quad x_4 \geq 4$

(2) $\quad 6\cdot x_1 - 2\cdot x_2 - \quad x_3 - 4\cdot x_4 \geq 0$

(3) $\quad -2\cdot x_1 + \quad x_2 - 3\cdot x_3 - 2\cdot x_4 \geq -3$

$\quad x_1, x_2, x_3, x_4 = 0 \vee 1$.

According to step 5 we can set $x_4 = 0$ (the coefficients of this variable are all negative); now the problem reads as follows :

P: $\min \pi = x_1 + 2\cdot x_2 + 3\cdot x_3 - 3$

(1) $\quad x_1 + 5\cdot x_2 + 3\cdot x_3 \geq 4$

(2) $\quad 6\cdot x_1 - 2\cdot x_2 - \quad x_3 \geq 0$

$$(3) \quad -2 \cdot x_1 + x_2 - 3 \cdot x_3 \geq -3$$
$$x_1, x_2, x_3 = 0 \vee 1$$

$l = 0$; $k = 0$; $Q = \emptyset$; $M^{(0)} = \{x^{(01)}\}$; $x^{(01)} = (0;0;0)$;

$x^{(0)} = x^{(01)} = (0;0;0)$; $R^{(0)} = (-4;0;3)$;

$M^{(0)} = M^{(0)} - \{x^{(0)}\} = \emptyset$; $k = 1$; $l = 1$;

$M^{(1)} = \{x^{(11)}; x^{(12)}; x^{(13)}\}$; $x^{(11)} = (1;0;0)$; $x^{(12)} = (0;1;0)$;

$x^{(13)} = (0;0;1)$; $x^{(1)} = x^{(11)} = (1;0;0)$; $R^{(1)} = (-3;6;1)$;

$M^{(1)} = M^{(1)} - \{x^{(1)}\} = \{x^{(12)}; x^{(13)}\}$; $k = 2$;

$x^{(2)} = x^{(12)} = (0;1;0)$; $R^{(2)} = (1;-2;4)$;

$M^{(1)} = M^{(1)} - \{x^{(2)}\} = \{x^{(13)}\}$; $k = 3$;

$x^{(3)} = x^{(13)} = (0;0;1)$; $R^{(3)} = (-1;-1;0)$;

$M^{(1)} = M^{(1)} - \{x^{(3)}\} = \emptyset$; $k = 4$; $l = 2$;

$M^{(2)} = \{x^{(21)}; x^{(22)}; x^{(23)}\}$; $x^{(21)} = (1;1;0)$; $x^{(22)} = (1;0;1)$;

$x^{(23)} = (0;1;1)$; $x^{(4)} = x^{(21)} = (1;1;0)$;

$R^{(4)} = (2;4;2) \overset{!}{\geq} \Theta$; $J^{(4)} = (1;2)$; $J^{(22)} = (1;3)$; $J^{(23)} = (2;3)$;

$Q = \{x^{(4)}\}$; $M^{(2)} = M^{(2)} - \{x^{(4)}\} - \{x^{(22)}; x^{(23)}\} = \emptyset$;

$l = 3$; $k = 5$; $M^{(3)} = \{x^{(31)}\} - \{x^{(31)}\} \overset{!}{=} \emptyset$;

$x^{(31)} = (1;1;1)$; $Q \neq \emptyset \rightarrow x^{(r)} = x^{(4)} = (1;1;0)$;

$x_1^{(4)} = 1 \rightarrow \tilde{x}_3^{(4)} = 1$; $\qquad x_3^{(4)} = 0 \rightarrow \tilde{x}_2^{(4)} = 1$;

$x_2^{(4)} = 1 \rightarrow \tilde{x}_1^{(4)} = 1$; $\qquad x_4^{(4)} = 0 \rightarrow \tilde{x}_4^{(4)} = 0$;

the optimal solution of $\tilde{P}$ is :

$\tilde{x} = (1;1;1;0)$; the value of the objective function is $\tilde{\pi} = \tilde{c} \cdot \tilde{x} = 0$.

134 *Integer Programming*

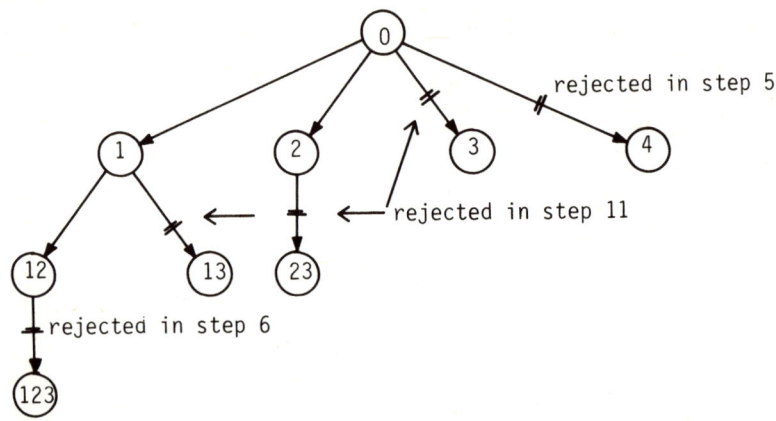

Note: In the sketch above, the numbers in the bounds are the indices j, for which $x_j^{(|i)} = 1$.

2.3 Primal–Dual Methods

2.3.1 A Partitioning Procedure for Mixed Integer Problems (Benders)

<u>Hypotheses</u>

Given the following problem P_1:

P_1: min π = c·x + g·w
    A·x + D·w $\geq$ b
         x $\geq$ 0
         w $\in$ $\mathbb{N}^{n-k}$,

where $c : c_{[1 \times k]}$; $g : g_{[1 \times (n-k)]}$;

$A : A_{[m \times k]}$; $D : D_{[m \times (n-k)]}$;

$b : b_{[m \times 1]}$; $x : x_{[k \times 1]}$;

$w : w_{[(n-k) \times 1]}$.

Note: The Partitioning Procedure may only be applied to mixed-integer problems !

Principle

The mixed-integer problem is partitioned into a continuous problem and a pure integer programming problem by this method. Both the integer subproblem (apply the dual of the all-integer problem) and the continuous problem is solved. The solution of the original problem is the union of the two sub-solutions.

Description

Step 1: Determine a feasible solution $\bar{u}$ of the problem P_2:
$$P_2: \quad u \cdot A \leq c$$
$$u \geq \theta$$
(no objective function!)

Step 2: Does at least one feasible solution $\bar{u}$ of P_2 exist ?
If yes: Set the running index $q := 1$. Go to step 3.
If no : Stop, P_1 has no finite solution.

Step 3: Compute $g_j^{(q)} := g_j - \bar{u} \cdot D_j \; \forall \; j; \quad z^{(q)} := \bar{u} \cdot b$,
where D_j is the j-th column of the matrix D.

Step 4: Solve the problem P_3:
$$P_3: \quad \min \; z$$
$$w \cdot g^{(p)} - z \leq -z^{(p)} \quad \forall \; p = 1, \ldots, q$$
$$w \in \mathbb{N}^{n-k}$$
$$z \in \mathbb{R}.$$
Let the solutions be given by $\bar{w}$ and $\bar{z}$. If $\bar{z}$ is unbounded from below, then set $\bar{z} := (-\infty)$.

Step 5: Calculate $\bar{b} := b - D_i \cdot \bar{w}$, where D_i is the i-th row of the matrix D.

Step 6: Solve the problem P_4 by the Dual Simplex-Algorithm:

P_4: max $u \cdot \bar{b}$
$u \cdot A \leq c$
$u \geq \theta$.

Let the optimal solution of P_4 be $\bar{u}$.

Step 7: Does P_4 have an unbounded solution ?
If yes: Formulate a new problem P_4 with the additional restriction $\sum_i u_i \leq C$, where C is a large positive constant. Go to step 6.
If no : Go to step 8.

Step 8: Is $\bar{z} = g \cdot \bar{w} + \bar{u} \cdot \bar{b}$?
If yes: Go to step 9.
If no : Set $q: = q + 1$. Go to step 3.

Step 9: Solve the problem P_5:
P_5: min $c \cdot x$
$A \cdot x \geq \bar{b}$
$x \geq \theta$.

Let the solution of P_5 be $\bar{x}$. Stop, $(\bar{x}; \bar{w})$ is an optimal mixed-integer solution of P_1 .

Example

Given the following problem P_1:

P_1: min $\pi = x + w$
(1) $4 \cdot x + 5 \cdot w \geq 20$
(2) $5 \cdot x + 3 \cdot w \geq 15$
$x \geq 0$
$w \in \mathbb{N}_0$

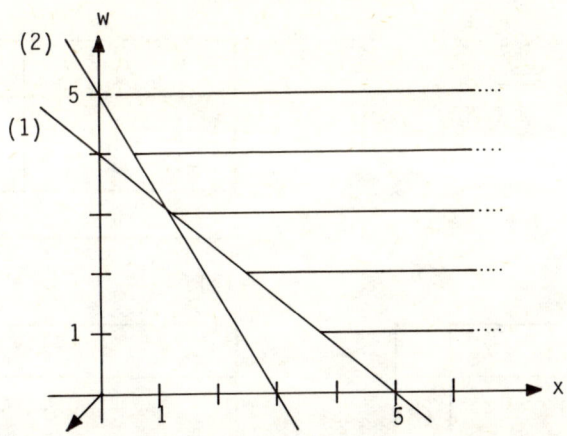

where: $c = 1$; $g = 1$; $A^T = (4;5)$; $D^T = (5;3)$; $b^T = (20;15)$;

P_2: $4 \cdot u_1 + 5 \cdot u_2 \leq 1$

$u_1, u_2 \geq 0$.

A feasible solution of P_2 is $\bar{u} = (1/4;0)$; $q = 1$;

$g_1^{(1)} = 1 - (1/4;0) \cdot \begin{pmatrix} 5 \\ 3 \end{pmatrix} = (-1/4)$; $z^{(1)} = (1/4;0) \cdot \begin{pmatrix} 20 \\ 15 \end{pmatrix} = 5$;

P_3: min z

$-1/4 \cdot w - z \leq -5$

$w \in \mathbb{N}_0$.

The solution of P_3 is $\bar{w} = 20$; $\bar{z} = 0$.
$\bar{b} = (20;15) - (5;3) \cdot 20 = (-80;-45)$;

P_4: max $(-80 \cdot u_1 - 45 \cdot u_2)$ \quad\quad min $80 \cdot u_1 + 45 \cdot u_2$

$4 \cdot u_1 + 5 \cdot u_2 \leq 1$ \quad $\rightarrow$ \quad $4 \cdot u_1 + 5 \cdot u_2 \leq 1$

$u_1, u_2 \geq 0$ \quad\quad\quad\quad $u_1, u_2 \geq 0$

The optimal solution of P_4 is $\bar{u} = (0;0)$;

$0 \stackrel{?}{=} 1 \cdot 20 + (0;0) \cdot \begin{pmatrix} -80 \\ -45 \end{pmatrix} = 20 \rightarrow 0 \neq 20$;

$q = 2$; $g^{(2)} = 1 - (0;0) \cdot \begin{pmatrix} 5 \\ 3 \end{pmatrix} = 1$; $z^{(2)} = (0;0) \cdot \begin{pmatrix} 20 \\ 15 \end{pmatrix} = 0$;

P_3: min z
  $-1/4 \cdot w - z \leq -5$
  $w - z \leq 0$
  $w \in \mathbb{N}_0$

$T^{(1)}$:

z	w	y_1	y_2	1
-1	1	1	0	0
-1	-1/4	0	1	-5
1	0	0	0	0

$T^{(2)}$:

z	w	y_1	y_2	1
-5	0	1	4	-20
4	1	0	-4	20
1	0	0	0	0

$T^{(3)}$:

z	w	y_1	y_2	1
1	0	-1/5	-4/5	4
0	1	4/5	-4/5	4
0	0	1/5	4/5	-4

The optimal solution of P_3 is $\bar{w} = 4$; $\bar{z} = 4$.

$\bar{b} = (20;15) - (5;3) \cdot 4 = (0;3);$

P_4: max $0 \cdot u_1 + 3 \cdot u_2$
  $4 \cdot u_1 + 5 \cdot u_2 \leq 1$
  $u_1, u_2 \geq 0$

$T^{(1)}$:

u_1	u_2	y	1
4	5	1	1
0	-3	0	0

$T^{(2)}$:

u_1	u_2	y	1
4/5	1	1/5	1/5
12/5	0	3/5	3/5

The optimal solution of P_4 is $\bar{u} = (0;1/5)$;

$4 \stackrel{?}{=} 1 \cdot 4 + (0;1/5) \cdot \begin{pmatrix} 0 \\ 3 \end{pmatrix} = 23/5 \to 4 \neq 23/5;$

$q = 3;$ $g^{(3)} = 1 - (0;1/5) \cdot \begin{pmatrix} 5 \\ 3 \end{pmatrix} = 2/5;$

$z^{(3)} = (0;1/5) \cdot \begin{pmatrix} 20 \\ 15 \end{pmatrix} = 3;$

P_3: min z
$-1/4 \cdot w - z \leq -5$
$w - z \leq 0$
$2/3 \cdot w - z \leq -3$
$w \in \mathbb{N}_0$

$T^{(1)}$:

z	w	y_1	y_2	y_3	1
-1	-1/4	1	0	0	-5
-1	1	0	1	0	0
-1	2/5	0	0	1	-3
1	0	0	0	0	0

$T^{(2)}$:

z	w	y_1	y_2	y_3	1
4	1	-4	0	0	20
-5	0	4	1	0	-20
-13/5	0	8/5	0	1	-11
1	0	0	0	0	0

$T^{(3)}$:

z	w	y_1	y_2	y_3	1
0	1	-4/5	4/5	0	4
1	0	-4/5	-1/5	0	4
0	0	-12/25	-13/25	1	-3/5
0	0	4/5	1/5	0	-4

$T^{(4)}$:

z	w	y_1	y_2	y_3	1
0	1	-20/13	0	20/13	40/13
1	0	-8/13	0	-5/13	55/13
0	0	12/13	1	-25/13	15/13
0	0	8/13	0	5/13	-55/13

Since $w \notin \mathbb{N}_0$, a Gomory-I-cut is applied:
$-6/13 \cdot y_1 - 7/13 \cdot y_3 + y_1^* = -1/13$

$T^{(5)}$:

z	w	y_1	y_2	y_3	y_1^*	1
0	1	-20/13	0	20/13	0	40/13
1	0	-8/13	0	-5/13	0	55/13
0	0	12/13	1	-25/13	0	15/13
0	0	-6/13	0	-7/13	1	-1/13
0	0	8/13	0	5/13	0	-55/13

$T^{(6)}$:

z	w	y_1	y_2	y_3	y_1^*	1
0	1	-20/7	0	0	20/7	20/7
1	0	-2/7	0	0	-5/7	30/7
0	0	18/7	1	0	-25/7	10/7
0	0	6/7	0	1	-13/7	1/7
0	0	2/7	0	0	5/7	-30/7

Since $w \notin \mathbb{N}_0$, a Gomory-I-cut is applied:
$- 1/7 \cdot y_1 - 6/7 \cdot y_1^* + y_2^* = -6/7$

$T^{(7)}$:

z	w	y_1	y_2	y_3	y_1^*	y_2^*	1
0	1	-20/7	0	0	20/7	0	20/7
1	0	-2/7	0	0	-5/7	0	30/7
0	0	18/7	1	0	-25/7	0	10/7
0	0	6/7	0	1	-13/7	0	1/7
0	0	-1/7	0	0	- 6/7	1	-6/7
0	0	2/7	0	0	5/7	0	-30/7

$T^{(8)}$:

z	w	y_1	y_2	y_3	y_1^*	y_2^*	1
0	1	-7/3	0	0	0	10/3	0
1	0	-1/6	0	0	0	-5/6	5
0	0	133/42	1	0	0	-25/6	5
0	0	7/6	0	1	0	-13/6	2
0	0	1/6	0	0	1	-7/6	1
0	0	4/21	0	0	0	5/6	-5

The optimal solution of P_3 is $\bar{w} = 0$; $\bar{z} = 5$;
$\bar{b} = (20;15) - (5;3) \cdot 0 = (20;15)$;

P_4: max $20 \cdot u_1 + 15 \cdot u_2$
    $4 \cdot u_1 + 5 \cdot u_2 \leq 1$
    $u_1, u_2 \geq 0$

$T^{(1)}$:

u_1	u_2	y	1
4	5	1	1
-20	-15	0	0

$T^{(2)}$:

u_1	u_2	y	1
1	5/4	1/4	1/4
0	10	5	5

The optimal solution of P_4 is $\bar{u} = (1/4;0)$.
$5 \stackrel{?}{=} 1 \cdot 0 + (1/4;0) \cdot \begin{pmatrix} 20 \\ 15 \end{pmatrix} = 5 \rightarrow 5 \stackrel{!}{=} 5$;

P_5: min $1 \cdot x$
    $4 \cdot x \geq 20$
    $5 \cdot x \geq 15$
    $x \geq 0$

$T^{(1)}$:

x	y_1	y_2	1
-4	1	0	-20
-5	0	1	-15
1	0	0	0

$T^{(2)}$:

x	y_1	y_2	1
1	-1/4	0	5
0	5/4	1	10
0	1/4	0	5

The optimal solution of P_5 is $\bar{x} = 5$.

Thereby, an optimal mixed-integer solution of P_1 is $(\bar{x};\bar{w}) = (5;0)$; the value of the objective function is $\bar{\pi} = 5$.

3. Theory of Graphs

3.0.1 Definitions

Definition 1: A graph G is an ordered pair $G = (N,A)$,
with: $N = \{n_1, n_2, \ldots, n_n\}$: set of vertices (nodes),
$n < \infty$; $A = \{a_{ij}\} = \{(n_i, n_j)\} \subseteq N \times N$: set of all
arcs, directed or undirected. The beginning node
(source) is given by n_1, the terminal node (sink)
by n_n. Unless otherwise stated, we assume from now
on that all graphs are directed (digraphs).

Definition 2: A loop is defined as an arc (n_i, n_j) with $i = j$.

Definition 3: Two arcs (n_i, n_j) and (n_k, n_l) are called parallel,
if $i = k$ and $j = l$.

Definition 4: A graph is called simple, if it contains neither
loops nor parallel arcs. Unless otherwise stated,
we assume from now on that all graphs are simple.

Definition 5: The directed arc (n_i, n_j) is called the inverted arc
of (n_k, n_l), if $i = l$ and $j = k$.

Definition 6: A node n_i is called isolated, if neither an arc
(n_i, n_j) nor an arc (n_j, n_i) exists with $n_j \in N - \{n_i\}$.

Definition 7: A path p is an alternating sequence of nodes and
arcs, with $p = (n_i, a_{ij}, n_j, \ldots, n_k, a_{kl}, n_l)$, or, in the
case of simple graphs $p = (n_i, n_j, \ldots, n_k, n_l) = (a_{ij}, \ldots, a_{kl})$.

Definition 8: A <u>cycle</u> is a path p, for which $n_i = n_1$, i.e. the beginning node of the path is equal to the end node of the path.

Definition 9: A cycle is called <u>elementary</u>, if all nodes on the cycle appear only once.

Definition 10: The set of <u>successor nodes</u> of n_i is defined as
$$N^S(n_i) := \{n_j \mid a_{ij} \in A; i \neq j\};$$ the set of <u>predecessor nodes</u> of n_i is defined as
$$P_N(n_i) := \{n_j \mid a_{ji} \in A; i \neq j\}.$$

Definition 11: The set of <u>adjacent nodes</u> of a given node n_i is defined as $P_N S := P_N \cup N^S$.

Definition 12: The <u>indegree</u> of a given node n_i is defined as
$$d^-(n_i) := |P_N(n_i)|;$$
the <u>outdegree</u> of a given node n_i is defined as
$$d^+(n_i) := |N^S(n_i)|.$$

Definition 13: A <u>valued graph</u> is an ordered triplet $\mathcal{N} = (N, A, \alpha)$, with $\alpha = (\alpha^{(1)}, \ldots, \alpha^{(k)})$: valuation of arcs; $\alpha^{(1)} = (\alpha_{ij}^{(1)})$. In general $\alpha_{ij}^{(1)} \in \mathbb{N}_0$.

In the following a valued graph is also called a <u>network</u>.

Definition 14: The <u>length of a path</u> $l(p)$ in a valued graph is defined as $$l(p) := \sum_{a_{ij} \in p} \alpha_{ij}.$$

Definition 15: A digraph G is called (<u>strongly</u>) <u>connected</u>, if there exists at least one path of finite length between all pairs of nodes $n_i, n_j \in N; i \neq j$.

Definition 16: A path from n_i to n_j is called <u>costminimal</u>, if its (cost-)length is minimal compared to the length of all other paths from n_i to n_j.

Definition 17: Let $N_\beta, N_\gamma \subset N$, so that $(N_\beta \cap N_\gamma = \emptyset) \wedge (N_\beta \cup N_\gamma = N)$, then $C(N_\beta, N_\gamma)$ is called a <u>cut</u> between N_β und N_γ.
If in a valued graph $\mathcal{N} = (N, A, \alpha)$ α_{ij} is defined as the flow capacity of the arc (n_i, n_j), then
$$C(N_\beta, N_\gamma) := \sum_{i,j} \alpha_{ij} \mid n_i \in N_\beta; n_j \in N_\gamma .$$

Definition 18: With cp_r we mean the <u>r-th critical path cp</u>.

Definition 19: $f^{opt}(\mathcal{N})$ is the <u>optimal</u>(maximal or minimal cost) <u>flow</u> through a valued graph $\mathcal{N}$.

Definition 20: The <u>adjacency matrix</u> $A^*(G):A^*(G)_{[n \times n]}$ of a given graph G is defined as
$$a^*_{ij} := \begin{cases} 1, & \text{if } a_{ij} \in A \\ 0, & \text{if } a_{ij} \notin A . \end{cases}$$

Definition 21: The <u>incidence matrix</u> $I^*(G):I^*(G)_{[n \times m]}$ of a given digraph G is defined as
$$\left. \begin{array}{l} i^*_{k,kl} = 1 \\ i^*_{l,kl} = (-1) \\ i^*_{s,kl} = 0 \, \forall s \neq k,l \end{array} \right\} \quad \forall \, a_{kl} \in A .$$

Here, m is the number of directed arcs in G.

3.0.2 The Determination of Rank in Graphs

Hypotheses
Given a finite graph G, the set Q is defined as Q: = N, the running index k: = 0.

Principle
In each iteration of this method, a set of numbers is assigned to each node using the adjacency matrix of the given graph. If the numbers are finite, then the graph contains no cycles.

Description
Step 1: Set up the adjacency matrix $A^*(G)$.

Step 2: Determine $P_{N_Q}(n_i) := \{n_j | (n_j \in Q) \wedge (a^*_{ji} = 1)\} \forall n_i \in Q$;

$M := \{n_i | P_{N_Q}(n_i) = \emptyset\}$.

Step 3: Is $M = \emptyset$?
If yes: Set $rk(n_i) := \infty \; \forall n_i \in Q$ and stop.
If no : Go to step 4.

Step 4: Set $rk(n_i) := k \; \forall n_i \in M$; $Q := Q - M$.

Step 5: Is $Q = \emptyset$?
If yes: Stop, the rank $rk(n_i)$ of each node n_i has been determined.
If no : Set k: = k + 1. Go to step 2.

Example 1
Given the following graph G:

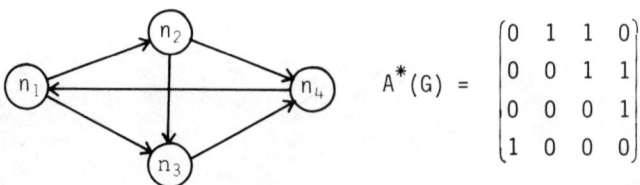

$$A^*(G) = \begin{pmatrix} 0 & 1 & 1 & 0 \\ 0 & 0 & 1 & 1 \\ 0 & 0 & 0 & 1 \\ 1 & 0 & 0 & 0 \end{pmatrix}$$

$k = 0$; $Q = \{n_1; n_2; n_3; n_4\}$;

$^PN_Q(n_1) = \{n_4\}$; $^PN_Q(n_2) = \{n_1\}$; $^PN_Q(n_3) = \{n_1; n_2\}$;

$^PN_Q(n_4) = \{n_2; n_3\}$; $M \stackrel{!}{=} \emptyset \rightarrow rk(n_1) = rk(n_2) = rk(n_3) = rk(n_4) = \infty$,

i.e. the graph G containes cycles.

Example 2

Given the following graph G :

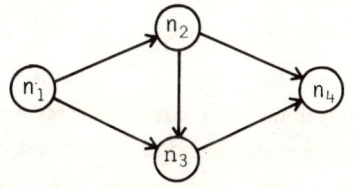

$A^*(G) = \begin{pmatrix} 0 & 1 & 1 & 0 \\ 0 & 1 & 1 & 0 \\ 0 & 0 & 0 & 1 \\ 0 & 0 & 0 & 0 \end{pmatrix}$

$k = 0$; $Q = \{n_1; n_2; n_3; n_4\}$;

$^PN_Q(n_1) = \emptyset$; $^PN_Q(n_2) = \{n_1\}$; $^PN_Q(n_3) = \{n_1; n_2\}$; $^PN_Q(n_4) = \{n_2, n_3\}$;

$M = \{n_1\}$; $rk(n_1) = 0$; $Q = \{n_2; n_3; n_4\}$; $k = 1$;

$^PN_Q(n_2) = \emptyset$; $^PN_Q(n_3) = \{n_2\}$; $^PN_Q(n_4) = \{n_2; n_3\}$;

$M = \{n_2\}$; $rk(n_2) = 1$; $Q = \{n_3; n_4\}$; $k = 2$;

$^PN_Q(n_3) = \emptyset$; $^PN_Q(n_4) = \{n_3\}$;

$M = \{n_3\}$; $rk(n_3) = 2$; $Q = \{n_4\}$; $k = 3$;

$^PN_Q(n_4) = \emptyset$; $M = \{n_4\}$; $rk(n_4) = 3$; $Q \stackrel{!}{=} \emptyset \rightarrow$ Stop,

$rk(n_1) = 0$; $rk(n_2) = 1$; $rk(n_3) = 2$; $rk(n_4) = 3$.

148 *Theory of Graphs*

3.0.3 The Number of Paths in a Graph

Hypotheses
Given a simple digraph, find out the number of paths from the source n_1 to the terminal n_n. (With slight modifications the algorithm is applicable for finding the numbers of paths between any two nodes in the graph.) Let $M(n_i)$ be the label assigned to the node n_i and N_α be the set of labelled nodes.

Principle
Starting with the label "1" for the beginning node, an iterative process is used to assign a label to each node. The label of the terminal is the number of different paths in the graph.

Description
Step 1: Set $M(n_1) := 1$; $N_\alpha := \{n_1\}$.

Step 2: Determine $M(n_r) := \sum_{j: n_j \in {}^PN(n_r)} M(n_j)$ for any $n_r \notin N_\alpha$, so that $\exists\ M(n_j)\ \forall\ n_j \in {}^PN(n_r)$; set $N_\alpha := N_\alpha \cup \{n_r\}$.

Step 3: Is $n_n \in N_\alpha$?
   If yes: Stop, $M(n_n)$ is the number of paths from the first node to the terminal.
   If no : Go to step 2.

Example
Given the following graph G :

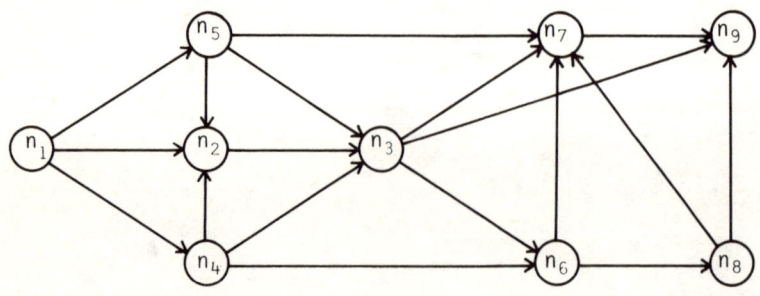

Let n_1 be the beginning node and n_9 the terminal.

$M(n_1) = 1$; $N_\alpha = \{n_1\}$;

$M(n_5) = M(n_1) = 1$; $N_\alpha = \{n_1; n_5\}$;

$M(n_4) = M(n_1) = 1$; $N_\alpha = \{n_1; n_4; n_5\}$;

$M(n_2) = M(n_1) + M(n_4) + M(n_5) = 1 + 1 + 1 = 3$; $N_\alpha = \{n_1; n_2; n_4; n_5\}$;

$M(n_3) = M(n_5) + M(n_2) + M(n_4) = 1 + 3 + 1 = 5$; $N_\alpha = \{n_1; n_2; n_3; n_4; n_5\}$;

$M(n_6) = M(n_3) + M(n_4) = 5 + 1 = 6$; $N_\alpha = \{n_1; n_2; n_3; n_4; n_5; n_6\}$;

$M(n_8) = M(n_6) = 6$; $N_\alpha = \{n_1; n_2; n_3; n_4; n_5; n_6; n_8\}$;

$M(n_7) = M(n_5) + M(n_3) + M(n_6) + M(n_8) = 1 + 5 + 6 + 6 = 18$;

$\qquad N_\alpha = \{n_1; n_2; n_3; n_4; n_5; n_6; n_7; n_8\}$;

$M(n_9) = M(n_7) + M(n_3) + M(n_8) = 18 + 5 + 6 = 29$; $N_\alpha = N = \{n_1; \ldots; n_9\}$;

$n_9 \in N_\alpha \to$ Stop, there are 29 different paths from n_1 to n_9.

3.0.4 The Determination of the Strongly Connected Components of a Graph

<u>Hypotheses</u>
Given the adjacency matrix $A^*(G)$ of a simple graph G, find the matrix $\hat{A}(G)$, which shows, what node n_j can be reached from a given node n_i, $\forall\ i,j=1,\ldots,n$. Set the running index $k := 1$.

<u>Principle</u>
In this method subsets of nodes are determined, that are connected by a cycle of directed arcs. In constructing the reduced graph each of these sets of nodes is contracted to one node.

<u>Description</u>
Step 1: Set up the matrix $A^{(o)}$, so that $A^{(o)} := A^*(G) + I$.

Step 2: Compute the matrix $A^{(k)} := (a_{ij}^{(k)})$, so that

$$a_{ij}^{(k)} := \max \{a_{ij}^{(k-1)}; a_{ik}^{(k-1)} \cdot a_{kj}^{(k-1)}\}.$$

150 *Theory of Graphs*

Step 3: Is k = n ?
   If yes: Go to step 4.
   If no : Set k: = k + 1. Go to step 2.

Step 4: Let $\hat{A}(G) := A^{(n)}$. Collect equivalent rows and columns of the matrix $\hat{A}(G)$ and group the corresponding nodes together in one new node n_i'.
   Result: A new matrix $\hat{A}(G'): \hat{A}(G')_{[1 \times 1]}$, with $1 \leq n$.
   Determine the matrix $A^*(G') := \hat{A}(G') - I$.
   Result: The adjacency matrix $A^*(G')$ of the reduced graph G'.

Example
Given the following graph G :

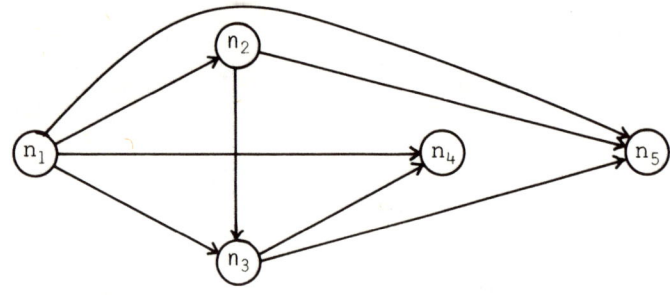

$$A^{(0)} = A^*(G) + I = \begin{pmatrix} 1 & 1 & 1 & 1 & 1 \\ 0 & 1 & 1 & 0 & 1 \\ 0 & 0 & 1 & 1 & 1 \\ 0 & 1 & 0 & 1 & 1 \\ 0 & 0 & 0 & 0 & 1 \end{pmatrix}$$

$$A^{(1)} = A^{(0)} = \begin{pmatrix} 1 & 1 & 1 & 1 & 1 \\ 0 & 1 & 1 & 0 & 1 \\ 0 & 0 & 1 & 1 & 1 \\ 0 & 1 & 0 & 1 & 1 \\ 0 & 0 & 0 & 0 & 1 \end{pmatrix} \quad A^{(2)} = \begin{pmatrix} 1 & 1 & 1 & 1 & 1 \\ 0 & 1 & 1 & 0 & 1 \\ 0 & 0 & 1 & 1 & 1 \\ 0 & 1 & 1 & 1 & 1 \\ 0 & 0 & 0 & 0 & 1 \end{pmatrix}$$

$$A^{(3)} = \begin{pmatrix} 1 & 1 & 1 & 1 & 1 \\ 0 & 1 & 1 & 1 & 1 \\ 0 & 0 & 1 & 1 & 1 \\ 0 & 1 & 1 & 1 & 1 \\ 0 & 0 & 0 & 0 & 1 \end{pmatrix} \quad A^{(4)} = \begin{pmatrix} 1 & 1 & 1 & 1 & 1 \\ 0 & 1 & 1 & 1 & 1 \\ 0 & 1 & 1 & 1 & 1 \\ 0 & 1 & 1 & 1 & 1 \\ 0 & 0 & 0 & 0 & 1 \end{pmatrix}$$

$$A^{(5)} = A^{(n)} = \hat{A}(G) = \begin{pmatrix} 1 & 1 & 1 & 1 & 1 \\ 0 & 1 & 1 & 1 & 1 \\ 0 & 1 & 1 & 1 & 1 \\ 0 & 1 & 1 & 1 & 1 \\ 0 & 0 & 0 & 0 & 1 \end{pmatrix} \quad \begin{aligned} n_1 &\to n_1' \\ (n_2; n_3; n_4) &\to n_2' \\ n_5 &\to n_3' \end{aligned}$$

$$\hat{A}(G') = \begin{pmatrix} 1 & 1 & 1 \\ 0 & 1 & 1 \\ 0 & 0 & 1 \end{pmatrix} \quad A^*(G') = \hat{A}(G') - I = \begin{pmatrix} 0 & 1 & 1 \\ 0 & 0 & 1 \\ 0 & 0 & 0 \end{pmatrix}$$

G':
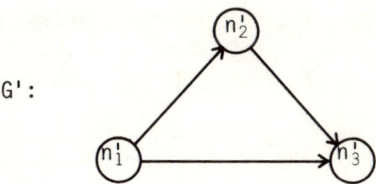

3.1 Shortest Paths in Graphs
3.1.1 The Algorithm of DIJKSTRA

<u>Hypotheses</u>
Given a connected valued graph $\mathcal{N}$ where the nonnegative arc-values c represent length or cost of the arcs, find the shortest or cost-minimal path from n_r to n_s. The set M is defined as M: = ∅ .

Principle

Starting with the beginning node n_r, the shortest paths from this node to arbitrary other nodes are determined successively until the shortest path to the terminal n_s is found.

Description

Step 1: Determine all paths from n_r to all successor nodes n_k and calculate each length c_{rk}.

Step 2: Let p_1 and p_2 be two different paths from n_r to n_k.
Is $l(p_1) \neq l(p_2)$?
If yes: Go to step 3.
If no : Go to step 4.

Step 3: Determine $l(p_g) := \max_{} \{l(p_1); l(p_2)\}$; set $M := M \cup \{p_g\}$.

Step 4: Determine $l(p_p) := \min_{t} \{l(p_t) | p_t \notin M\}$.

Step 5: $\exists\ l(p_p)$?
If yes: Go to step 6.
If no : Stop, there is no cost-minimal path from n_r to n_s.

Step 6: Let $l(p_p)$ be the length of the path $p_p = (n_r, \ldots, n_i)$.
Is $n_i = n_s$?
If yes: Stop, the shortest or cost-minimal path from n_r to n_s is given by p_p with length $l(p_p)$.
If no : Go to step 7.

Step 7: Determine all paths from n_i to each successor node n_{j_q}, which does not lie on the path p_p and calculate its length $l(n_r, \ldots, n_i, n_{j_q}) := l(p_p) + c_{ij_q}$.
Set $M := M \cup \{p_p\}$; go to step 2.

Note: If the given graph is not a digraph, then transform it into one.

Section 3.1.1   153

Example
Given the following graph $\mathcal{N}$,

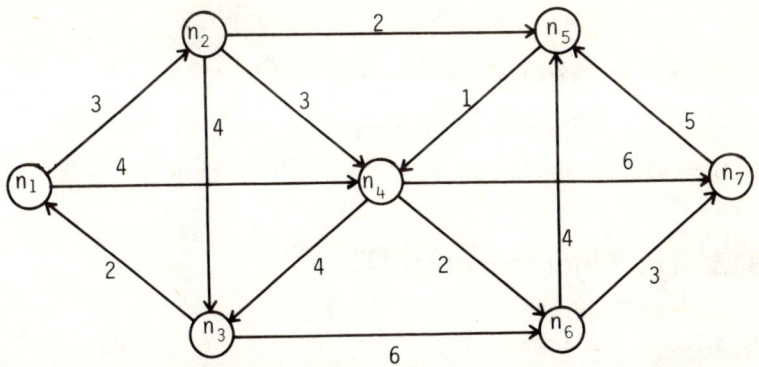

find out the shortest path from n_1 to n_7.

$M = \emptyset$;
$p_1 = (n_1; n_2)$; $l(p_1) = 3$; $p_2 = (n_1; n_4)$; $l(p_2) = 4$;
$p_p = p_1$;
$p_3 = (n_1; n_2; n_3)$; $l(p_3) = 3 + 4 = 7$;
$p_4 = (n_1; n_2; n_4)$; $l(p_4) = 3 + 3 = 6$;
$p_5 = (n_1; n_2; n_5)$; $l(p_5) = 3 + 2 = 5$;
$M = \{p_1; p_4\}$; $p_p = p_2$;

$p_6 = (n_1; n_4; n_3)$; $l(p_6) = 4 + 4 = 8$;
$p_7 = (n_1; n_4; n_6)$; $l(p_7) = 4 + 2 = 6$;
$p_8 = (n_1; n_4; n_7)$; $l(p_8) = 4 + 6 = 10$;
$M = \{p_1; p_2; p_4; p_6\}$; $p_p = p_5$;

$p_9 = (n_1; n_2; n_5; n_4)$; $l(p_9) = 5 + 1 = 6$;
$M = \{p_1; p_2; p_4; p_5; p_6; p_9\}$; $p_p = p_7$;

$p_{10} = (n_1; n_4; n_6; n_5)$; $l(p_{10}) = 6 + 4 = 10$;
$p_{11} = (n_1; n_4; n_6; n_7)$; $l(p_{11}) = 6 + 3 = 9$;

$M = \{p_1; p_2; p_4; p_5; p_6; p_7; p_8; p_{10}\}$; $P_p = p_3$;

$p_{12} = (n_1; n_2; n_3; n_6)$; $l(p_{12}) = 7 + 6 = 13$;

$M = \{p_1; p_2; p_3; p_4; p_5; p_6; p_7; p_8; p_{10}; p_{12}\}$; $P_p = p_{11}$;

$n_7 \overset{!}{\in} p_{11} \rightarrow$ Stop, $p_{11} = (n_1; n_4; n_6; n_7)$

is the shortest path from n_1 to n_7 with length $l(p_{11}) = 9$.

3.1.2 The Algorithm of DANTZIG

__Hypotheses__
The same as 3.1.1; initially define the set M as M: = $\{n_r\}$.

__Principle__   See 3.1.1

__Description__

Step 1: Formulate the following table, which will be utilized in the calculations displaying the results at each step:

n_j		n_1	n_2	. . .	n_n
$p_j = (n_r, \ldots, n_j)$					
$l(p_j)$					
$N^s(n_j)$					

Step 2: Determine all nodes $n_{k_{(j)}} \in N^s(n_j)$, so that
$n_{k_{(j)}} \notin M \;\forall\; n_j \neq n_s$.

(The index (j) refers to the predecessor node n_j.)

Step 3: Determine all paths $p_{k_{(j)}} = (p_j, n_{k_{(j)}}) \;\forall\; n_j \in M$
and their lengths $l(p_{k_{(j)}}) := l(p_j) + c_{jk_{(j)}}$.

Step 4: Determine $l(p_{q(p)}) := \min_{k,j} \{l(p_{k(j)})\}$ and
write the path $p_q = (n_r, \ldots, n_p, n_q)$ as well as its length
$l(p_q)$ in the above table under node n_q. Set $M := M \cup \{n_q\}$.

Step 5: Is $n_q = n_s$?
If yes: Stop, the shortest path from n_r to n_s is given by
p_q with length $l(p_q)$.

If no : Eliminate node $n_q \in N^S(n_j) \; \forall n_j \neq n_s$.
Go to step 6.

Step 6: Is $N^S(n_j) = \emptyset \; \forall n_j$?
If yes: Stop, there is no cost-minimal path from n_r to n_s.
If no : Go to step 3.

Note: If the given graph is not a digraph, then transform it into one.

Example
Given the following graph $\mathcal{N}$,

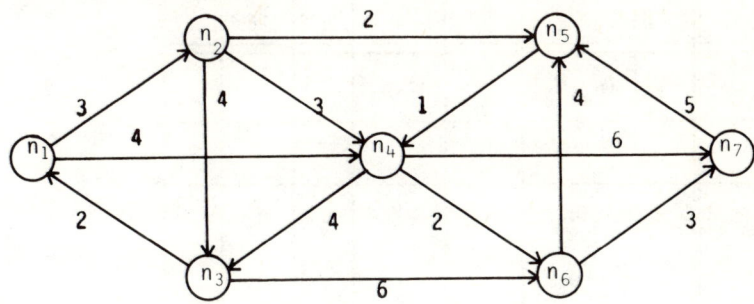

find out the shortest path from n_1 to n_7.

$M = \{n_1\}$;

$p_{2(1)} = (n_1; n_2); \; l(p_{2(1)}) = 3 =: l(p_{q(p)})$;

$p_{4(1)} = (n_1; n_4); \; l(p_{4(1)}) = 4$;

n_j	n_1	n_2	n_3	n_4	n_5	n_6	n_7
p_j		$(n_1;n_2)$					
$l(p_j)$		3					
$N^s(n_j)$	n_2 n_4	n_3 n_4 n_5	n_6	n_3 n_6 n_7	n_4	n_5 n_7	

$M = \{n_1;n_2\}$;

$p_{4_{(1)}} = (n_1;n_4); \; l(p_{4_{(1)}}) = 4 =: l(p_{q_{(p)}})$;

$p_{3_{(2)}} = (n_1;n_2;n_3); \; l(p_{3_{(2)}}) = 7$;

$p_{4_{(2)}} = (n_1;n_2;n_4); \; l(p_{2_{(4)}}) = 6$;

$p_{5_{(2)}} = (n_1;n_2;n_5); \; l(p_{5_{(2)}}) = 5$;

n_j	n_1	n_2	n_3	n_4	n_5	n_6	n_7
p_j		$(n_1;n_2)$		$(n_1;n_4)$			
$l(p_j)$		3		4			
$N^s(n_j)$	n_4	n_3 n_4 n_5	n_6	n_3 n_6 n_7	n_4	n_5 n_7	

$M = \{n_1;n_2;n_4\}$;

$p_{3_{(2)}} = (n_1;n_2;n_3); \; l(p_{3_{(2)}}) = 7$;

$p_{5_{(2)}} = (n_1;n_2;n_5); \quad 1(p_{5_{(2)}}) = 5 =: 1(p_{q_{(p)}})$;

$p_{3_{(4)}} = (n_1;n_4;n_3); \quad 1(p_{3_{(4)}}) = 8;$

$p_{6_{(4)}} = (n_1;n_4;n_6); \quad 1(p_{6_{(4)}}) = 6;$

$p_{7_{(4)}} = (n_1;n_4;n_7); \quad 1(p_{7_{(4)}}) = 10;$

n_j	n_1	n_2	n_3	n_4	n_5	n_6	n_7
p_j		$(n_1;n_2)$		$(n_1;n_4)$	$(n_1;n_2;n_5)$		
$1(p_j)$		3		4	5		
$N^s(n_j)$		n_3 n_5	n_6	n_3 n_6 n_7		n_5 n_7	

$M = \{n_1;n_2;n_4;n_5\}$;

$p_{3_{(2)}} = (n_1;n_2;n_3); \quad 1(p_{3_{(2)}}) = 7; \quad p_{3_{(4)}} = (n_1;n_4;n_3); \quad 1(p_{3_{(4)}}) = 8;$

$p_{6_{(4)}} = (n_1;n_4;n_6); \quad 1(p_{6_{(4)}}) = 6 =: 1(p_{q_{(p)}});$

$p_{7_{(4)}} = (n_1;n_4;n_7); \quad 1(p_{7_{(4)}}) = 10;$

n_j	n_1	n_2	n_3	n_4	n_5	n_6	n_7
p_j		$(n_1;n_2)$		$(n_1;n_4)$	$(n_1;n_2;n_5)$	$(n_1;n_4;n_6)$	
$1(p_j)$		3		4	5	6	
$N^s(n_j)$		n_3	n_6	n_3 n_6 n_7		n_7	

158 *Theory of Graphs*

$M = \{n_1; n_2; n_4; n_5; n_6\}$;

$p_{3_{(2)}} = (n_1; n_2; n_3); \quad l(p_{3_{(2)}}) = 7 =: l(p_{q_{(p)}})$;

$p_{3_{(4)}} = (n_1; n_4; n_3); \quad l(p_{3_{(4)}}) = 8$;

$p_{7_{(4)}} = (n_1; n_4; n_7); \quad l(p_{7_{(4)}}) = 10$;

$p_{7_{(6)}} = (n_1; n_4; n_6; n_7); \quad l(p_{7_{(6)}}) = 9$;

n_j	n_1	n_2	n_3	n_4	n_5	n_6	n_7
p_j		$(n_1;n_2)$	$(n_1;n_2;n_3)$	$(n_1;n_4)$	$(n_1;n_2;n_5)$	$(n_1;n_4;n_6)$	
$l(p_j)$		3	7	4	5	6	
$N^s(n_j)$		n_3		n_3 n_7		n_7	

$M = \{n_1; n_2; n_3; n_4; n_5; n_6\}$;

$p_{7_{(4)}} = (n_1; n_4; n_7); \quad l(p_{7_{(4)}}) = 10$;

$p_{7_{(6)}} = (n_1; n_4; n_6; n_7); \quad l(p_{7_{(6)}}) = 9 =: l(p_{q_{(p)}})$;

n_j	n_1	n_2	n_3	n_4	n_5	n_6	n_7
p_j		$(n_1;n_2)$	$(n_1;n_2;n_3)$	$(n_1;n_4)$	$(n_1;n_2;n_5)$	$(n_1;n_4;n_6)$	$(n_1;n_4;n_6;n_7)$
$l(p_j)$		3	7	4	5	6	9
$N^s(n_j)$				n_7		n_7	

$n_q \stackrel{!}{=} n_7 \rightarrow$ Stop, the shortest path from n_1 to n_7 is given by
$p_7 = (n_1; n_4; n_6; n_7)$ with length $l(p_7) = 9$.

3.1.3 The FORD Algorithm I (shortest path(s))

Hypotheses
Given a valued graph $\mathscr{N}$ with nonnegative arc-values $c_{ij} < \infty$ $\forall$ $a_{ij} \in A$, determine the length of the shortest path from n_r to each other node n_j, $\forall j \neq r$.

Principle
The algorithm determines in an iterative process a sequence of labels, whose value is monotonic non-increasing, to all nodes, until no new labels can be assigned. The label of a node n_j is then equal to the length of the shortest path from the beginning node n_r to this node n_j.

Description
Step 1: Define the labels $M(n_r) := 0$; $M(n_j) := \infty$ $\forall j \neq r$.

Step 2: Determine $Q := \{a_{ij} \mid M(n_j) - M(n_i) > c_{ij}\}$.

Step 3: Is $Q = \emptyset$?
  If yes: Stop, $M(n_j)$ is the length of the shortest path from n_r to n_j $\forall j$.
  If no : Go to step 4.

Step 4: Select one arc $a_{st} \in Q$ and define
  $M(n_t) := M(n_s) + c_{st}$. Go to step 2.

Hint: In step 4 select first the arcs a_{st} with indices s and t having values similar to r.

Example
Given the following graph $\mathscr{N}$,

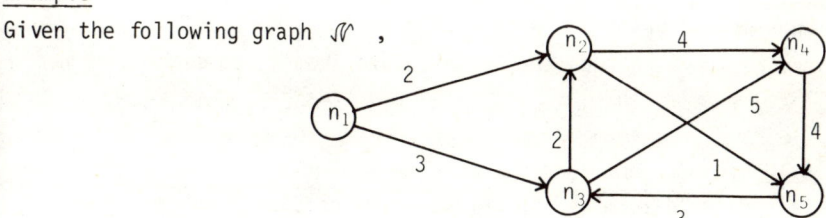

determine the length of the shortest paths from n_1 to all other nodes.

$M(n_1) = 0$; $M(n_2) = M(n_3) = M(n_4) = M(n_5) = \infty$;

$Q = \{a_{12}; a_{13}\}$; $a_{st} = a_{12}$;

$M(n_2) = 0 + 2 = 2$; $M(n_1) = 0$; $M(n_3) = M(n_4) = M(n_5) = \infty$;

$Q = \{a_{13}; a_{24}; a_{25}\}$; $a_{st} = a_{13}$;

$M(n_3) = 0 + 3 = 3$; $M(n_1) = 0$; $M(n_2) = 2$; $M(n_4) = M(n_5) = \infty$;

$Q = \{a_{24}; a_{25}; a_{34}\}$; $a_{st} = a_{34}$;

$M(n_4) = 3 + 5 = 8$; $M(n_1) = 0$; $M(n_2) = 2$; $M(n_3) = 3$; $M(n_5) = \infty$;

$Q = \{a_{24}; a_{25}; a_{45}\}$; $a_{st} = a_{45}$;

$M(n_5) = 8 + 4 = 12$; $M(n_1) = 0$; $M(n_2) = 2$; $M(n_3) = 3$; $M(n_4) = 8$;

$Q = \{a_{24}; a_{25}\}$; $a_{st} = a_{24}$;

$M(n_4) = 2 + 4 = 6$; $M(n_1) = 0$; $M(n_2) = 2$; $M(n_3) = 3$; $M(n_5) = 12$;

$Q = \{a_{25}; a_{45}\}$; $a_{st} = a_{25}$;

$M(n_5) = 2 + 1 = 3$; $M(n_1) = 0$; $M(n_2) = 2$; $M(n_3) = 3$; $M(n_4) = 6$;

$Q \stackrel{!}{=} \emptyset \rightarrow$ Stop.

3.1.4 The FORD Algorithm II (longest path(s))

Hypotheses
Given a valued graph $\mathcal{N}$ with arc-values $c_{ij} < \infty \; \forall \; a_{ij} \in A$, determine the length of the longest path from n_r to each other node n_j, $\forall \; j \neq r$.

Principle
The algorithm determines in an iterative process a sequence of

Section 3.1.4 161

labels, whose value is monotonic non-decreasing, to all nodes, until no new labels can be assigned. The label of a node n_j is then the length of the longest path from the beginning node n_r to this node n_j.

Description

Step 1: Define the labels $M(n_j) := 0 \ \forall \ j = 1,\ldots,n$.

Step 2: Determine $Q := \{a_{ij} | M(n_j) - M(n_i) < c_{ij}\}$.

Step 3: Is $Q = \emptyset$?
  If yes: Stop, $M(n_j)$ is the length of the longest path from n_r to $n_j \ \forall \ j$.
  If no : Go to step 4.

Step 4: Select one arc $a_{st} \in Q$ and define $M(n_t) := M(n_s) + c_{st}$. Go to step 2.

Hint: In step 4 select first the arcs a_{st} with indices s and t having values similar to r.

Example

Given the following graph $\mathcal{N}$,

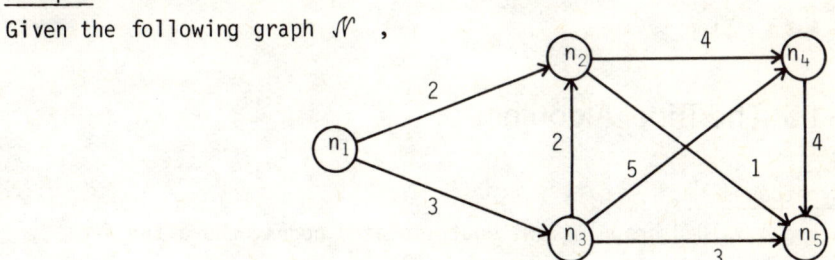

determine the length of the longest paths from n_1 to all other nodes.

$M(n_1) = M(n_2) = M(n_3) = M(n_4) = M(n_5) = 0$;

$Q = \{a_{12}; a_{13}; a_{24}; a_{25}; a_{32}; a_{34}; a_{35}; a_{45}\}$; $a_{st} = a_{12}$;

$M(n_2) = 0 + 2 = 2$; $M(n_1) = M(n_3) = M(n_4) = M(n_5) = 0$;

$Q = \{a_{13}; a_{24}; a_{25}; a_{34}; a_{35}; a_{45}\}$; $a_{st} = a_{24}$;
$M(n_4) = 2 + 4 = 6$; $M(n_1) = 0$; $M(n_2) = 2$; $M(n_3) = M(n_5) = 0$;

$Q = \{a_{13}; a_{25}; a_{34}; a_{35}; a_{45}\}$; $a_{st} = a_{45}$;
$M(n_5) = 6 + 4 = 10$; $M(n_1) = 0$; $M(n_2) = 2$; $M(n_3) = 0$; $M(n_4) = 6$;

$Q = \{a_{13}\}$; $a_{st} = a_{13}$;
$M(n_3) = 0 + 3 = 3$; $M(n_1) = 0$; $M(n_2) = 2$; $M(n_4) = 6$; $M(n_5) = 10$;

$Q = \{a_{32}; a_{34}\}$; $a_{st} = a_{32}$;
$M(n_2) = 3 + 2 = 5$; $M(n_1) = 0$; $M(n_3) = 3$; $M(n_4) = 6$; $M(n_5) = 10$;

$Q = \{a_{24}; a_{34}\}$; $a_{st} = a_{24}$;
$M(n_4) = 5 + 4 = 9$; $M(n_1) = 0$; $M(n_2) = 5$; $M(n_3) = 3$; $M(n_5) = 10$;

$Q = \{a_{45}\}$; $a_{st} = a_{45}$;
$M(n_5) = 9 + 4 = 13$; $M(n_1) = 0$; $M(n_2) = 5$; $M(n_3) = 3$; $M(n_4) = 9$;

$Q \stackrel{!}{=} \emptyset \rightarrow$ Stop.

3.1.5 The Tripel Algorithm

<u>Hypotheses</u>
Given a valued graph $\mathcal{N}$ without isolated nodes, where the arc-values c_{ij} represent the distance (cost) from n_i to n_j, determine the shortest path from each node n_i to each other node n_j $\forall$ $j \neq i$. This graph can be represented by a distance matrix $C^{(0)}$: $C^{(0)}_{[n \times n]}$ with:

$c^{(0)}_{ii} = 0$ $\forall i=1,\ldots,n$; $c^{(0)}_{ij} = \infty$, if $a_{ij} \notin A$; $c^{(0)}_{ij} = c_{ij}$, otherwise.

Principle

In this algorithm a sequence of n matrices is constructed. In each iteration one node is considered as a detour-node.

Description

Step 1: Set up the "detour-matrix" $B^{(0)}$: $B^{(0)}$ [n x n] , belonging to $C^{(0)}$, with:

$$b_{ij}^{(0)} := \begin{cases} 0, & \text{if } c_{ij}^{(0)} = \infty \\ j, & \text{otherwise} \end{cases} \quad \forall \, i,j \, .$$

Initially set the running index k: = 1 .

Step 2: Determine the matrices $C^{(k)}$ and $B^{(k)}$, so that

$$c_{ik}^{(k)} := c_{ik}^{(k-1)} \quad \forall \, i = 1,\ldots,n$$

$$c_{kj}^{(k)} := c_{kj}^{(k-1)} \quad \forall \, j = 1,\ldots,n$$

$$c_{ij}^{(k)} := \min \{c_{ij}^{(k-1)}; \, c_{ik}^{(k-1)} + c_{kj}^{(k-1)}\} \quad \forall \, i \neq k; \, j \neq k$$

$$b_{ij}^{(k)} := \begin{cases} b_{ij}^{(k-1)}, & \text{if } c_{ij}^{(k)} = c_{ij}^{(k-1)} \\ k, & \text{otherwise} \end{cases} \quad \forall \, i,j \, .$$

Step 3: Is k = n ?
If yes: Go to step 4.
If no : Set k: = k + 1. Go to step 2.

Step 4: The length of the shortest path from each node n_i to each other node n_j is given by the matrix $C^{(n)}$; each path is given by the matrix $B^{(n)}$ as follows:
Take for example the shortest path from n_r to n_s; the r-th element in the column vector $B_s^{(n)}$ defines the detour-node n_1. At this point the search continues identically. As soon as $n_1 = n_s$, then the shortest path from n_r to n_s has been determined.

164 *Theory of Graphs*

Note: Determining the matrix $C^{(n)}$, the maximal number of elementary operations is approximately $2 \cdot n^3 + 8 \cdot n$.

<u>Example</u>

Given the following graph $\mathcal{N}$,

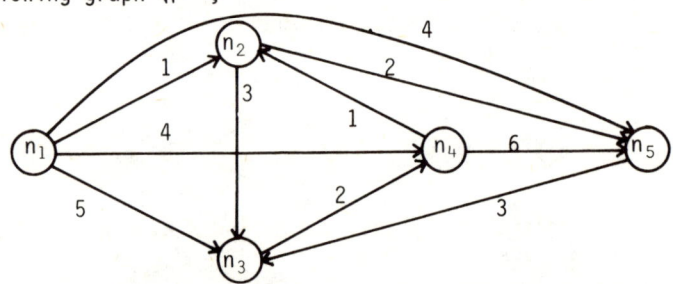

determine the shortest paths from each node n_i to each other node n_j, $\forall\ j \neq i$.

$C^{(0)}$:

	n_1	n_2	n_3	n_4	n_5
n_1	0	1	5	4	4
n_2	∞	0	3	∞	2
n_3	∞	∞	0	2	∞
n_4	∞	1	∞	0	6
n_5	∞	∞	3	∞	0

$B^{(0)}$:

	n_1	n_2	n_3	n_4	n_5
n_1	1	2	3	4	5
n_2	0	2	3	0	5
n_3	0	0	3	4	0
n_4	0	2	0	4	5
n_5	0	0	3	0	5

$C^{(1)}$:

	n_1	n_2	n_3	n_4	n_5
n_1	0	1	5	4	4
n_2	∞	0	3	∞	2
n_3	∞	∞	0	2	∞
n_4	∞	1	∞	0	6
n_5	∞	∞	3	∞	0

$B^{(1)}$:

	n_1	n_2	n_3	n_4	n_5
n_1	1	2	3	4	5
n_2	0	2	3	0	5
n_3	0	0	3	4	0
n_4	0	2	0	4	5
n_5	0	0	3	0	5

$C^{(2)}$:

	n_1	n_2	n_3	n_4	n_5
n_1	0	1	4	4	3
n_2	∞	0	3	∞	2
n_3	∞	∞	0	2	∞
n_4	∞	1	4	0	3
n_5	∞	∞	3	∞	0

$B^{(2)}$:

	n_1	n_2	n_3	n_4	n_5
n_1	1	2	2	4	2
n_2	0	2	3	0	5
n_3	0	0	3	4	0
n_4	0	2	2	4	2
n_5	0	0	3	0	5

$C^{(3)}$:

	n_1	n_2	n_3	n_4	n_5
n_1	0	1	4	4	3
n_2	∞	0	3	5	2
n_3	∞	∞	0	2	∞
n_4	∞	1	4	0	3
n_5	∞	∞	3	5	0

$B^{(3)}$:

	n_1	n_2	n_3	n_4	n_5
n_1	1	2	2	4	2
n_2	0	2	3	3	5
n_3	0	0	3	4	0
n_4	0	2	2	4	2
n_5	0	0	3	3	5

$C^{(4)}$:

	n_1	n_2	n_3	n_4	n_5
n_1	0	1	4	4	3
n_2	∞	0	3	5	2
n_3	∞	3	0	2	5
n_4	∞	1	4	0	3
n_5	∞	6	3	5	0

$B^{(4)}$:

	n_1	n_2	n_3	n_4	n_5
n_1	1	2	2	4	2
n_2	0	2	3	3	5
n_3	0	4	3	4	4
n_4	0	2	2	4	2
n_5	0	4	3	3	5

166 *Theory of Graphs*

$C^{(5)}$:

	n_1	n_2	n_3	n_4	n_5
n_1	0	1	4	4	3
n_2	∞	0	3	5	2
n_3	∞	3	0	2	5
n_4	∞	1	4	0	3
n_5	∞	6	3	5	0

$B^{(5)}$:

	n_1	n_2	n_3	n_4	n_5
n_1	1	2	2	4	2
n_2	0	2	3	3	5
n_3	0	4	3	4	4
n_4	0	2	2	4	2
n_5	0	4	3	3	5

For example the shortest path from n_3 to n_5 is sought. In $C^{(5)}$ the length is found to be 5 units.

In $B^{(5)}$ we find: 1st detour-node : n_4

2nd detour-node : n_2.

Result: The path $n_3 \rightarrow n_4 \rightarrow n_2 \rightarrow n_5$ is the shortest path from n_3 to n_5.

3.1.6 The HASSE Algorithm

Hypotheses

See 3.1.5 . Let the distance matrix be $C^{(1)}$; initially set the running index k: = 1 .

Principle

The algorithm consists of a construction of a sequence of matrices. In each iteration the length of the direct path between two nodes is compared with the previously determined length of all paths between these two nodes, which include exactly one detour-node.

Description

Step 1: Compute the matrix $C^{(k+1)}$: = $(c_{ij}^{(k+1)})$, so that

$$c_{ij}^{(k+1)} := \min_{l} \{c_{il}^{(k)} + c_{lj}^{(1)}\} \; \forall \; i,j \, .$$

Note: In general it is more efficient to modify the above formula as follows:

$$C^{(2k)} := (c_{ij}^{(2k)}), \quad \text{where}$$

$$c_{ij}^{(2k)} := \min_{l} \{c_{il}^{(k)} + c_{lj}^{(k)}\} \quad \forall\, i,j .$$

Step 2: Is $C^{(k+1)} = C^{(k)}$ (or $C^{(2k)} = C^{(k)}$) ?
   If yes: Stop, the total-distance matrix $C^{(k+1)}$ (or $C^{(2k)}$) has been found.
   If no : Set $k := k + 1$ (or $k := 2k$). Go to step 1 .

Note: Determining the matrix $C^{(\kappa)}$ that terminates the algorithm, the maximal number of elementary operations is approximately $2 \cdot n^4$.

Example
Given the following distance matrix $C^{(1)}$, determine the total-distance matrix.

$$C^{(1)} = \begin{pmatrix} 0 & 1 & 5 & 4 & 4 \\ \infty & 0 & 3 & \infty & 2 \\ \infty & \infty & 0 & 2 & \infty \\ \infty & 1 & \infty & 0 & 6 \\ \infty & \infty & 3 & \infty & 0 \end{pmatrix} \quad C^{(2)} = \begin{pmatrix} 0 & 1 & 4 & 4 & 3 \\ \infty & 0 & 3 & 5 & 2 \\ \infty & 3 & 0 & 2 & 8 \\ \infty & 1 & 4 & 0 & 3 \\ \infty & \infty & 3 & 4 & 0 \end{pmatrix}$$

$$C^{(3)} = \begin{pmatrix} 0 & 1 & 4 & 4 & 3 \\ \infty & 0 & 3 & 5 & 2 \\ \infty & 3 & 0 & 2 & 5 \\ \infty & 1 & 4 & 0 & 3 \\ \infty & 6 & 3 & 5 & 0 \end{pmatrix} \quad C^{(4)} = C^{(3)} = \begin{pmatrix} 0 & 1 & 4 & 4 & 3 \\ \infty & 0 & 3 & 5 & 2 \\ \infty & 3 & 0 & 2 & 5 \\ \infty & 1 & 4 & 0 & 3 \\ \infty & 6 & 3 & 5 & 0 \end{pmatrix} \text{ Stop !}$$

(In this case the modified algorithm (see note in step 1) is as efficient as the other one, because $C^{(8)} = C^{(4)}$.)

3.1.7 The Cascade Algorithm

Hypotheses See 3.1.5

Principle
See 3.1.6 . Here however, the results of an iteration are utilized as soon as possible for the calculations within the same iteration.

Description
Step 1: Compute the matrix $C^{(1)}$. The elements of this matrix are determined in the following sequence:

$$c_{11}^{(1)}, c_{12}^{(1)}, \ldots, c_{1n}^{(1)}, c_{21}^{(1)}, \ldots, c_{2n}^{(1)}, \ldots, c_{n1}^{(1)}, \ldots, c_{nn}^{(1)},$$

so that $c_{ij}^{(1)} := \min_{l} \{c_{il}^{(r)} + c_{lj}^{(s)}\}$, where

$$r = \begin{cases} 0, & \text{if } l \geq j \\ 1, & \text{if } l < j \end{cases} \; ; \quad s = \begin{cases} 0, & \text{if } l \geq i \\ 1, & \text{if } l < i \end{cases} \; ;$$

i.e. the upper indices r and s indicate that the elements $c_{ij}^{(1)}$, which have already been computed, are the ones used in determining the later elements.

Step 2: Compute the matrix $C^{(2)}$. The elements of this matrix are determined in the following sequence:

$$c_{nn}^{(2)}, c_{n,n-1}^{(2)}, \ldots, c_{n1}^{(2)}, c_{n-1,n}^{(2)}, \ldots, c_{n-1,1}^{(2)}, \ldots, c_{1n}^{(2)}, \ldots, c_{11}^{(2)},$$

so that $c_{ij}^{(2)} := \min_{l} \{c_{il}^{(r)} + c_{lj}^{(s)}\}$, where

$$r = \begin{cases} 1, & \text{if } l \leq j \\ 2, & \text{if } l > j \end{cases} \; ; \quad s = \begin{cases} 1, & \text{if } l \leq i \\ 2, & \text{if } l > i \end{cases} \; .$$

Result: The matrix $C^{(2)}$ includes the length of the shortest paths between all nodes n_i and n_j.

Note 1: Determining the matrix $C^{(2)}$, the maximal number of elementary operations is approximately $4 \cdot n^3$.

Note 2: If the original distance matrix $C^{(0)}$ is symmetric, then $C^{(1)}$ and $C^{(2)}$ are symmetric too.

Note 3: If a topologically sorted graph without cycles is given, then $C^{(2)} \equiv C^{(1)}$.

Example

Given the following distance matrix $C^{(0)}$, determine the total-distance matrix $C^{(2)}$.

$$C^{(0)} = \begin{pmatrix} 0 & 1 & 5 & 4 & 4 \\ \infty & 0 & 3 & \infty & 2 \\ \infty & \infty & 0 & 2 & \infty \\ \infty & 1 & \infty & 0 & 6 \\ \infty & \infty & 3 & \infty & 0 \end{pmatrix} \qquad C^{(1)} = C^{(2)} = \begin{pmatrix} 0 & 1 & 4 & 4 & 3 \\ \infty & 0 & 3 & 5 & 2 \\ \infty & 3 & 0 & 2 & 5 \\ \infty & 1 & 4 & 0 & 3 \\ \infty & 6 & 3 & 5 & 0 \end{pmatrix}$$

3.1.8 The Algorithm of LITTLE

Hypotheses

Given a strongly connected simple graph $\mathcal{N}$ and the matrix $\bar{C}$ consisting of the arc-values $\bar{c}_{ij} \geq 0 \ \forall \ i,j = 1,\ldots,n$, representing the distances (costs) between the nodes ($\bar{c}_{ii} = \infty \ \forall \ i = 1,\ldots,n$), determine a cost-minimal tour which includes all nodes only once (Hamiltonian circuit).

Principle

In this algorithm pairs of node-indices (direct connections) are successively determined so that the minimal detour, not using this arc is as long as possible. Allowing no subtours, more index-pairs are determined, beginning with the partial-path with minimal costs, as long as a sequence of index-pairs can be found which forms a Hamiltonian circuit and has the minimal cost.

Description

Step 1: Use FLOOD's Technique on the matrix $\bar{C}$.
Result: A matrix C and the reduction-constant r_0.
The index-set Q is defined as:
$Q := \emptyset$, furthermore let $R := r_0$.

Step 2: Let $c_{ij} = 0$. Assign to this element an index u, so that

$$u := (\min_{k \neq j} \{c_{ik}\} + \min_{l \neq i} \{c_{lj}\}).$$

Do this step for all elements $c_{ij} = 0$.

Step 3: Determine the element c_{rs}, so that

$$c_{rs} := (c_{ij}^{(v)} \mid v = \max \{u\}).$$

Step 4: (inclusion of (r,s))
Set $Q := Q \cup \{r\}$, where Q is the current index-set. Eliminate row r and column s and set up matrix $\hat{C}_1$, so that

$\hat{c}_{sq} := \infty \quad \forall \ q \in Q; \quad \hat{c}_{ij} := c_{ij} \quad$ otherwise.

Step 5: Use FLOOD's Technique on the matrix $\hat{C}_1$.
Result: A matrix $\hat{C}$ and the reduction-constant r_1.
Set $\hat{R} := R + r_1$; eliminate all assigned indices u.

Step 6: $((r,s)$ is not included $\equiv (\overline{r,s}))$
Set $c_{rs} := \infty$ in the matrix C from step 1 and then use FLOOD's Technique on the matrix C.
Result: A matrix $\tilde{C}$ and the reduction-constant r_2.
Set $\tilde{R} := R + r_2$.
Result: Two more end-nodes of the solution tree with the values $\hat{R}$ and $\tilde{R}$.

Section 3.1.8   171

Step 7: Select that node from all end-nodes of the solution tree
which has the minimal value R (either $\hat{R}$ or $\tilde{R}$).
For the index-set Q at <u>this</u> node is $|Q| = (n - 1)$?
If yes: Stop, the optimal tour has been found, the n-th
assignment is uniquely determined by the current
matrix $\hat{C}$.
If no : Let the approriate matrix $\hat{C}$ or $\tilde{C}$ be the new
matrix C, and R be the current node-value.
Go to step 2 .

Note 1 : If a maximum problem is given, then the matrix $\bar{C}$ has to
be transformed using the "constant"matrix as explained in 1.2.2. The
algorithm then begins with the transformed matrix.

Note 2 : The algorithm is also applicable to problems concerning the
working-sequence of products on machines, assuming cyclic production.

Example

Given the matrix $\bar{C}$, it describes completely the corresponding
graph $\mathcal{N}$.

$\bar{C} =$

from \ to	1	2	3	4	5	$c^{(r)}$
1	∞	3	9	8	2	2
2	2	∞	9	4	5	2
3	1	7	∞	9	5	1
4	3	5	6	∞	4	3
5	1	6	8	4	∞	1

$C' =$

from \ to	1	2	3	4	5
1	∞	1	7	6	0
2	0	∞	7	2	3
3	0	6	∞	8	4
4	0	2	3	∞	1
5	0	5	7	3	∞
$c^{(c)}$:	0	1	3	2	0

$C =$

from \ to	1	2	3	4	5
1	∞	0^1	4	4	0^1
2	0^0	∞	4	0^1	3
3	0^4	5	∞	6	4
4	0^0	1	0^4	∞	1
5	0^1	4	4	1	∞

reduction-constant $r_o = 15$

R: $= r_o = 15$

$v = 4, \to c_{rs} = c_{31}; Q = \emptyset$

172 *Theory of Graphs*

stage 1

$\tilde{C} = $

to from	1	2	3	4	5
1	∞	0	4	4	0
2	0	∞	4	0	3
3	∞	1	∞	2	0
4	0	1	0	∞	1
5	0	4	4	1	∞

$\hat{C}_1 = $

to from	2	3	4	5
1	0	∞	4	0
2	∞	4	0	3
4	1	0	∞	1
5	4	4	1	∞

$Q = \emptyset$; $r_2 = 4$; $\tilde{R} = 19$

$\hat{C} = $

to from	2	3	4	5
1	0^1	∞	4	0^1
2	∞	4	0^3	3
4	1	0^4	∞	1
5	3	3	0^3	∞

$Q = \{3\}$; $r_1 = 1$; $\hat{R} = 16$;
$v = 4$; → $c_{rs} = c_{43}$

stage 2

$\tilde{C} = $

to from	2	3	4	5
1	0	∞	4	0
2	∞	1	0	3
4	0	∞	∞	0
5	3	0	0	∞

$\hat{C}_1 = \hat{C} = $

to from	2	4	5
1	0^3	4	0^3
2	∞	0^3	3
5	3	0^3	∞

$Q = \{3\}$; $r_2 = 4$; $\tilde{R} = 20$;

$Q = \{3;4\}$; $r_1 = 0$;
$\hat{R} = 16$; $v = 3$;
→ $c_{rs} = c_{12}$

Section 3.1.8 173

stage 3

$\tilde{C} = $

to from	2	4	5
1	∞	4	0^7
2	∞	0^3	3
5	0^∞	0^0	∞

$\hat{C}_1 = $

to from	4	5
2	∞	3
5	0	∞

$Q = \{3;4\}$; $r_2 = 3$;
$\tilde{R} = 19$; $v = $; $\to c_{rs} = c_{52}$

$\hat{C} = $

to from	4	5
2	∞	0^∞
5	0^∞	∞

$Q = \{1;3;4\}$; $r_1 = 3$; $\hat{R} = 19$; $v = \infty$;
$\to c_{rs} = c_{25}$

stage 4
Now select $\tilde{C}$ from stage 3

$\tilde{C} = $

to from	2	4	5
1	0	4	0
2	0	0	3
5	0	0	∞

$Q = \{3;4\}$; $r_2 = \infty$;
$\tilde{R} = \infty$;

$\hat{C}_1 = $

to from	4	5
1	4	0
2	∞	3

$\hat{C} = $

to from	4	5
1	0	0
2	∞	0

$Q = \{3;4\}$; $r_1 = 7$; $\hat{R} = 26$;

stage 5
Now select $\hat{C}$ from stage 3

$\tilde{C} = $

to from	4	5
2	0	0
5	0	∞

$Q = \{1;3;4\}$; $r_2 = \infty$; $\tilde{R} = \infty$;

$\hat{C} = $

to from	4
5	0

$Q = \{1;2;3;4\}$; $\hat{R} = 19$;
Optimal !

The optimal tour is given by (3,1);(4,3);(1,2);(2,5);(5,4), i.e.:
$n_1 \to n_2 \to n_5 \to n_4 \to n_3 \to n_1$.

The associated cost is 19 m.u.

The solution tree developed the following form during the course of the algorithm:

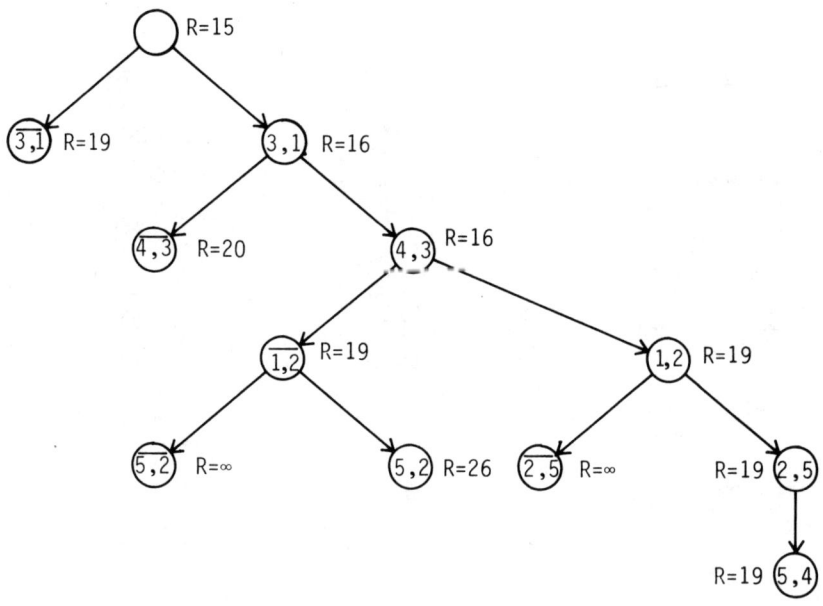

3.1.9 The Method of EASTMAN

<u>Hypotheses</u>

See 3.1.8 . Set the running index t: = 1 and furthermore $\pi^* := \infty$.

<u>Principle</u>

First a set of assignments is determined. If at least two subtours are produced, successively eliminate those index-pairs belonging to a subtour and determine new assignments. The method terminates when a tour is produced, using all nodes only once.

Section 3.1.9   175

Description

Step 1: Completely solve the problem with the Hungarian Method.
Result: An assignment $\bar{z}$ with assignment costs $\bar{\pi}$.

Step 2: Does the assignment $\bar{z}$ form a tour, in which all n nodes are included ?
If yes: Stop, $\bar{z}$ is an optimal tour with costs $\bar{\pi}$.
If no : Go to step 3.

Step 3: At this point there are k subtours S_i, $i=1,\ldots,k$. Each subtour S_i can be represented by an ordered set of r_i index-pairs:
$$S_i := ((j_1,j_2),(j_2,j_3),\ldots(j_{r_i},j_1)).$$
Set up the problems:
$$z_{il}^{(t)} := \bar{z} - \{(j_1,j_{l+1})\} \;\forall\; i=1,\ldots,k;\; l=1,\ldots,r_i;$$
where $(r_i+1) := 1$, with the appropriate matrices $C_{il}^{(t)}$, setting the element $c_{j_1,j_{l+1}} := \infty$ in the current matrix $\tilde{C}$.

Step 4: Use the Hungarian Method on the matrices $C_{il}^{(t)}$.
Result: Assignments $\bar{z}_{il}^{(t)}$ with assignment costs $\bar{\pi}_{il}^{(t)}$.

Step 5: Define $M := \{\bar{z}_{il}^{(\tau)} \mid \bar{z}_{il}^{(\tau)}$ forms a tour with all n nodes, for all $\tau = 1,\ldots,t\}$.

Step 6: Is $M = \emptyset$?
If yes: Go to step 8.
If no : Go to step 7.

Step 7: Determine $\bar{\pi}_{pq}^{(\tau)} := \min\{\bar{\pi}_{il}^{(\tau)} \mid \bar{z}_{il}^{(\tau)} \in M\}$
and set $\pi^* := \bar{\pi}_{pq}^{(\tau)}$; $z^* := \bar{z}_{pq}^{(\tau)}$.

176 Theory of Graphs

Step 8: Define $M := \{\bar{z}_{il}^{(t)} \mid (\bar{z}_{il}^{(t)} \in M) \wedge (\pi_{il}^{(t)} \leq \pi^*)\}$.

Step 9: Is $M = \emptyset$?
If yes: Stop, z^* is an optimal tour with costs π^*.
If no : Set $t := t+1$. Go to step 10.

Step 10: Now, there are k subtours S_i, $i=1,\ldots,k$, in each assignment $\bar{z}_{gh}^{(t-1)} \in \tilde{M}$.

Each of these subtours S_i can be represented by an ordered set of r_i index-pairs:

$S_i := ((j_1, j_2), (j_2, j_3), \ldots, (j_{r_i}, j_1))$.

Set up the problems:

$z_{il}^{(t)} := \bar{z}_{gh}^{(t-1)} - \{(j_1, j_{l+1})\} \forall \bar{z}_{gh}^{(t-1)} \in \tilde{M}$;

$i=1,\ldots,k$; $l=1,\ldots,r_i$; where $(r_i + 1) := 1$, with the appropriate matrices $C_{il}^{(t)}$, setting the element

$\tilde{c}_{j_1, j_{l+1}} := \infty$ in the current matrix $\tilde{C}_{gh}^{(t-1)}$. Go to step 4.

Example

Given the matrix $\bar{C}$, it describes completely the corresponding graph $\mathscr{N}$:

$\bar{C} =$

from \ to	1	2	3	4
1	∞	1	5	6
2	2	∞	6	4
3	4	7	∞	3
4	5	5	1	∞

→ $\tilde{C} =$

from \ to	1	2	3	4
1	∞	0	4	5
2	0	∞	4	2
3	1	4	∞	0
4	4	4	0	∞

$t=1$; $\pi^* = \infty$; with the Hungarian Method the following assignment $\bar{z}$ is found: $\bar{z} = ((1,2),(2,1),(3,4),(4,3))$; $\bar{\pi} = 7$; $\exists$ two subtours: $S_1 = ((1,2),(2,1))$; $S_2 = ((3,4),(4,3))$; $z^{(1)} = \bar{z} - \{(1,2)\}$;

$z_{12}^{(1)} = \bar{z} - \{(2,1)\}$; $z_{21}^{(1)} = \bar{z} - \{(3,4)\}$; $z_{22}^{(1)} = \bar{z} - \{(4,3)\}$;

$c_{11}^{(1)} =$

from\to	1	2	3	4
1	∞	∞	4	5
2	0	∞	4	2
3	1	4	∞	0
4	4	4	0	∞

→ $\tilde{c}_{11}^{(1)} =$

from\to	1	2	3	4
1	∞	∞	0	1
2	0	∞	4	2
3	1	0	∞	0
4	4	0	0	∞

$\bar{z}_{11}^{(1)} = ((1,3),(3,4),(4,2),(2,1))$; $\bar{\pi}_{11}^{(1)} = 15$;

$c_{12}^{(1)} =$

from\to	1	2	3	4
1	∞	0	4	5
2	∞	∞	4	2
3	1	4	∞	0
4	4	4	0	∞

→ $\tilde{c}_{12}^{(1)} =$

from\to	1	2	3	4
1	∞	0	4	5
2	∞	∞	2	0
3	0	4	∞	0
4	3	4	0	∞

$\bar{z}_{12}^{(1)} = ((1,2),(2,4),(4,3),(3,1))$; $\bar{\pi}_{12}^{(1)} = 10$;

$c_{21}^{(1)} =$

from\to	1	2	3	4
1	∞	0	4	5
2	0	∞	4	2
3	1	4	∞	∞
4	4	4	0	∞

→ $\tilde{c}_{21}^{(1)} =$

from\to	1	2	3	4
1	∞	0	4	3
2	0	∞	4	0
3	0	3	∞	∞
4	4	4	0	∞

$\bar{z}_{21}^{(1)} = ((1,2),(2,4),(4,3),(3,1))$; $\bar{\pi}_{21}^{(1)} = 10$;

$c_{22}^{(1)} =$

from\to	1	2	3	4
1	∞	0	4	5
2	0	∞	4	2
3	1	4	∞	0
4	4	4	∞	∞

→ $\tilde{c}_{22}^{(1)} =$

from\to	1	2	3	4
1	∞	0	0	5
2	0	∞	0	2
3	1	4	∞	0
4	0	0	∞	∞

$\bar{z}_{22}^{(1)} = ((1,2),(2,3),(3,4),(4,1))$; $\bar{\pi}_{22}^{(1)} = 15$;

$M = \{\bar{z}_{11}^{(1)}; \bar{z}_{12}^{(1)}; \bar{z}_{21}^{(1)}; \bar{z}_{22}^{(1)}\}; \bar{\pi}_{pq}^{(\tau)} = \min\{15;10;10;15\} = 10;$

$\bar{\pi}_{pq}^{(\tau)} := \bar{\pi}_{12}^{(1)} = \bar{\pi}_{21}^{(1)} = 10; \quad \pi^*: = 10; \quad z^* = ((1,2),(2,4),(4,3),(3,1));$

$\hat{M} \stackrel{!}{=} \emptyset \quad \rightarrow \quad$ Stop, z^* is an optimal tour with costs $\pi^* = 10$ m.u.

3.2 Flows in Networks
3.2.1 The Algorithm of FORD and FULKERSON

Hypotheses
Given a simple graph $\mathcal{N} = (G, \kappa)$, find the maximal flow from the source n_1 to the sink n_n. The arc values $\kappa_{ij} \in \mathbb{N}_0$ define the capacities of the arcs $a_{ij} \in A$. Let x_{ij} be the flow in arc a_{ij}. The problem can be stated as one in Linear Programming as follows:

$$\max \pi = \sum_j x_{1j} = \sum_j x_{jn}$$

$$\sum_j x_{ij} - \sum_j x_{ji} = 0 \quad \forall\, i = 2,\ldots,n-1$$

$$\sum_j x_{1j} - \sum_j x_{jn} = 0$$

$$\left.\begin{array}{l} x_{ij} \leq \kappa_{ij} \\ x_{ij} \in \mathbb{N}_0 \end{array}\right\} \quad \forall\, i,j$$

Initially set $x = (x_{ij}) = (0,\ldots,0)$, furthermore let $f^*: = 0$.

Principle
The algorithm determines in each iteration a set of paths from source to sink and the corresponding capacity. Superimposing the flows over the various paths, a flow pattern with maximal flow is determined.

Description

Step 1: Define $T := \emptyset$; $R^+ := R^- := \emptyset$; $P := \{n_1\}$; $M(n_1) := (n_1, \infty)$.

Step 2: Define $I_p := \{i \mid n_i \in P\}$,
determine r: $\max \{i \mid (i \in I_p) \wedge (i \notin T)\}$
and set $T := T \cup \{r\}$.

Step 3: (labelling phase)
Determine $P^+ := \{n_j \mid (n_j \notin P) \wedge (\kappa_{rj} > x_{rj})\}$ and set
$M(n_j) := (n_r, \varepsilon_j) \; \forall \, n_j \in P^+$; $R^+ := R^+ \cup \{a_{rj} \mid n_j \in P^+\}$,
where $\varepsilon_j := \min \{\varepsilon_r; \kappa_{rj} - x_{rj}\}$.

Step 4: Determine $P^- := \{n_j \mid (n_j \notin P) \wedge (\kappa_{jr} > 0)\}$ and set
$M(n_j) := (n_r, \varepsilon_j) \; \forall \, n_j \in P^-$; $R^- := R^- \cup \{a_{jr} \mid n_j \in P^-\}$,
where $\varepsilon_j := \min \{\varepsilon_r; x_{jr}\}$.

Step 5: Is $(P^+ \cup P^- = \phi) \wedge (|T| = |P|)$?
If yes: Stop, $x = (x_{ij})$ is the maximal flow in $\mathcal{N}$, its
value is $f^{max}(\mathcal{N}) := f^*$.
If no : Set $P := P \cup P^+ \cup P^-$. Go to step 6.

Step 6: Is $n_n \in P$?
If yes: Go to step 7.
If no : Go to step 2.

Step 7: Beginning at node n_n determine a path
$p = (n_1, \ldots, n_i, n_j, \ldots, n_n)$ by retracing the labels $M(n_j)$.

Step 8: (flow-alteration phase) FAP
Set $\varepsilon := \varepsilon_n$ and determine

$$x_{ij} := \begin{cases} x_{ij} + \varepsilon, & \text{if } (a_{ij} \in p) \wedge (a_{ij} \in R^+) \\ x_{ij} - \varepsilon, & \text{if } (a_{ij} \in p) \wedge (a_{ij} \in R^-) \\ x_{ij}, & \text{otherwise} \end{cases}$$

$f^* := f^* + \varepsilon$.

180 *Theory of Graphs*

Eliminate all labels $M(n_j)$. Go to step 1.

Example

Given the following network $\mathcal{N}$, find the maximal flow in $\mathcal{N}$.

Legend: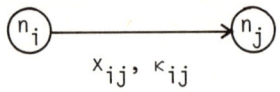
x_{ij}, κ_{ij}

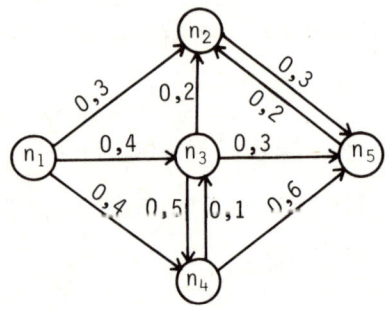

$f^* = 0; R^+ = R^- = \emptyset; P = \{n_1\}; M(n_1) = (n_1, \infty); I_p = \{1\}; r = 1; T=\{1\};$
$P^+ = \{n_2; n_3; n_3\}; M(n_2)=(n_1,3); M(n_3)=(n_1,4); M(n_4) = (n_1,4);$
$R^+ = \{a_{12}; a_{13}; a_{14}\}; P^- = \emptyset; R^- = \emptyset; (P^+ \cup P^-) \neq \emptyset \rightarrow P=\{n_1;n_2;n_3;n_4\};$
$n_5 \notin P; I_p=\{1;2;3;4\}; r=4; T=\{1;4\}; P^+=\{n_5\}; M(n_5) = (n_4,4);$
$R^+=\{a_{12};a_{13};a_{14};a_{45}\}; P^-=\emptyset; R^-=\emptyset; (P^+ \cup P^-) \neq \emptyset \rightarrow$
$P=\{n_1;n_2;n_3;n_4;n_5\}; n_5 \in P; p=(n_1;n_4;n_5); \varepsilon = \varepsilon_5 = 4;$
$x_{14} = 4; x_{45} = 4; f^* = 4;$

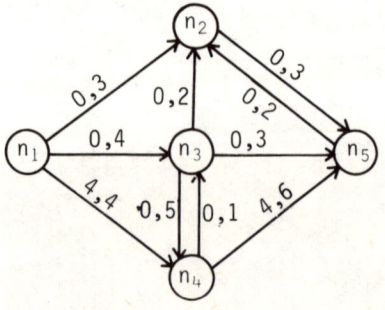

$T = \emptyset$; $R^+ = R^- = \emptyset$; $P = \{n_1\}$; $M(n_1) = (n_1; \infty)$; $I_P = \{1\}$; $r=1$; $T = \{1\}$;
$P^+ = \{n_2; n_3\}$; $M(n_2) = (n_1, 3)$; $M(n_3) = (n_1, 4)$; $R^+ = \{a_{12}; a_{13}\}$;
$P^- = \emptyset$; $R^- = \emptyset$; $(P^+ \cup P^-) \neq 0 \rightarrow P = \{n_1; n_2; n_3\}$; $n_5 \notin P$; $I_P = \{1; 2; 3\}$;
$r=3$; $T = \{1; 3\}$; $P^+ = \{n_4; n_5\}$; $M(n_4) = (n_3, 4)$; $M(n_5) = (n_3, 3)$;
$R^+ = \{a_{12}; a_{13}; a_{34}; a_{35}\}$; $P^- = \emptyset$; $R^- = \emptyset$; $(P^+ \cup P^-) \neq \emptyset \rightarrow$
$P = \{n_1; n_2; n_3; n_4; n_5\}$; $n_5 \overset{!}{\in} P$; $p = (n_1; n_3; n_5)$; $\varepsilon = \varepsilon_5 = 3$;
$x_{13} = 3$; $x_{35} = 3$; $f^* = 7$;

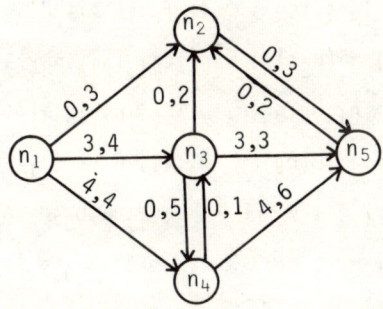

$T = \emptyset$; $R^+ = R^- = \emptyset$; $P = \{n_1\}$; $M(n_1) = (n_1, \infty)$; $I_P = \{1\}$; $r=1$; $T = \{1\}$;
$P^+ = \{n_2\}$; $M(n_2) = (n_1; 3)$; $R^+ = \{a_{12}\}$; $P^- = \emptyset$; $(P^+ \cup P^-) \neq \emptyset \rightarrow$
$P = \{n_1; n_2\}$; $n_5 \notin P$; $I_P = \{1; 2\}$; $r = 2$; $T = \{1; 2\}$; $P^+ = \{n_5\}$;
$M(n_5) = (n_2; 3)$; $R^+ = \{a_{12}; a_{25}\}$; $P^- = \emptyset$; $R^- = \emptyset$; $(P^+ \cup P^-) \neq \emptyset \rightarrow$
$P = \{n_1; n_2; n_5\}$; $n_5 \overset{!}{\in} P$; $p = (n_1; n_2; n_5)$; $\varepsilon = \varepsilon_5 = 3$;
$x_{12} = 3$; $x_{25} = 3$; $f^* = 10$;

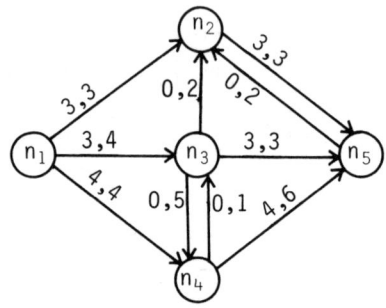

$T = \emptyset$; $R^+ = R^- = \emptyset$; $P = \{n_1\}$; $M(n_1) = (n_1, \infty)$; $I_P=\{1\}$; $r=1$; $T=\{1\}$;
$P^+ =\{n_3\}$; $M(n_3) = (n_1;1)$; $R^+ = \{a_{13}\}$; $P^- = \emptyset$; $R^- = \emptyset$; $(P^+ \cup P^-) \neq \emptyset \rightarrow$
$P = \{n_1;n_3\}$; $n_5 \notin P$; $I_P =\{1;3\}$; $r=3$; $T =\{1;3\}$; $P^+ = \{n_2;n_4\}$;
$M(n_2) = (n_3;1)$ $M(n_4) = (n_3;1)$; $R^+ = \{a_{13};a_{32};a_{34}\}$; $P^- = \emptyset$; $R^- = \emptyset$;
$(P^+ \cup P^-) \neq \emptyset \rightarrow P = \{n_1;n_2;n_3;n_4\}$; $n_5 \notin P$; $I_P =\{1;2;3;4\}$; $r = 4$;
$T = \{1;3;4\}$; $P^+ = \{n_5\}$; $M(n_5) = (n_4;1)$; $R^+ = \{a_{13};a_{32};a_{34};a_{45}\}$;
$P^- = \emptyset$; $R^- = \emptyset;(P^+ \cup P^-) \neq \emptyset \rightarrow P = \{n_1;n_2;n_3;n_4;n_5\}$; $n_5 \overset{!}{\in} P$;
$p = (n_1;n_3;n_4;n_5)$; $\varepsilon = \varepsilon_5 = 1$; $x_{13} = 4$; $x_{34} = 1$; $x_{45} = 5$; $f^* = 11$;

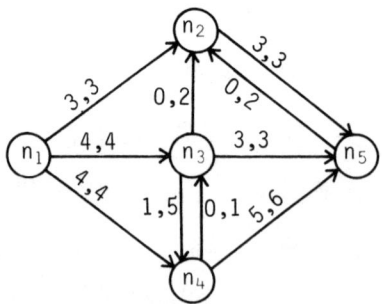

$T = \emptyset$; $R^+ = R^- = \emptyset$; $P = \{n_1\}$; $M(n_1) = (n_1, \infty)$; $I_P=\{1\}$; $r=1$; $T=\{1\}$;
$P^+ = \emptyset$; $R^+ = \emptyset$; $R^- = \emptyset$; $(P^+ \cup P^- \overset{!}{=} \emptyset) \wedge (|T| \overset{!}{=} |P|) \rightarrow$ Stop,
$f^{max}(\mathcal{N}) = f^* = 11$ is the value of the maximal flow in $\mathcal{N}$,
the corresponding flow pattern is shown above .

3.2.2 The Algorithm of BUSACKER and GOWEN

Hypotheses
Given a simple graph $\mathcal{N} = (G,\kappa,c)$, find a minimal-cost flow from the source n_1 to the sink n_n with a given flow-value $\bar{f}$. The values $\kappa_{ij} \in \mathbb{N}$ define the arc-capacities and the values $c_{ij} \in \mathbb{R}_+$ are the transportation costs per unit from n_i to n_j, $\forall\, a_{ij} \in A$. Let x_{ij} be the flow in arc a_{ij}. The problem can be stated as one in Linear Programming as follows:

$$\min\ \pi = \sum_{a_{ij} \in A} c_{ij} \cdot x_{ij}$$

$$\sum_j x_{ij} - \sum_j x_{ji} = 0 \quad \forall\, i = 2,\ldots,n-1$$

$$\sum_j x_{1j} = \bar{f}$$

$$\left.\begin{array}{r} x_{ij} \leq \kappa_{ij} \\ x_{ij} \in \mathbb{N}_0 \end{array}\right\} \quad \forall\, i,j$$

Initially set $x = (x_{ij}) = (0,\ldots,0)$, furthermore let $f^* := 0$.

Principle
In the algorithm a sequence of incremental graphs is constructed in which the shortest paths are determined. For each path the maximum feasible flow is determined. Superimposing the appropriate flow over the corresponding path yields the required flow. The algorithm terminates, when the required flow-value is reached or when no feasible flows exist.

Description
Step 1: Construct an incremental graph $I(\mathcal{N}) = (N,A^*,\gamma)$, which is associated to the current flow pattern x, so that $A^* := A' \cup A''$, where

$$x_{ij} < \kappa_{ij} \Leftrightarrow a^*_{ij} \in A'$$

184 *Theory of Graphs*

$$x_{ij} > 0 \Leftrightarrow a^*_{ji} \in A'' \;;$$

$$\gamma_{ij} := \begin{cases} c_{ij}, & \text{if } a_{ij} \in A' \\ -c_{ji}, & \text{if } a_{ij} \in A'' \end{cases}.$$

Step 2: Determine a shortest path $p = (n_1, \ldots, n_n)$ in $I(\mathcal{N})$ with respect to cost γ_{ij}.

Step 3: Are there any shortest paths $p = (n_1, \ldots, n_n)$ in $I(\mathcal{N})$ with respect to cost ?
If yes: Go to step 4.
If no : Stop, a minimal-cost flow with the given flow-value $\bar{f}$ does not exist.

Step 4: Let $\delta' := \delta'' := \infty$, determine
$$\delta' := \min \{\kappa_{ij} - x_{ij} \mid (a^*_{ij} \in p) \wedge (a^*_{ij} \in A')\}$$
$$\delta'' := \min \{x_{ji} \mid (a^*_{ij} \in p) \wedge (a^*_{ij} \in A'')\}$$
$$\varepsilon := \min \{\delta'; \delta''; \bar{f} - f^*\}$$

$$x_{ij} := \begin{cases} x_{ij} + \varepsilon, & \text{if } a^*_{ij} \in p \\ x_{ij} - \varepsilon, & \text{if } a^*_{ji} \in p \\ x_{ij}, & \text{otherwise} \end{cases},$$

set $f^* := f^* + \varepsilon$.

Step 5: Is $f^* = \bar{f}$?
If yes: Stop, $x = (x_{ij})$ is the minimal-cost flow pattern with the flow-value $f^* = \bar{f}$.
If no : Go to step 1.

Example
Given the following network $\mathcal{N}$, find a minimal-cost flow with flow-value $\bar{f} = 10$. Let $f^* = 0$.

Legend: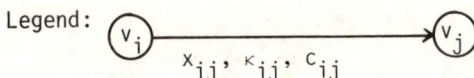

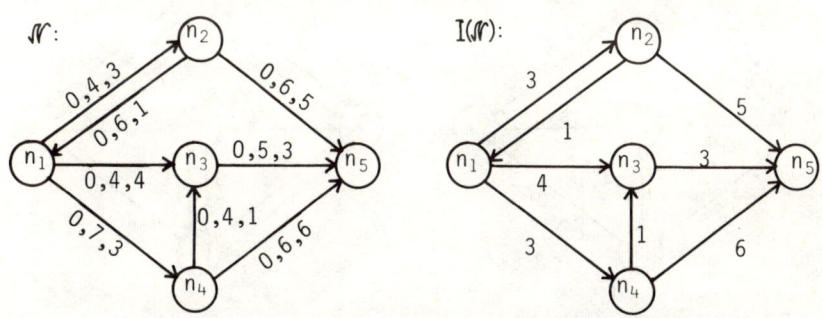

$p = (n_1, n_3, n_5);\ \varepsilon = \min \{4; \infty; 10\} = 4;\ f^* = 4$

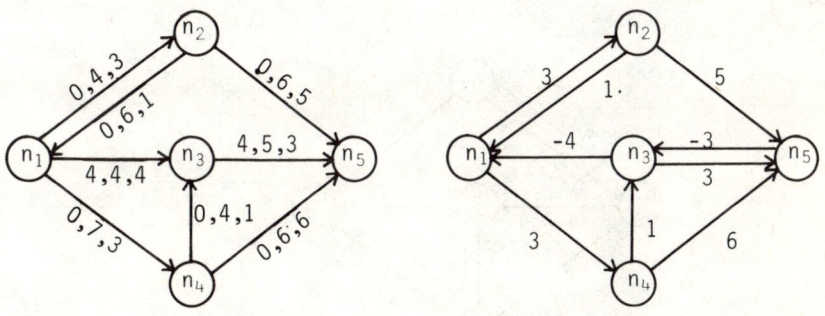

$p = (n_1, n_4, n_3, n_5);\ \varepsilon = \min \{1; \infty; 6\} = 1;\ f^* = 5$

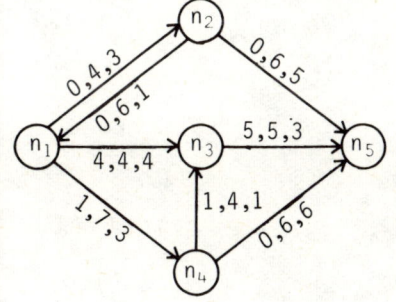
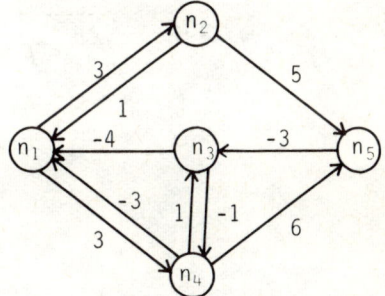

$p = (n_1, n_2, n_5)$; $\varepsilon = \min \{4; \infty; 5\} = 4$; $f^* = 9$

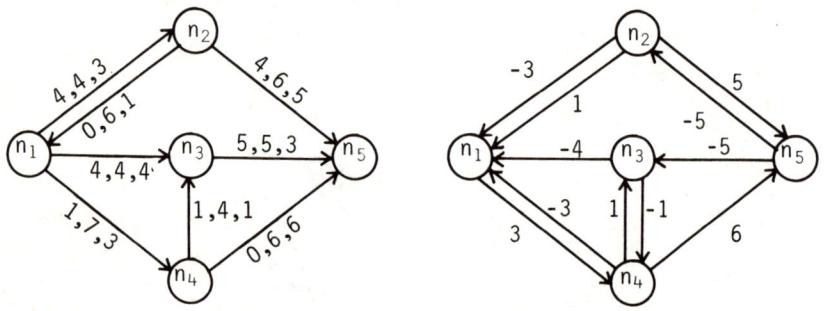

$p = (n_1, n_4, n_5)$; $\varepsilon = \min \{6; \infty; 1\} = 1$; $f^* = 10$

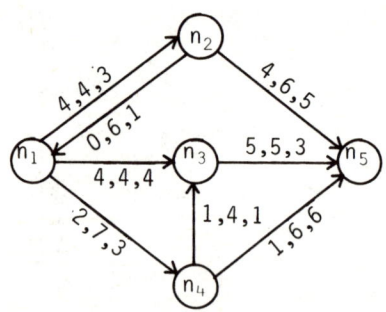

Stop, $f^* = \bar{f} = 10$; $\pi = \sum_{a_{ij} \in A} c_{ij} \cdot x_{ij} = 76$ m.u.

Section 3.2.2 187

3.2.3 The Method of KLEIN

Hypotheses   See 3.2.2

Principle
After determination of a feasible flow (if one exists), incremental graphs are constructed in which cycles of negative length are sought. Along these tours the flow is altered as much as possible, effecting a reduction in total costs. The method terminates producing a minimal-cost flow, when there are no tours of negative length in the current incremental graph.

Description

Step 1: Determine a feasible flow pattern $x = (x_{ij})$ with the flow-value $\bar{f}$ (e.g. with the Maximal-Flow Algorithm of FORD and FULKERSON).

Step 2: Are there any flow patterns $x = (x_{ij})$ with value $\bar{f}$?
If yes: Go to step 3.
If no : Stop, a feasible flow pattern with value $\bar{f}$ does not exist.

Step 3: Form the incremental graph $I(\mathcal{N}) = (N, A^*, \gamma)$ associated to the current flow pattern x, so that
$A^* := A' \cup A''$, where

$(x_{ij} < \kappa_{ij}) \wedge (x_{ji} = 0) \Leftrightarrow a^*_{ij} \in A'$
$x_{ij} > 0 \Leftrightarrow a^*_{ji} \in A''$;

$$\gamma_{ij} := \begin{cases} c_{ij} , & \text{if } a^*_{ij} \in A' \\ -c_{ji} , & \text{if } a^*_{ij} \in A'' \end{cases}.$$

Step 4: Set up the matrix $\bar{C} : \bar{C}_{[n \times n]}$, so that

$$\bar{c}_{ij} := \begin{cases} \gamma_{ij} & \forall\ a^*_{ij} \in A^* \\ \infty & , \text{otherwise} \end{cases}$$

$$\qquad\qquad\qquad\qquad\qquad\qquad\qquad\qquad (1)\quad (2)$$

If parallel arcs a^*_{ij} exist, set $\bar{c}_{ij} := \min\{\gamma^{(1)}_{ij}\ ;\ \gamma^{(2)}_{ij}\}$.

Step 5: Execute the Tripel-Algorithm using the matrix $\bar{C}$.
Result: A matrix $\tilde{C}$.

Step 6: $\exists\ \tilde{c}_{ii} < 0$?
If yes: Go to step 7.
If no : Stop, the flow pattern $x = (x_{ij})$ is a minimal-cost flow with flow-value $\bar{f}$.

Step 7: Determine $\tilde{c}_{rr} := \min\{\tilde{c}_{ii}\,|\,\tilde{c}_{ii} < 0\}$, herewith the node n_r is given.

Step 8: Define the labels $M(n_i) := (\tilde{n}_i, \varepsilon_i)$, where $\tilde{n}_i := n_i$;

$$\varepsilon_i := \begin{cases} 0 & , \text{if } i = r \\ \infty & , \text{otherwise} \end{cases} \quad \text{set } P := \{n_r\};\ T := \emptyset\ .$$

Step 9: Define $I_p := \{i\,|\,n_i \in P\}$,
determine $s := \max\{i\,|\,(i \in I_p) \wedge (i \notin T)\}$
and set $T := T \cup \{s\}$.

Step 10: Determine
$$P^* := \{n_i\,|\,(a^*_{si} \in A^*) \wedge (n_i \neq \tilde{n}_k\,|\,M(n_s) = (\tilde{n}_k, \varepsilon_s)) \wedge$$
$$\wedge\ (\varepsilon_s + \gamma_{si} \leq \varepsilon_i)\}\ .$$

Step 11: Is $P^* = \emptyset$?
If yes: Go to step 9.
If no : Go to step 12.

Step 12: Define the labels $M(n_i) := (n_s, \varepsilon_s + \gamma_{si}) \forall\ n_i \in P^*$ and set $P := P \cup P^*$.

Step 13: Is $n_r \in P^*$?
If yes: Go to step 14.
If no : Go to step 9.

Step 14: Beginning with node n_r determine a directed cycle
$p = (n_r,\ldots,n_i, n_j,\ldots,n_r)$ of negative length by retracing
the labels $M(n_j)$.

Step 15: Let $\delta' := \delta'' := \infty$, determine
$$\delta' := \min \{\kappa_{ij} - x_{ij} \mid (a^*_{ij} \in p) \wedge (a^*_{ij} \in A')\}$$
$$\delta'' := \min \{x_{ji} \mid (a^*_{ij} \in p) \wedge (a^*_{ij} \in A'')\}$$
$$\varepsilon := \min \{\delta'; \delta''\}$$

$$x_{ij} := \begin{cases} x_{ij} + \varepsilon, & \text{if } a^*_{ij} \in p \\ x_{ij} - \varepsilon, & \text{if } a^*_{ji} \in p \\ x_{ij} & \text{, otherwise} \end{cases}$$

Go to step 3.

Example
Given the following network $\mathcal{N}$, find a minimal-cost flow pattern
with value $\bar{f} = 12$.

Legend: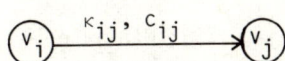

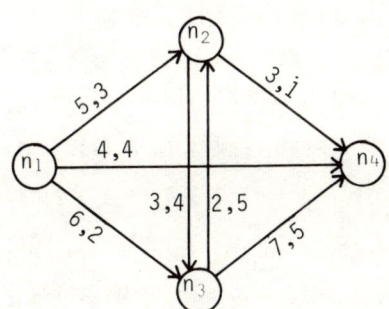

190 *Theory of Graphs*

Initial feasible flow pattern:

Legend: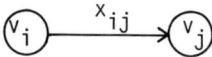

$\mathcal{N}$: 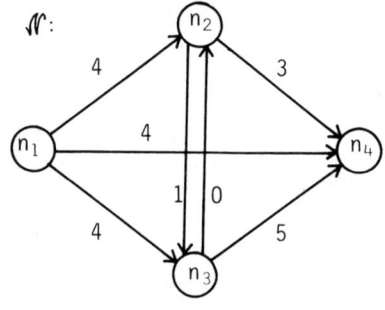 $I(\mathcal{N})$: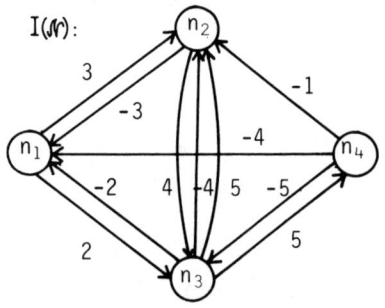

$$\bar{C} := \begin{pmatrix} \infty & 3 & 2 & \infty \\ -3 & \infty & 4 & \infty \\ -2 & -4 & \infty & 5 \\ -4 & -1 & -5 & \infty \end{pmatrix} \qquad \tilde{C} := \begin{pmatrix} 5 & -2 & 2 & 7 \\ -8 & -5 & -1 & 4 \\ -7 & -4 & -5 & 5 \\ -12 & -9 & -5 & 0 \end{pmatrix}$$

$$\tilde{c}_{rr} = \tilde{c}_{11}$$

$M(n_1)=(n_1;0)$; $M(n_2)=(n_2,\infty)$; $M(n_3)=(n_3,\infty)$; $M(n_4)=(n_4,\infty)$
$P = \{n_1\}$; $T = \emptyset$; $I_P = \{1\}$; $s = 1$; $T = \{1\}$; $P^* = \{n_2;n_3\}$

$M(n_1)=(n_1,0)$; $M(n_2)=(n_1,3)$; $M(n_3)=(n_1,2)$; $M(n_4)=(n_4,\infty)$
$P = \{n_1;n_2;n_3\}$; $n_r = n_1 \notin P^*$; $I_P = \{1;2;3\}$; $s = 3$; $T = \{1;3\}$;
$P^* = \{n_2;n_4\}$

$M(n_1)=(n_1,0)$; $M(n_2)=(n_3,-2)$; $M(n_3)=(n_1,2)$; $M(n_4)=(n_3,7)$
$P = \{n_1;n_2;n_3;n_4\}$; $n_r = n_1 \notin P^*$; $I_P = \{1;2;3;4\}$; $s = 4$;
$T = \{1;3;4\}$; $P^* = \emptyset$; $I_P = \{1;2;3;4\}$; $s = 2$; $T = \{1;2;3;4\}$; $P^*=\{1\}$;
$M(n_1)=(n_2,-5)$; $P = \{n_1;n_2;n_3;n_4\}$; $n_r = n_1 \overset{!}{\in} P^*$;
$p = (n_1;n_3;n_2;n_1)$; $\delta' = 2$; $\delta'' = \min\{1;4\} = 1$; $\varepsilon = \min\{2;1\} = 1$

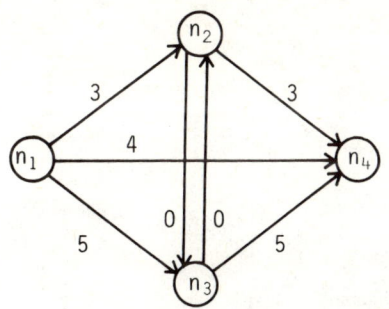 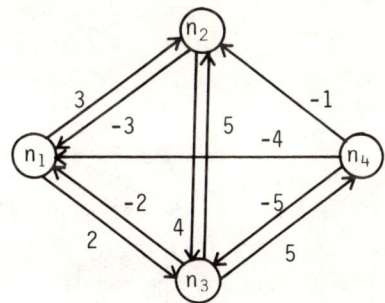

$$\bar{C} = \begin{pmatrix} \infty & 3 & 2 & \infty \\ -3 & \infty & 4 & \infty \\ -2 & 5 & \infty & 5 \\ -4 & -1 & -5 & \infty \end{pmatrix} \qquad \tilde{C} = \begin{pmatrix} 0 & 3 & 2 & 7 \\ -3 & 0 & -1 & 4 \\ -2 & 1 & 0 & 5 \\ -7 & -4 & -5 & 0 \end{pmatrix}$$

$\not\exists \; \tilde{c}_{ii} < 0 \rightarrow$ Stop, $x = (x_{ij})$ is a minimal-cost flow with value $\bar{f} = 12$.

The costs are: $\pi = \sum\limits_{a_{ij} \in A} c_{ij} \cdot x_{ij} = 63$ m.u.

3.2.4 The Out-of-Kilter Algorithm (Ford; Fulkerson)

<u>Hypotheses</u>

Given a simple graph $\mathcal{N} = (G, \lambda, \kappa, c)$, find a feasible, minimal-cost circulation. Define $\lambda_{ij} \in \mathbb{N}_0$, $\kappa_{ij} \in \mathbb{N}$, $c_{ij} \in \mathbb{Z}$ as follows:

λ_{ij} : lower bound on the arc $a_{ij} \in A$

κ_{ij} : upper bound (capacity) on the arc $a_{ij} \in A$

c_{ij} : transportation costs per unit from n_i to n_j

$a \leq b < \infty$.

If $a_{n1} \notin A$, establish an artificial arc a_{n1}, so that $\lambda_{n1} = 0$; $\kappa_{n1} \gg 0$; $c_{n1} = 0$ (or set $\lambda_{n1} = \kappa_{n1} = \bar{f}$, if a flow value $\bar{f}$ is

given). Let x_{ij} be the flow in arc a_{ij}.
The problem can be stated as one in Linear Programming as follows:

$$\min \pi = c \cdot x$$
$$I^*(G) \cdot x = \Theta$$
$$x \geq \lambda$$
$$x \leq \kappa$$

Let $u_i \in \mathbb{N}_0$ be the values of the nodes n_i, $i=1,\ldots,n$. Initially let $u_i := 0 \; \forall \; i=1,\ldots,n$; $x := (x_{ij}) := (0,\ldots,0)$; $f^* := 0$ and set the running index $t := 1$.

Note: The algorithm may also start with an arbitrary $u_i \in \mathbb{Z}$ and an arbitrary circulation x with the value $f^* < \infty$.

Principle

The algorithm determines tours using a labeling procedure, beginning with any circulation. If a tour exists, the flow is altered, if not, the "dual variables" (node-values) are altered and a test for optimality according to the complementary-slackness-theorem is performed.

Description

Step 1: Set up the network $\mathcal{N}^{(t)} = (G, x, d, z_\nu^{sign}, k, u)$,
where $d_{ij} := c_{ij} + u_i - u_j$

$$z_\nu^{sign} := \begin{cases} z_1^+ \\ z_2^+ \\ z_3^+ \\ z_1^- \\ z_2^- \\ z_3^- \\ z_4^- \\ z_5^- \\ z_6^- \end{cases} \text{if} \begin{cases} (d_{ij} > 0) \wedge (x_{ij} = \lambda_{ij}) \\ (d_{ij} < 0) \wedge (x_{ij} = \kappa_{ij}) \\ (d_{ij} = 0) \wedge (x_{ij} \in [\lambda_{ij}; \kappa_{ij}]) \\ (d_{ij} > 0) \wedge (x_{ij} < \lambda_{ij}) \\ (d_{ij} > 0) \wedge (x_{ij} > \lambda_{ij}) \\ (d_{ij} < 0) \wedge (x_{ij} < \kappa_{ij}) \\ (d_{ij} < 0) \wedge (x_{ij} > \kappa_{ij}) \\ (d_{ij} = 0) \wedge (x_{ij} < \lambda_{ij}) \\ (d_{ij} = 0) \wedge (x_{ij} > \kappa_{ij}) \end{cases}$$

$$k_{ij} := \begin{cases} d_{ij} \cdot (x_{ij} - \lambda_{ij}) = 0 \\ d_{ij} \cdot (x_{ij} - \kappa_{ij}) = 0 \\ d_{ij} \cdot (x_{ij} - \kappa_{ij}) = 0 \end{cases} \text{if} \begin{cases} r_{ij} : z_1^+ \\ r_{ij} : z_2^+ \\ r_{ij} : z_3^+ \end{cases} \\ \begin{cases} \lambda_{ij} - x_{ij} \\ d_{ij} \cdot (x_{ij} - \lambda_{ij}) \\ d_{ij} \cdot (x_{ij} - \kappa_{ij}) \end{cases} \text{if} \begin{cases} r_{ij} : z_1^- \\ r_{ij} : z_2^- \\ r_{ij} : z_3^- \end{cases} \\ \begin{cases} x_{ij} - \kappa_{ij} \\ \lambda_{ij} - x_{ij} \\ x_{ij} - \kappa_{ij} \end{cases} \text{if} \begin{cases} r_{ij} : z_4^- \\ r_{ij} : z_5^- \\ r_{ij} : z_6^- \end{cases} \end{cases}$$

Step 2: Is $k_{ij} = 0 \ \forall a_{ij} \in A$?
If yes: Stop, $x = (x_{ij})$ is an optimal flow pattern with the value $f^{opt}(\mathcal{N}) = f^*$.
If no : Go to step 3.

Step 3: Determine an arc $a_{kl} \in \{a_{ij} \mid k_{ij} > 0\}$.

Step 4: Let $R^+ := R^- := T := \emptyset$.
Label the node n_k or n_l according to the following rule :

$z_\nu^{sign}(a_{kl})$	to be labelled	label $M(n.)$	where	and set
z_1^-	n_l	(n_k, ε_l)	$\varepsilon_l = \lambda_{kl} - x_{kl}$	$P := \{n_l\}; R^+ := R^+ \cup \{a_{kl}\}$
z_3^-, z_5^-	n_l	(n_k, ε_l)	$\varepsilon_l = \kappa_{kl} - x_{kl}$	$P := \{n_l\}; R^+ := R^+ \cup \{a_{kl}\}$
z_2^-, z_6^-	n_k	(n_l, ε_k)	$\varepsilon_k = x_{kl} - \lambda_{kl}$	$P := \{n_k\}; R^- := R^- \cup \{a_{kl}\}$
z_4^-	n_k	(n_l, ε_k)	$\varepsilon_k = x_{kl} - \kappa_{kl}$	$P := \{n_k\}; R^- := R^- \cup \{a_{kl}\}$

Step 5: Define $I_P := \{i \mid n_i \in P\}$,
determine $s := \max \{i \mid (i \in I_P) \land (i \notin T)\}$

and set $T := T \cup \{s\}$.

Step 6: Consider all nodes n_j, for which $(n_j \in {}^{P}N^{S}(n_s)) \wedge (n_j \notin P)$ holds and label these nodes according to the following rule:

condition for labelling n_j		label $M(n_j)$	where	and set
$n_j \in N^{S}(n_s)$	$(d_{sj}>0) \wedge (x_{sj}<\lambda_{sj})$	(n_s, ε_j)	$\varepsilon_j = \min\{\varepsilon_s; \lambda_{sj} - x_{sj}\}$	$R^+ := R^+ \cup \{a_{sj}\}$
	$(d_{sj} \leq 0) \wedge (x_{sj}<\kappa_{sj})$	(n_s, ε_j)	$\varepsilon_j = \min\{\varepsilon_s; \kappa_{sj} - x_{sj}\}$	$R^+ := R^+ \cup \{a_{sj}\}$
$n_j \in {}^{P}N(n_s)$	$(d_{js} \geq 0) \wedge (x_{js}>\lambda_{js})$	(n_s, ε_j)	$\varepsilon_j = \min\{\varepsilon_s; x_{js} - \lambda_{js}\}$	$R^- := R^- \cup \{a_{js}\}$
	$(d_{js}<0) \wedge (x_{js}>\kappa_{js})$	(n_s, ε_j)	$\varepsilon_j = \min\{\varepsilon_s; x_{js} - \kappa_{js}\}$	$R^- := R^- \cup \{a_{js}\}$

Define $P^* := \{n_j \mid M(n_j) = (n_s, \varepsilon_j)\}$.

Step 7: Is $((n_k \in P) \wedge (n_l \in P^*)) \vee ((n_k \in P^*) \wedge (n_l \in P))$?
If yes: Go to step 12.
If no : Go to step 8.

Step 8: Is $(P^* = \emptyset) \wedge (|P| = |T|)$?
If yes: Go to step 9.
If no : Set $P := P \cup P^*$. Go to step 5.

Step 9: Let $\delta_1 := \delta_2 := \infty$, define
$R_1 := \{a_{ij} \mid (n_i \in P) \wedge (n_j \notin P) \wedge (d_{ij} > 0) \wedge (x_{ij} \leq \kappa_{ij})\}$
$R_2 := \{a_{ij} \mid (n_i \notin P) \wedge (n_j \in P) \wedge (d_{ij} < 0) \wedge (x_{ij} \geq \lambda_{ij})\}$
and determine
$\delta_1 := \min \{d_{ij} \mid a_{ij} \in R_1\}$
$\delta_2 := \min \{-d_{ij} \mid a_{ij} \in R_2\}$
$\delta := \min \{\delta_1; \delta_2\}$.

Step 10: Is $\delta = \infty$?

If yes: Stop, there is no feasible circulation in the network $\mathcal{N}$.

If no : Go to step 11.

Step 11: Define
$$u_i := \begin{cases} u_i + \delta, & \text{if } n_i \notin P \\ u_i, & \text{otherwise} \end{cases}$$
and set t: = t+1 . Eliminate all labels $M(n_j)$; go to step 1.

Step 12: Let $n_1(n_k)$ be the first node to be labelled in step 4. Beginning with $n_k(n_1)$ determine the cycle
$$p=(n_1,\ldots,n_i,n_j,\ldots,n_k,n_1) \text{ [or } p=(n_k,\ldots,n_i,n_j,\ldots,n_1,n_k)]$$
by retracing the labels $M(n_j)$.

Step 13: Determine $\varepsilon := \min\{\varepsilon_j \mid n_j \in p\}$
$$x_{ij} := \begin{cases} x_{ij} + \varepsilon, & \text{if } (a_{ij} \in p) \wedge (a_{ij} \in R^+) \\ x_{ij} - \varepsilon, & \text{if } (a_{ij} \in p) \wedge (a_{ij} \in R^-) \\ x_{ij}, & \text{otherwise} \end{cases}$$
set $f^* := f^* + \varepsilon$; t: = t+1 .
Eliminate all labels $M(n_j)$; go to step 1 .

Example

Given the following network $\mathcal{N}$, find a minimal-cost flow pattern with value $\bar{f} = 5$.

Legend:

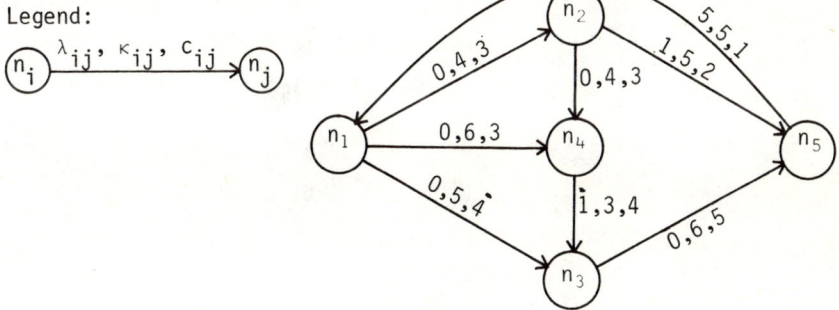

196 *Theory of Graphs*

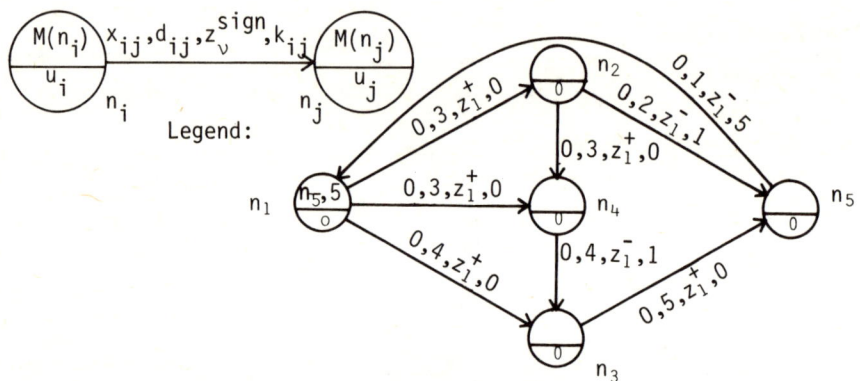

$f^* = 0$; $t = 1$; $R^+ = R^- = \emptyset$; $T = \emptyset$; $a_{kl} = a_{51}$; $P = \{n_1\}$; $R^+ = \{a_{51}\}$
$I_p = \{1\}$; $s = 1$; $T = \{1\}$; $(P^* \stackrel{!}{=} \emptyset) \wedge (|P| \stackrel{!}{=} |T|)$; $n_5 \notin P \rightarrow \delta_1 = \delta_2 = \infty$;
$R_1 = \{a_{12}; a_{13}; a_{14}\}$; $R_2 = \emptyset$; $\delta_1 = \min \{3;4;3\} = 3$;
$\delta = \min \{3;\infty\} = 3$; $t = 2$;

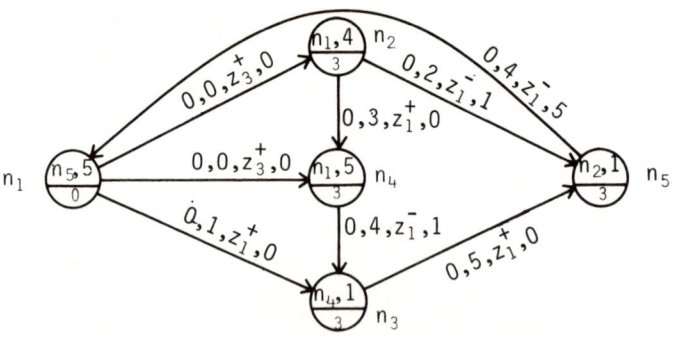

$R^+ = R^- = \emptyset$; $T = \emptyset$; $a_{kl} = a_{51}$; $P = \{n_1\}$; $R^+ = \{a_{51}\}$; $I_p = \{1\}$;
$s = 1$; $T = \{1\}$; $P^* = \{n_2; n_4\}$; $R^+ = \{a_{12}; a_{14}; a_{51}\}$; $P^* \neq \emptyset \rightarrow$
$P = \{n_1; n_2; n_4\}$; $I_p = \{1;2;4\}$; $s = 4$; $T = \{1;4\}$; $P^* = \{n_3\}$;
$R^+ = \{a_{12}; a_{14}; a_{43}; a_{51}\}$; $P^* \neq \emptyset \rightarrow P = \{n_1; n_2; n_3; n_4\}$;
$I_p = \{1;2;3;4\}$; $s = 3$; $T = \{1;3;4\}$; $P^* \stackrel{!}{=} \emptyset$; $|P| \neq |T|$ $\rightarrow$
$P = \{n_1; n_2; n_3; n_4\}$; $I_p = \{1;2;3;4\}$; $s = 2$; $T = \{1;2;3;4\}$;

Section 3.2.4   197

$P^* = \{n_5\}$; $R^+ = \{a_{12}; a_{14}; a_{25}; a_{34}; a_{51}\}$; $P^* \neq \emptyset \to P = \{n_1; n_2; n_3; n_4; n_5\}$;
$I_p = \{1;2;3;4;5\}$; $s = 5$; $T = \{1;2;3;4;5\}$; $(P^* \stackrel{!}{=} \emptyset) \wedge (|P| \stackrel{!}{=} |T|)$
$(n_1 \stackrel{!}{\in} P)$ $(n_5 \stackrel{!}{\in} P) \to$ $p = (n_1, n_2, n_5, n_1)$; $\varepsilon = 1$; $f^* = 1$; $t = 3$;

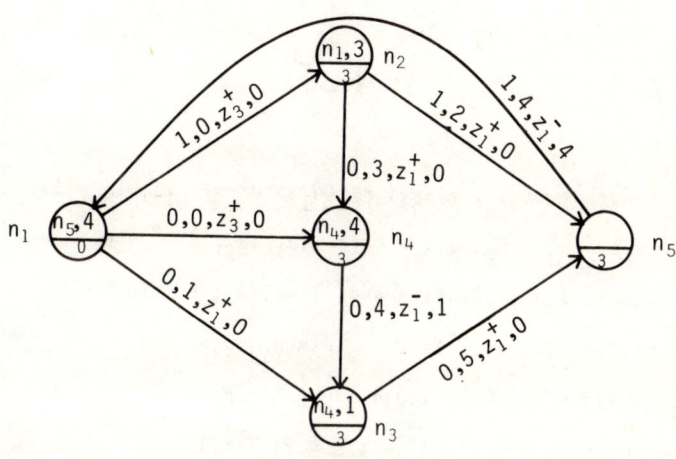

$R^+ = R^- = \emptyset$; $T = \emptyset$; $a_{k1} = a_{51}$; $P = \{n_1\}$; $R^+ = \{a_{51}\}$; $I_p = \{1\}$;
$s = 1$; $T = \{1\}$; $P^* = \{n_2; n_4\}$; $R^+ = \{a_{12}; a_{14}; a_{51}\}$; $P^+ \neq \emptyset \to$
$P = \{n_1; n_2; n_4\}$; $I_p = \{1;2;4\}$; $s = 4$; $T = \{1;4\}$; $P^* = \{n_3\}$;
$R^+ = \{a_{12}; a_{14}; a_{43}; a_{51}\}$; $P^* \neq \emptyset \to P = \{n_1; n_2; n_3; n_4\}$;
$I_p = \{1;2;3;4\}$; $s = 3$; $T = \{1;3;4\}$; $P^* \stackrel{!}{=} \emptyset$; $|P| \neq |T| \to$
$P = \{n_1; n_2; n_3; n_4\}$; $I_p = \{1;2;3;4\}$; $s = 2$; $T = \{1;2;3;4\}$;
$(P^* \stackrel{!}{=} \emptyset) \wedge (|P| \stackrel{!}{=} |T|)$; $n_5 \notin P \to \delta_1 = \delta_2 = \infty$; $R_1 = \{a_{25}; a_{35}\}$;
$R_2 = \emptyset$; $\delta_1 = \min \{2;5\} = 2$; $\delta = \min \{2; \infty\} = 2$; $t = 4$;

198 *Theory of Graphs*

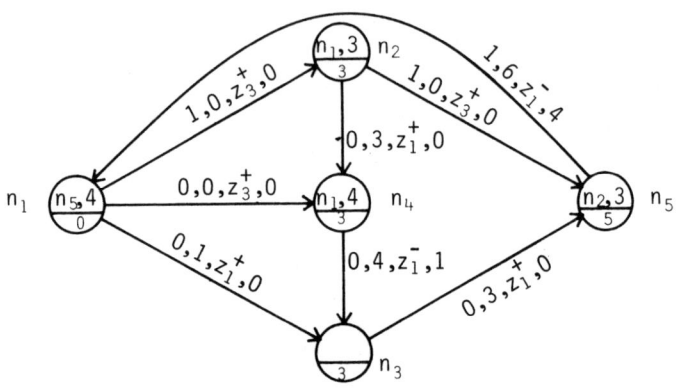

$R^+ = R^- = \emptyset$; $T = \emptyset$; $a_{k1} = a_{51}$; $P = \{n_1\}$; $R^+ = \{a_{51}\}$; $I_p = \{1\}$;
$s = 1$; $T = \{1\}$; $P^* = \{n_2; n_4\}$; $R^+ = \{a_{12}; a_{14}; a_{51}\}$; $P^* \stackrel{!}{=} \emptyset \rightarrow$
$P = \{n_1; n_2; n_4\}$; $I_p = \{1; 2; 4\}$; $s = 4$; $T = \{1; 4\}$; $P^* = \{n_3\}$;
$R^+ = \{a_{12}; a_{14}; a_{43}; a_{51}\}$; $P^* \neq \emptyset \rightarrow P = \{n_1; n_2; n_3; n_4\}$;
$I_p = \{1; 2; 3; 4\}$; $s = 3$; $T = \{1; 3; 4\}$; $P \stackrel{!}{=} \emptyset$; $|P| \neq |T| \rightarrow$
$P = \{n_1; n_2; n_3; n_4\}$; $I_p = \{1; 2; 3; 4\}$; $s = 2$; $T = \{1; 2; 3; 4\}$;
$P^* = \{n_5\}$; $R^+ = \{a_{12}; a_{14}; a_{25}; a_{43}; a_{51}\}$; $P^* \stackrel{!}{=} \emptyset \rightarrow P = \{n_1; n_2; n_3; n_4; n_5\}$
$I_p = \{1; 2; 3; 4; 5\}$; $s = 5$; $T = \{1; 2; 3; 4; 5\}$; $(P^* \stackrel{!}{=} \emptyset) \wedge (|P| \stackrel{!}{=} |T|)$;
$(n_1 \stackrel{!}{\in} P) \wedge (n_5 \stackrel{!}{\in} P) \rightarrow p = (n_1, n_2, n_5, n_1)$; $\varepsilon = 3$; $f^* = 4$; $t = 5$;

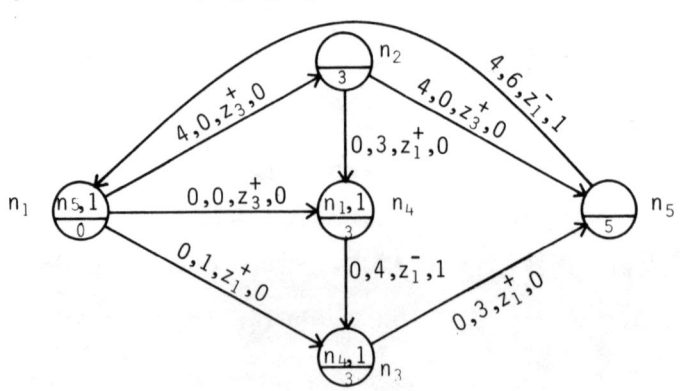

Section 3.2.4   199

$R^+ = R^- = \emptyset$; $T = \emptyset$; $a_{kl} = a_{51}$; $P = \{n_1\}$; $R^+ = \{a_{51}\}$; $I_p = \{1\}$;
$s = 1$; $T = 1$; $P^* = \{n_4\}$; $R^+ = \{a_{14}; a_{51}\}$; $P^* \neq \emptyset \to P = \{n_1; n_4\}$;
$I_p = \{1; 4\}$; $s = 4$; $T = \{1; 4\}$; $P^* = \{n_3\}$; $R^+ = \{a_{14}; a_{43}; a_{51}\}$; $P^* \neq \emptyset \to$
$P = \{n_1; n_3; n_4\}$; $I_p = \{1; 3; 4\}$; $s = 3$; $T = \{1; 3; 4\}$;
$(P^* \stackrel{!}{=} \emptyset) \wedge (|P| \stackrel{!}{=} |T|)$; $n_5 \notin P \to \delta_1 = \delta_2 = \infty$; $R_1 = \{a_{35}\}$; $R_2 = \emptyset$;
$\delta_1 = 3$; $\delta = \min\{3; \infty\} = 3$; $t = 6$;

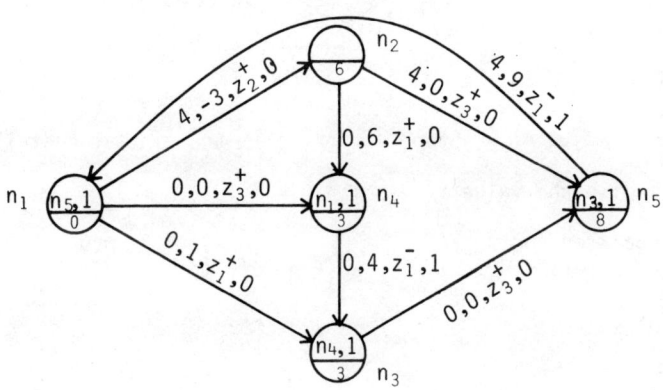

$R^+ = R^- = \emptyset$; $T = \emptyset$; $a_{kl} = a_{51}$; $P = \{n_1\}$; $R^+ = \{a_{51}\}$; $I_p = \{1\}$;
$s = 1$; $T = \{1\}$; $P^* = \{n_4\}$; $R^+ = \{a_{14}; a_{51}\}$; $P^* \neq \emptyset \to P = \{n_1; n_4\}$;
$I_p = \{1; 4\}$; $s = 4$; $T = \{1; 4\}$; $P^* = \{n_3\}$; $R^+ = \{a_{14}; a_{43}; a_{51}\}$;
$P^* \neq \emptyset$; $\to P = \{n_1; n_3; n_4\}$; $I_p = \{1; 3; 4\}$; $s = 3$; $T = \{1; 3; 4\}$;
$P^* = \{n_5\}$; $R^+ = \{a_{14}; a_{35}; a_{43}; a_{51}\}$; $P^* \neq \emptyset \to P = \{n_1; n_3; n_4; n_5\}$;
$I_p = \{1; 3; 4; 5\}$; $s = 5$; $T = \{1; 3; 4; 5\}$; $P^* = \{n_2\}$; $R^- = \{a_{25}\}$;
$P^* \neq \emptyset \to P = \{n_1; n_2; n_3; n_4; n_5\}$; $I_p = \{1; 2; 3; 4; 5\}$; $s = 2$;
$T = \{1; 2; 3; 4; 5\}$; $(P^* \stackrel{!}{=} \emptyset) \wedge (|P| \stackrel{!}{=} |T|)$; $(n_1 \stackrel{!}{\in} P) \wedge (n_5 \stackrel{!}{\in} P) \longrightarrow$
$p = (n_1; n_4; n_3; n_5; n_1)$ $\varepsilon = 1$; $f^* = \bar{f} = 5$; $t = 7$;

200 *Theory of Graphs*

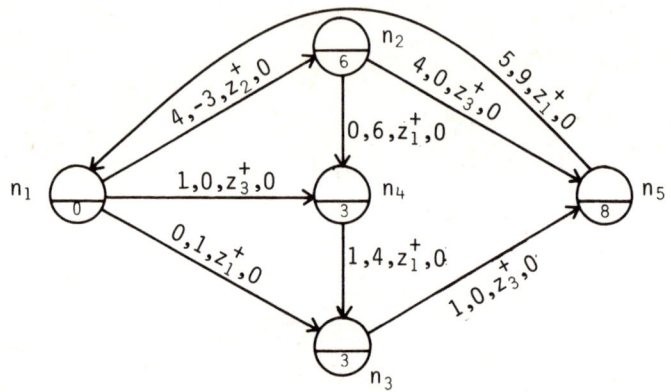

$k_{ij} \stackrel{!}{=} 0 \ \forall \ a_{ij} \in A \rightarrow$ Stop, $x = (x_{ij})$ is the minimal-cost flow pattern with the value $f^{opt}(\mathcal{N}) = \bar{f} = 5$.

The associated costs are $\pi = \sum_{a_{ij} \in A} c_{ij} \cdot x_{ij} = 37$ m.u.

3.3 Shortest Spanning Subtrees of a Graph

3.3.1 The Method of KRUSKAL

Hypotheses
Given a valued undirected graph $\mathcal{N} = (N,A,c)$ without isolated nodes, find a subgraph $\mathcal{N}' = (N,A',c)$, so that each node n_i can be reached from each other node n_j and that $\sum_{a_{ij} \in A'} c_{ij}$ is minimal. The so constructed subgraph contains exactly (n-1) arcs. Define $\tilde{A} := \{a_{ij}\}$; $A' := \emptyset$; set the running index $t := 1$.

Principle
A sequence of graphs is formed, in which the cost-minimal arcs, which form no cycle in the current spanning tree, are successively added to the given set of nodes. The method terminates when a

Section 3.3.1   201

connected graph $\mathcal{N}'$ has been constructed.

Description

Step 1: Determine $c_{kl} := \min\{c_{ij} | a_{ij} \in \tilde{A}\}$.

Step 2: Does the arc a_{kl} complete a cycle in the current subgraph?
If yes: Set $\tilde{A} := \tilde{A} - \{a_{kl}\}$; go to step 1.
If no : Go to step 3.

Step 3: Include the arc a_{kl} in the current subgraph. Set
$\tilde{A} := \tilde{A} - \{a_{kl}\}$; $A' := A' \cup \{a_{kl}\}$; $t := t + 1$.

Step 4: Is $t = n$?
If yes: Stop, the shortest spanning subtree $\mathcal{N}'$ has been
found, its value is given by $\sum_{a_{ij} \in A'} c_{ij}$.
If no : Go to step 1.

Example

Given the following graph $\mathcal{N}$, find the shortest spanning subtree $\mathcal{N}'$.

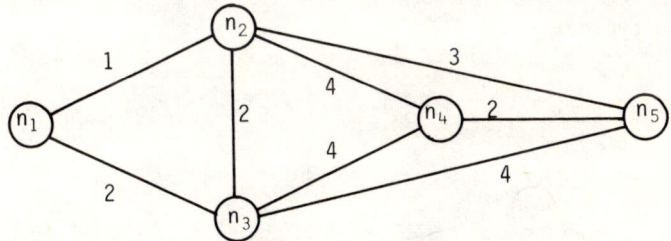

$\tilde{A} = \{a_{12}; a_{13}; a_{23}; a_{24}; a_{25}; a_{34}; a_{35}; a_{45}\}$; $A' = \emptyset$; $t=1$; $a_{kl} = a_{12}$
$\tilde{A} = \{a_{13}; a_{23}; a_{24}; a_{25}; a_{34}; a_{35}; a_{45}\}$; $A' = \{a_{12}\}$; $t=2$;

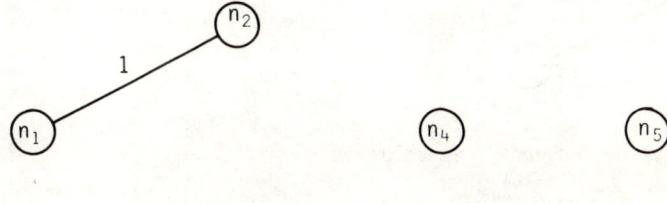

202 *Theory of Graphs*

$a_{kl}=a_{13}$; $\tilde{A}=\{a_{23};a_{24};a_{25};a_{34};a_{35};a_{45}\}$; $A'=\{a_{12};a_{13}\}$; $t=3$;

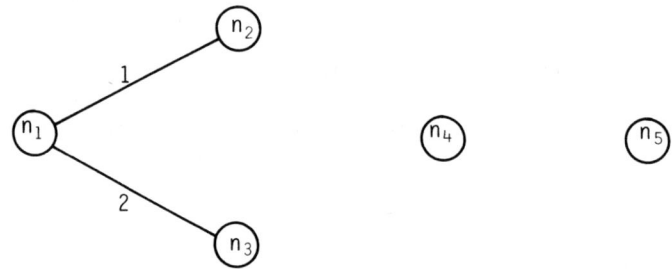

$a_{kl}=a_{45}$; $\tilde{A}=\{a_{23};a_{24};a_{25};a_{34};a_{35}\}$; $A'=\{a_{12};a_{13};a_{45}\}$; $t=4$;

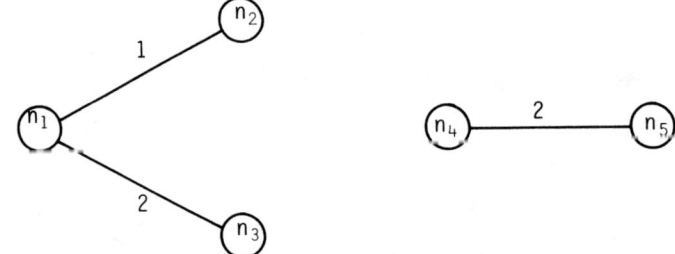

The arc a_{23} may not be added to the current subgraph, because it would form a cycle.

$\tilde{A}=\{a_{24};a_{25};a_{34};a_{35}\}$; $a_{kl}=a_{25}$; $\tilde{A}=\{a_{24};a_{34};a_{35}\}$; $A'=\{a_{12};a_{13};a_{25};a_{45}\}$; $t=n=5$

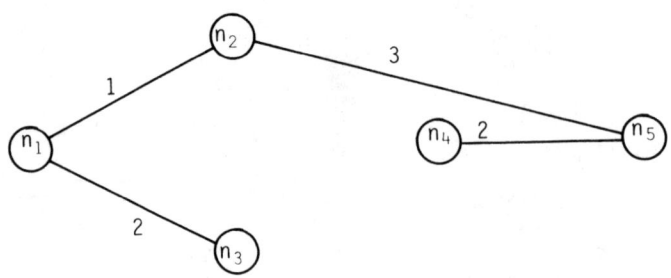

Stop, $\mathcal{N}'$ has been found; its value is

$$\sum_{a_{ij} \in A'} c_{ij} = 8 \text{ m.u.}$$

3.3.2 The Method of SOLLIN

Hypotheses
See 3.3.1 ; let $A':=\emptyset$; set the running index $k:=1$.

Principle
For each node the cost-minimal arc is determined which does not form a cycle in the current spanning tree nor has already been selected for another node. If the so determined subgraph is not connected, then adjoin the necessary arcs of minimum cost, which effect the desired property (a connected subgraph).

Description

Step 1: Determine $c_{kl} := \min_{j} \{b_{kj}\}$.

Step 2: Is $a_{kl} \in A'$?
    If yes: Go to step 5.
    If no : Go to step 3.

Step 3: Does the arc a_{kl} complete a cycle in the current subgraph?
    If yes: Go to step 5.
    If no : Go to step 4.

Step 4: Include the arc a_{kl} in the current subgraph; set $A' := A' \cup \{a_{kl}\}$.

Step 5: Is $k = n$?
    If yes: Go to step 6.
    If no : Set $k := k + 1$. Go to step 1.

Step 6: Is $|A'| = (n - 1)$?
    If yes: Stop, the shortest spanning subtree $\mathcal{N}'$ has been found, its value is given by $\sum_{a_{ij} \in A'} c_{ij}$.
    If no : Go to step 7.

204 *Theory of Graphs*

Step 7: Now there are at least two separate subgraphs. Determine the arc a_{rs} with minimal c_{rs}, so that two of those subgraphs are connected. Include this arc in the current subgraph. Set $A' := A' \cup \{a_{rs}\}$; go to step 6.

<u>Example</u>
Given the following graph $\mathcal{N}$, find the shortest spanning subtree $\mathcal{N}'$.

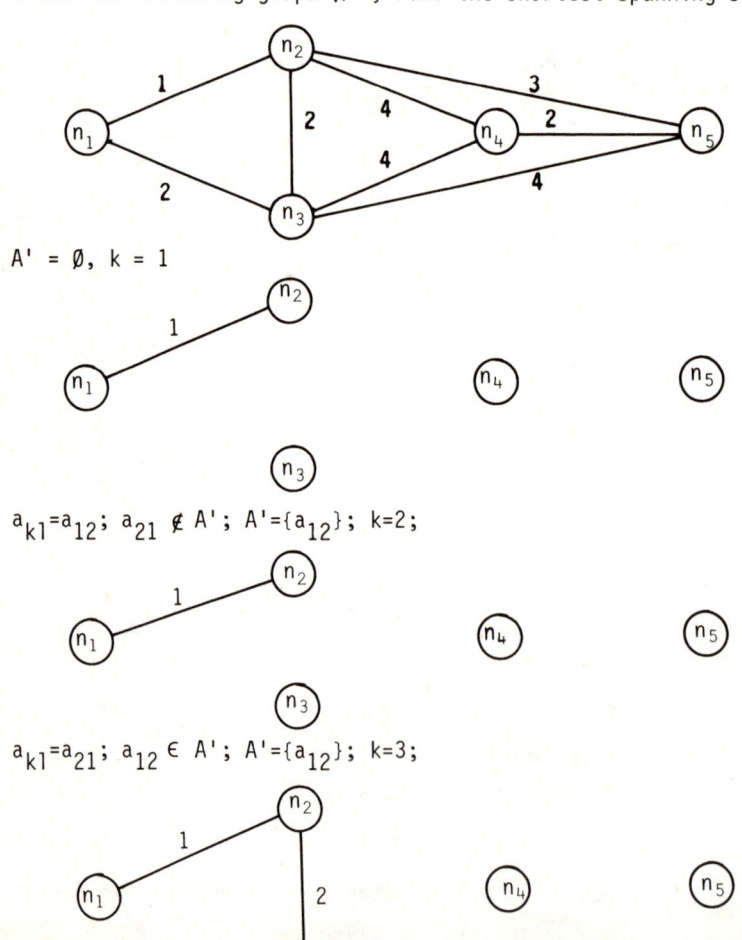

$A' = \emptyset$, k = 1

$a_{kl} = a_{12}$; $a_{21} \notin A'$; $A' = \{a_{12}\}$; k=2;

$a_{kl} = a_{21}$; $a_{12} \in A'$; $A' = \{a_{12}\}$; k=3;

$a_{kl} = a_{32}$; $a_{23} \notin A'$; $A' = \{a_{12}; a_{32}\}$; k=4;

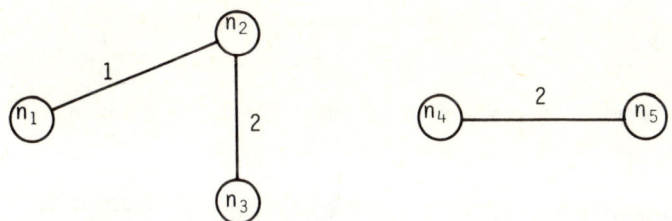

$a_{kl}=a_{45}$; $a_{54} \notin A'$; $A'=\{a_{12};a_{32};a_{45}\}$; $k=5$;

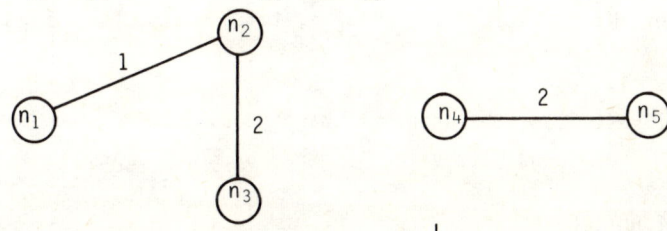

$a_{kl}=a_{54}$; $a_{45} \in A'$; $A'=\{a_{12};a_{32};a_{45}\}$; $k\stackrel{!}{=}n=5$; $|A'| \neq 4$;

$a_{rs}=a_{25}$ ist the arc with the least value c connecting the two subgraphs. $A'=\{a_{12};a_{25};a_{32};a_{45}\}$; $|A'| \stackrel{!}{=} 4$;

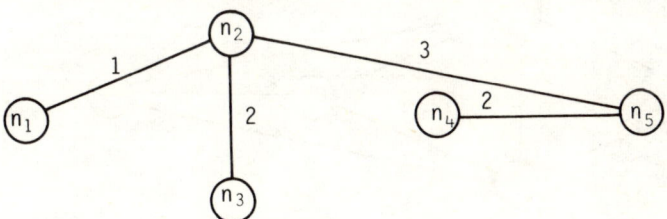

Stop, $\mathcal{N}'$ has been found, its value is $\sum\limits_{a_{ij} \in A'} c_{ij} = 8$ m.u.

3.3.3 The Method of WOOLSEY

<u>Hypotheses</u>
See 3.3.1 . Define $\tilde{N}:=A':=\emptyset$; set the running index $t:=1$.

206 *Theory of Graphs*

Principle   See 3.3.1

Description

Step 1: Determine $c_{kl} := \min\{c_{ij}\}$ and connect the nodes n_k and n_l. Set $\tilde{N} := \tilde{N} \cup \{n_k; n_l\}$; $A' := A' \cup \{a_{kl}\}$.

Step 2: Determine $c_{kl} := \min\{c_{ij} | (n_i \in \tilde{N}) \wedge (n_j \notin \tilde{N})\}$ and connect the nodes n_k und n_l.
Set $\tilde{N} := \tilde{N} \cup \{n_l\}$; $A' := A' \cup \{a_{kl}\}$; $t := t + 1$.

Step 3: Is $t = (n - 1)$?
If yes: Stop, the shortest spanning subtree $\mathcal{N}'$ has been found, its value is given by $\sum_{a_{ij} \in A'} c_{ij}$.
If no : Go to step 2.

Example

Given the following graph $\mathcal{N}$, find the shortest spanning subtree $\mathcal{N}'$.

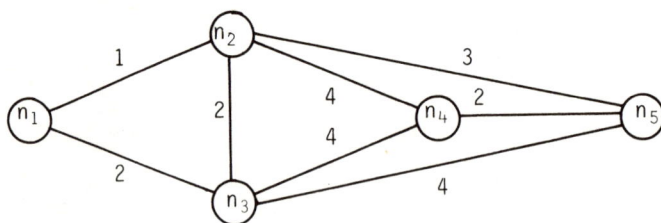

$\tilde{N} = A' = \emptyset$; $t = 1$;

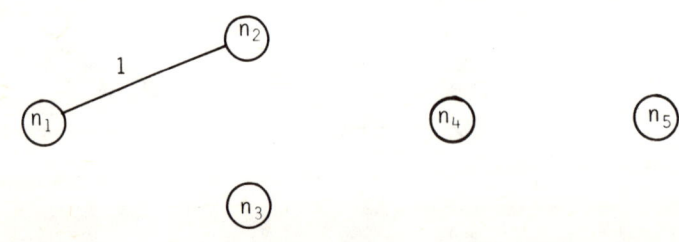

$c_{kl} = c_{12}$; $\tilde{N} = \{n_1; n_2\}$; $A' = \{a_{12}\}$;

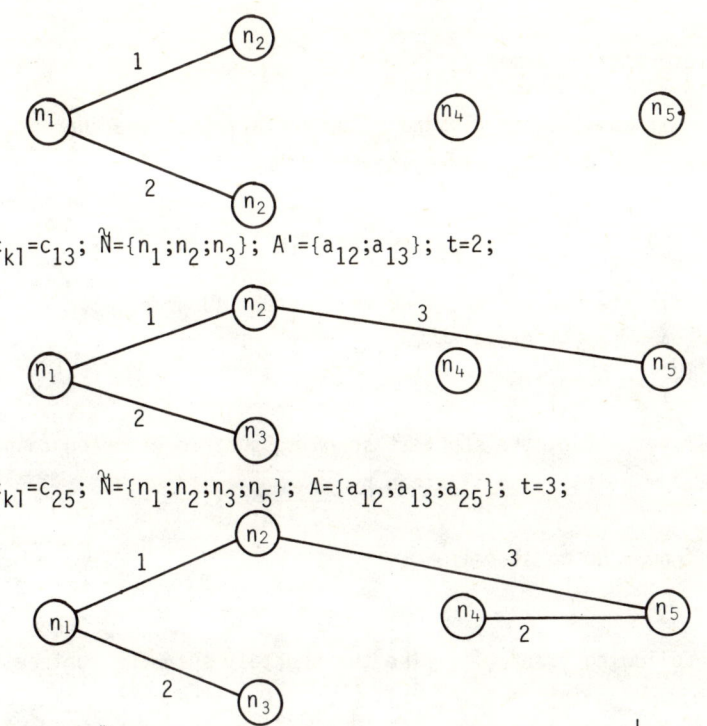

$c_{kl}=c_{13}$; $\hat{N}=\{n_1;n_2;n_3\}$; $A'=\{a_{12};a_{13}\}$; $t=2$;

$c_{kl}=c_{25}$; $\hat{N}=\{n_1;n_2;n_3;n_5\}$; $A=\{a_{12};a_{13};a_{25}\}$; $t=3$;

$c_{kl}=c_{45}$; $\hat{N}=\{n_1;n_2;n_3;n_4;n_5\}$; $A'=\{a_{12};a_{13};a_{25};a_{45}\}$; $t\stackrel{!}{=}(n-1)=4$;

Stop, $\mathcal{N}'$ has been found, its value is $\sum_{a_{ij}\in A'} c_{ij} = 8$ m.u.

3.3.4 The Method of BERGE

<u>Hypotheses</u>
See 3.3.1 . Define $\hat{A}:=A':=\{a_{ij}\}$ and set the running index $t:=|A|$.

<u>Principle</u>
Cost-maximal arcs are eliminated from the given graph as long as the subgraph remains connected.

Description

Step 1: Determine $c_{kl} := \max\{c_{ij} | a_{ij} \in \tilde{A}\}$.

Step 2: Is it possible to eliminate the arc a_{kl} from the current subgraph without destroying its connectivity?
If yes: Go to step 3.
If no : Set $\tilde{A} := \tilde{A} - \{a_{kl}\}$. Go to step 1.

Step 3: Eliminate the arc a_{kl} from the current subgraph. Set $\tilde{A} := \tilde{A} - \{a_{kl}\}$; $A' := A' - \{a_{kl}\}$; $t := t - 1$.

Step 4: Is $t = (n - 1)$?
If yes: Stop, the shortest spanning subtree has been found, its value is given by $\sum_{a_{ij} \in A'} c_{ij}$.
If no : Go to step 1.

Example

Given the following graph $\mathcal{N}$, find the shortest spanning subtree $\mathcal{N}'$.

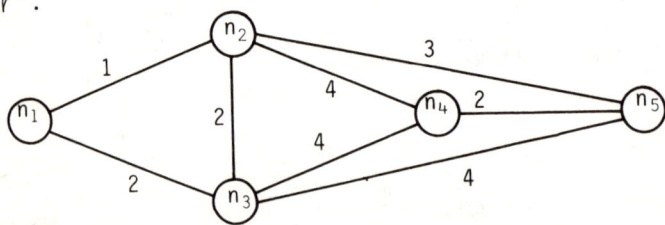

$t=8$; $\tilde{A}=A'=\{a_{12}; a_{13}; a_{23}; a_{24}; a_{25}; a_{34}; a_{35}; a_{45}\}$;

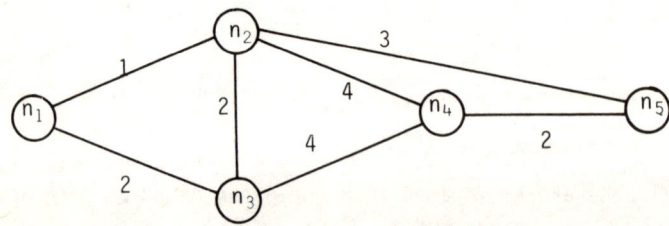

$a_{kl}=a_{35}$; $\tilde{A}=A' = \{a_{12}; a_{13}; a_{23}; a_{24}; a_{25}; a_{34}; a_{45}\}$; $t=7$;

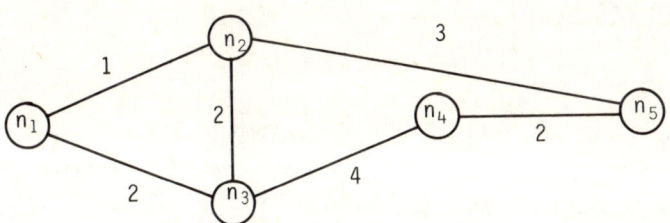

$a_{kl}=a_{24}$; $\tilde{A}:=A'=\{a_{12};a_{13};a_{23};a_{25};a_{34};a_{45}\}$; t=6;

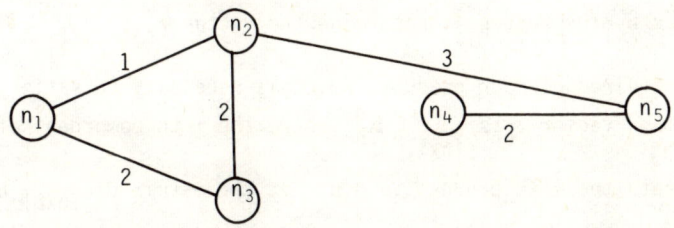

$a_{kl}=a_{34}$; $\tilde{A}=A'=\{a_{12};a_{13};a_{23};a_{25};a_{45}\}$; t=5;

$a_{kl}=a_{25}$; the arc a_{25} may not be eliminated, because the resulting subgraph would not be connected. $\tilde{A}=\{a_{12};a_{13};a_{23};a_{45}\}$; $A'=\{a_{12};a_{13};a_{23};a_{25};a_{45}\}$;

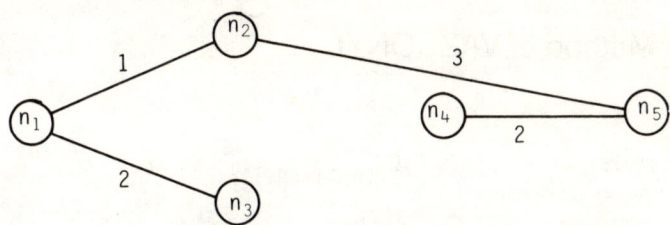

$a_{kl}=a_{23}$; $\tilde{A}=\{a_{12};a_{13};a_{45}\}$; $A'=\{a_{12};a_{13};a_{25};a_{45}\}$; $t\stackrel{!}{=}(n-1)=4$.

Stop, $\mathcal{N}'$ has been found, its value is $\sum_{a_{ij} \in A'} c_{ij}$ = 8 m.u.

3.4 Gozinto Graphs

Hypotheses
Given a simple, valued graph $\mathscr{N} = (N,A,c)$ without cycles so that $(n_i, n_j) \in A \rightarrow i < j$ (if necessary this can be reached by determining the rank and topological sorting) in which the arc values $c_{ij} \in \mathbb{N}$ indicate how many units of the i-th product (represented by the node n_i) are necessary to produce one unit of the j-th product (represented by the node n_j), determine the vector $z_s : z_{s_{[n \times 1]}} \in \mathbb{N}_0^n$, the total required of each product, which is necessary to satisfy a given demand vector $z_0 : z_{0_{[n \times 1]}} \in \mathbb{N}_0^n$ where the i-th component of z_0 represents the i-th product in q.u. Let the matrix $\bar{C} : \bar{C}_{[n \times n]}$ be the direct demand-matrix (also called the "next assembly quantity matrix"), so that

$$\bar{c}_{ij} := \begin{cases} c_{ij}, & \text{if } a_{ij} \in A \\ 0, & \text{if } a_{ij} \notin A \end{cases}.$$

3.4.1 The Method of VAZSONYI

Description
Step 1: Formulate the matrix $\tilde{C} : \tilde{C}_{[(n+1) \times (n+1)]}$, so that

$$\tilde{C} = \begin{array}{|c|c|} \hline \bar{C} & z_0 \\ \hline \Theta & 0 \\ \hline \end{array}.$$

Step 2: Determine (with an appropriate method (Gauss-Jordan; Gauss)) the matrix $D := (I - \tilde{C})^{-1}$.

Step 3: Compute the matrix $\hat{D} := D - I$.
Stop, the first n components of the (n+1)-th column vector of $\hat{D}$ form the vector z_s, the total requirements-vector

Section 3.4.1 211

Example

Given the following Gozinto-graph $\mathcal{N}$ and the demand vector $z_0^T = (0;80;0;490;720)$, find the vector z_s.

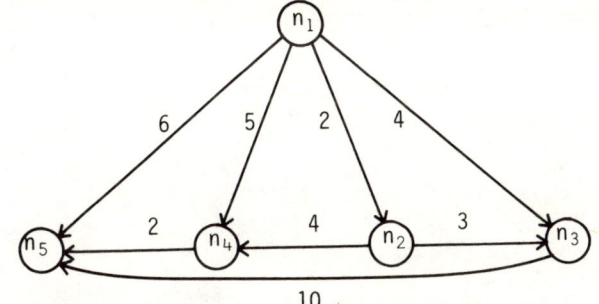

$$\bar{C} = \begin{pmatrix} 0 & 2 & 4 & 5 & 6 \\ 0 & 0 & 3 & 4 & 0 \\ 0 & 0 & 0 & 0 & 10 \\ 0 & 0 & 0 & 0 & 2 \\ 0 & 0 & 0 & 0 & 0 \end{pmatrix} \qquad (I-\tilde{C}) = \begin{pmatrix} 1 & -2 & -4 & -5 & -6 & 0 \\ 0 & 1 & -3 & -4 & 0 & -80 \\ 0 & 0 & 1 & 0 & -10 & 0 \\ 0 & 0 & 0 & 1 & -2 & -490 \\ 0 & 0 & 0 & 0 & 1 & -720 \\ 0 & 0 & 0 & 0 & 0 & 1 \end{pmatrix}$$

$$D = (I-\tilde{A})^{-1} = \begin{pmatrix} 1 & 2 & 10 & 13 & 132 & 101{,}570 \\ 0 & 1 & 3 & 4 & 38 & 29{,}400 \\ 0 & 0 & 1 & 0 & 10 & 7{,}200 \\ 0 & 0 & 0 & 1 & 2 & 1{,}930 \\ 0 & 0 & 0 & 0 & 1 & 720 \\ 0 & 0 & 0 & 0 & 0 & 1 \end{pmatrix}$$

$$\tilde{D} = (D-I) = \begin{pmatrix} 0 & 2 & 10 & 13 & 132 & 101{,}570 \\ 0 & 0 & 3 & 4 & 38 & 29{,}400 \\ 0 & 0 & 0 & 0 & 10 & 7{,}200 \\ 0 & 0 & 0 & 0 & 2 & 1{,}930 \\ 0 & 0 & 0 & 0 & 0 & 720 \\ 0 & 0 & 0 & 0 & 0 & 0 \end{pmatrix}$$

The total requirements-vector z_s is:

$z_s^T = (101{,}570;\ 29{,}400;\ 7{,}200;\ 1{,}930;\ 720)$.

3.4.2 The Method of TISCHER

Hypotheses
Set the running index k:=1.

Description
Step 1: Determine the vector z_k, so that $z_k := \bar{C} \cdot z_{k-1}$.

Step 2: Is $z_k = z_{k-1}$?
  If yes: Go to step 3.
  If no : Set k:=k+1. Go to step 1.

Step 3: Determine z_s, the total requirements-vector, so that
$$z_s := \sum_{r=0}^{k-1} z_r .$$

Example
Given the following Gozinto-graph $\mathcal{N}$ and the demand-vector $z_0^T = (0;80;0;490;720)$, find the vector z_s.

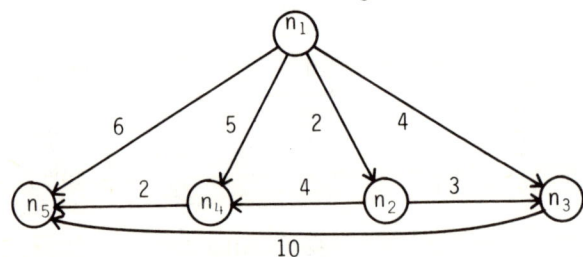

$z_1^T = \bar{C} \cdot z_0 = (6,930; 1,960; 7,200; 1,440; 0)$
$z_2^T = \bar{C} \cdot z_1 = (39,920; 27,360; 0; 0; 0)$
$z_3^T = \bar{C} \cdot z_2 = (54,720; 0; 0; 0; 0)$
$z_4^T = \bar{C} \cdot z_3 = (0; 0; 0; 0; 0)$
$z_5^T = \bar{C} \cdot z_4 = (0; 0; 0; 0; 0)$
$z_5 \stackrel{!}{=} z_4$; $z_s = \sum_{r=0}^{4} z_r$; $z_s^T = (101,570; 29,400; 7,200; 1,930; 720)$.

3.4.3 The Method of FLOYD

Hypotheses
Set the running index $k:=1$.

Description
Step 1: Formulate the matrix $C^{(o)}: C^{(o)}_{[(n+1)\times(n+1)]}$, so that

$$C^{(o)} = \begin{pmatrix} \bar{C} & z_o \\ \Theta & 0 \end{pmatrix}.$$

Step 2: Compute the matrix $C^{(k)}$, so that

$$c_{ik}^{(k)} := c_{ik}^{(k-1)} \qquad \forall\ i=1,\ldots,n+1$$

$$c_{kj}^{(k)} := c_{kj}^{(k-1)} \qquad \forall\ j=1,\ldots,n+1$$

$$c_{ij}^{(k)} := c_{ij}^{(k-1)} + (c_{ik}^{(k-1)} \cdot c_{kj}^{(k-1)}) \quad \forall\ i \neq k;\ j \neq k.$$

Step 3: Is $k = n$?
If yes: Stop, the first n components of the $(n+1)$-th column vector of $C^{(k)}$ form the vector z_s, the total requirements-vector.
If no: Set $k:=k+1$. Go to step 2.

Example
Given the following Gozinto-graph $\mathcal{N}$ and the demand-vector $z_o^T = (0;\ 80;\ 0;\ 490;\ 720)$, find the vector z_s.

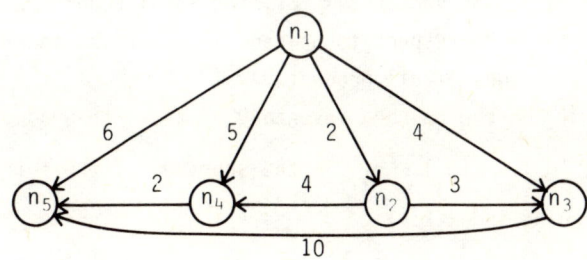

214 *Theory of Graphs*

$$C^{(0)} = C^{(1)} = \begin{pmatrix} 0 & 2 & 4 & 5 & 6 & 0 \\ 0 & 0 & 3 & 4 & 0 & 80 \\ 0 & 0 & 0 & 0 & 10 & 0 \\ 0 & 0 & 0 & 0 & 2 & 490 \\ 0 & 0 & 0 & 0 & 0 & 720 \\ 0 & 0 & 0 & 0 & 0 & 0 \end{pmatrix} \quad C^{(2)} = \begin{pmatrix} 0 & 2 & 10 & 13 & 6 & 160 \\ 0 & 0 & 3 & 4 & 0 & 80 \\ 0 & 0 & 0 & 0 & 10 & 0 \\ 0 & 0 & 0 & 0 & 2 & 490 \\ 0 & 0 & 0 & 0 & 0 & 720 \\ 0 & 0 & 0 & 0 & 0 & 0 \end{pmatrix}$$

$$C^{(3)} = \begin{pmatrix} 0 & 2 & 10 & 13 & 106 & 160 \\ 0 & 0 & 3 & 4 & 30 & 80 \\ 0 & 0 & 0 & 0 & 10 & 0 \\ 0 & 0 & 0 & 0 & 2 & 490 \\ 0 & 0 & 0 & 0 & 0 & 720 \\ 0 & 0 & 0 & 0 & 0 & 0 \end{pmatrix} \quad C^{(4)} = \begin{pmatrix} 0 & 2 & 10 & 13 & 132 & 6{,}530 \\ 0 & 0 & 3 & 4 & 38 & 2{,}040 \\ 0 & 0 & 0 & 0 & 10 & 0 \\ 0 & 0 & 0 & 0 & 2 & 490 \\ 0 & 0 & 0 & 0 & 0 & 720 \\ 0 & 0 & 0 & 0 & 0 & 0 \end{pmatrix}$$

$$C^{(5)} = \begin{pmatrix} 0 & 2 & 10 & 13 & 132 & 101{,}570 \\ 0 & 0 & 3 & 4 & 38 & 29{,}400 \\ 0 & 0 & 0 & 0 & 10 & 7{,}200 \\ 0 & 0 & 0 & 0 & 2 & 1{,}930 \\ 0 & 0 & 0 & 0 & 0 & 720 \\ 0 & 0 & 0 & 0 & 0 & 0 \end{pmatrix}$$

$k \stackrel{!}{=} n = 5$; $z_s^T = (101{,}570;\ 29{,}400;\ 7{,}200;\ 1{,}930;\ 720)$

3.4.4 The Gozinto List Method

<u>Hypotheses</u>

In this method it is also possible to consider warehouse inventories other than zero. The vector $z_s \in \mathbb{Z}^n$, the total requirements-vector, is sought; and with respect to on-hand inventory balances z_s defines the total shortages of all products.

Let $L(n_i) \in \mathbb{N}_0$ be the on-hand inventory of the i-th product, then $S_0(n_i) := z_0(n_i) - L(n_i) \ \forall \ n_i$ is the shortage of the i-th product. Set the running index $k := 1$.

Section 3.4.4   215

Description

Step 1: Construct the following table, which is expanded during the course of the procedure:

n_i	$d_0^+(n_i)$	$S_0(n_i)$	$j^{(1)}$		$j^{(k)}$		$j^{(n-1)}$	
			$d_1^+(n_i)$	$S_1(n_i)$ ... $d_k^+(n_i)$	$S_k(n_i)$ ... $d_{n-1}^+(n_i)$	$S_{n-1}(n_i)$		
n_1				...		...		
.								
.								
.								
n_n				...		...		

Note: $d^+(n_i)$ means the outdegree of n_i.

Step 2: Determine a node n_i, so that $d_{k-1}^+(n_i) = 0$; set $j^{(k)} := i$.

Step 3: Set

$$d_k^+(n_i) := \begin{cases} d_{k-1}^+(n_i) - 1, & \text{if } a_{ij(k)} \in A \\ d_{k-1}^+(n_i), & \text{if } a_{ij(k)} \notin A \\ (*), & \text{if } n_i = n_{j(k)} \end{cases}$$

Step 4: Compute

$$S_k(n_i) := \begin{cases} S_{k-1}(n_i) + \bar{c}_{ij(k)} \cdot S_{k-1}(n_{j(k)}), & \text{if } d_k^+(n_i) < d_{k-1}^+(n_i) \\ S_{k-1}(n_i), & \text{if } \begin{cases} d_k^+(n_i) = d_{k-1}^+(n_i) \\ d_k^+(n_i) = (*) \end{cases} \end{cases}$$

Step 5: Is $d_k^+(n_i) = 0 \lor (*) \; \forall \; n_i$?
   If yes: Stop, $z_s := S_k(n_i)$ ist the total requirements-vector.
   If no : Set $k := k + 1$. Go to step 2.

Example

Given the following Gozinto-graph $\mathcal{N}$ and the demand-vector $z_0^T = (0; 80; 0; 490; 720)$; in this case the on-hand balance of each product is zero, i.e. $L^T=(0;\ldots;0)$, find the vector z_s.

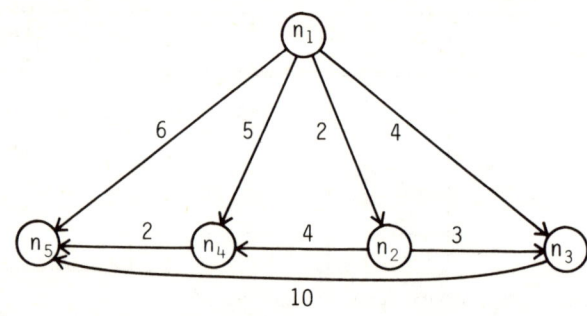

			$j^{(1)}=5$		$j^{(2)}=4$		$j^{(3)}=3$	
n_i	$d_0^+(n_i)$	$S_0(n_i)$	$d_1^+(n_i)$	$S_1(n_i)$	$d_2^+(n_i)$	$S_2(n_i)$	$d_3^+(n_i)$	$S_3(n_i)$
n_1	4	0	3	4,320	2	13,970	1	42,770
n_2	2	80	2	80	1	7,800	0	29,400
n_3	1	0	0	7,200	0	7,200	(*)	7,200
n_4	1	490	0	1,930	(*)	1,930	(*)	1,930
n_5	0	720	(*)	720	(*)	720	(*)	720

	$j^{(4)}=2$
$d_4^+(n_i)$	$S_4(n_i)$
0	101,570
(*)	29,400
(*)	7,200
(*)	1,930
(*)	720

$d_4^+(n_i) \stackrel{!}{=} 0 \lor (*) \; \forall \; n_i$; the vector of shortages is :

$z_s^T = S_4(n_i) = (101,570; 29,400; 7,200; 1,930; 720)$.

4. Planning Networks

4.0.1 The Critical Path Method (CPM)

<u>Hypotheses</u>
Given a simple network $\mathcal{N} = (N,A,d)$ without cycles, with one source n_1 and one sink n_n, where the arcs represent activities, determine the critical path(s) cp and the project duration. (If necessary add an artificial source and/or an artificial sink to the given network). The arc-values $d_{ij} \in \mathbb{N}_0$ denote the duration of the activities $(n_i, n_j) \in A \; \forall \; i,j=1,\ldots,n$. If the following information is given in a table like the one below, the corresponding network must be constructed.

Description of the activity	Number of the activity	Duration d_{ij}	Direct predecessor activity	Direct successor activity
.	.	.	.	.
.	.	.	.	.
.	.	.	.	.
.	.	.	.	.

<u>Principle</u>
In this method the longest paths from the source to the sink and, using the corresponding reverse arcs, the shortest paths from the sink to the source are determined. A "Covering" of these paths yields the critical path(s) cp.

<u>Description</u>
Step 1: Determine the longest paths from the source n_1 to all other nodes n_j, $j \neq 1$, with the FORD Algorithm II .

218 *Planning Networks*

Result: The labels $M(n_j) \; \forall j=1,\ldots,n$.

Step 2: Define $ES(n_j) := M(n_j) \; \forall \; j=1,\ldots,n$; $LS(n_n) := ES(n_n)$.
Here $ES(n_j)$ denotes the earliest possible time, the result n_j can be obtained (the earliest event start time) and $LS(n_j)$ denotes the latest acceptable time, the result n_j can be obtained (the latest allowable event start time).

Step 3: Determine
$$LS(n_j) := \min\{(LS(n_k)-d_{jk}) | (n_k \in N^S(n_j)) \land$$
$$(\exists LS(n_k) \forall n_k \in N^S(n_j))\} \; \forall j=1,\ldots,n-1.$$

Step 4: Define $M:=\{n_j | \; ES(n_j) = LS(n_j)\}$
and determine
$$\{cp\}:=\{cp_l=(n_1,\ldots,n_i,n_j,\ldots,n_n) | (n_i \in M) \land (n_j \in M) \land$$
$$(ES(n_j)-ES(n_i) = d_{ij}) \; \forall l=1,\ldots,q\}.$$

Here $q \in \mathbb{N}$ critical paths cp exist. The project duration is $ES(n_n)$ t.u.

Note: The slack of an activity (n_i,n_j) determines to what extent its beginning may be delayed without changing $ES(n_n)$. We distinguish between the following types of slack:

- total slack $S_{tot}(n_i,n_j) := LS(n_j)-(ES(n_i) + d_{ij})$

  it denotes the maximal increase of d_{ij}, assuming that all other activities begin at the most favorable moments with respect to this activity.

- free slack $S_f(n_i,n_j) := ES(n_j) - (LS(n_i) + d_{ij})$,

  it denotes the maximal increase of d_{ij}, assuming that all events n_k start at $ES(n_k)$.

- independent slack $S_{ind}(n_i,n_j) := \max \{0; \; ES(n_j)-(LS(n_i)+d_{ij})\}$,

  it denotes the maximal increase of d_{ij}, assuming that all other activities begin at the most unfavorable moments with respect to this activity.

- interferring slack $S_{int}(n_j) := LS(n_j) - ES(n_j)$,

it denotes the interval in which the event n_j can happen.

Example
Project: Renovation of an apartment

Description of the activity	Number of the activity	d_{ij}	Direct predecessor activity	Direct successor activity
clear out the apartment	A	3	-	C, D, E
buy wall papers	B	2	-	F, G
take-down curtains	C	1	A	F, G
mix the paint	D	6	A	H
lay out the rug	E	9	A	I
wash the curtains	F	4	B, C	H
paper the apartment	G	3	B, C	I
paint the ceiling	H	2	D, F	I
clean the apartment; return the furniture	I	8	E, H, G	-

The corresponding network $\mathcal{N}$ is:

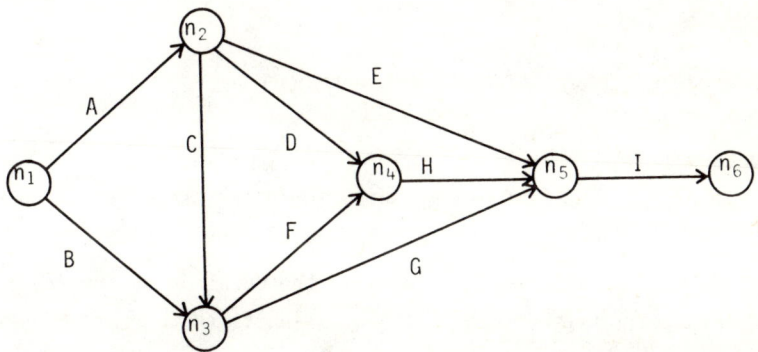

Legend: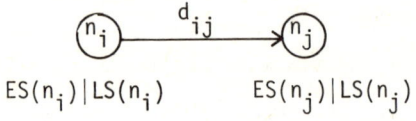

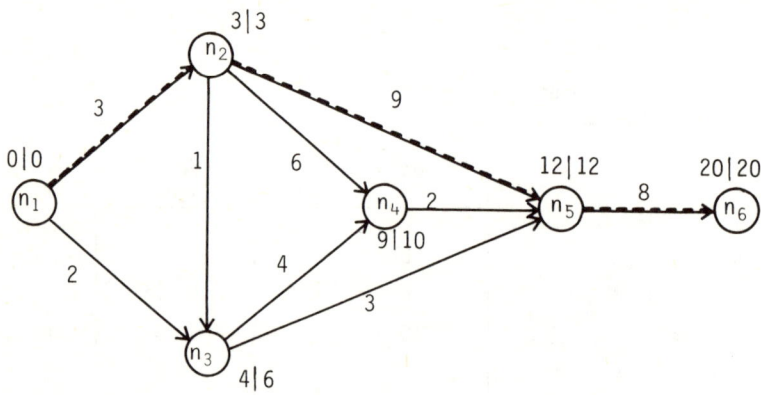

$M = \{n_1; n_2; n_5; n_6\}$; $\{cp\} = \{p_1\}$, where $p_1 = (n_1, n_2, n_5, n_6)$; i.e. the activities A, E and I form the only critical path cp.

Determine the slack for the activity F:

$S_{tot}(F) = 10 - (4 + 4) = 2$; $\quad S_f(F) = 9 - (4 + 4) = 1$;

$S_{ind}(F) = \max\{0; 9 - (5 + 4)\} = 0$; $\quad S_{int}(n_3) = 6 - 4 = 2$;

$S_{int}(n_4) = 10 - 9 = 1$.

4.0.2 The CPM Project Acceleration

Hypotheses

Given: (1) A CPM-Planning Network, for which the critical path(s) cp and the project duration $ES(n_n)$ have already been determined.

(2) Costs $\Delta c_{ij} > 0$, which occur when the duration d_{ij} of the activity (n_i, n_j) is shortened by one time-unit. (Note that the activity (n_i, n_j) needs at least $\lambda_{ij} \in \mathbb{N}_0$ time-units, so that $d_{ij} \geq \lambda_{ij}$). Accelerate the project completion with minimal costs, so that the moment of completion $x \leq ES(n_n)$ is realized. Define $C := 0$.

Principle

In this method a sequence of networks is constructed in which the duration of the activities, which have not reached their lower bound λ_{ij}, is reduced on all critical paths so that the reduction is cost-minimal.

Description

Step 1: Is $ES(n_n) = x$?
  If yes: Stop, the optimal solution has been obtained.
  If no : Go to step 2.

Step 2: Let $\{cp\} := \{cp_l | \ l=1,\ldots,q\}$ the set of the current critical paths. Determine the activity $(n_r,n_s)^{(l)} \ \forall \ l=1,\ldots,q$, so that
  (a) $d_{rs} > \lambda_{rs}$;
  (b) the number of the critical paths is not decreasing for $d_{rs}:=d_{rs}-1$;
  (c) Δc_{rs} is minimal with respect to (a) and (b).

Step 3: Is $|\{(n_r,n_s)^{(l)}\}| = q$?
  If yes: Go to step 4.
  If no : Stop, the moment of completion $ES(n_n) = x$ can not be realized.

Step 4: Define $d_{rs}^{(1)} := d_{rs}^{(1)} - 1 \ \forall \ l=1,\ldots,q$;

$$C := C + \sum_{l=1}^{q} \Delta c_{rs}^{(1)} ,$$

determine the new values $ES(n_j)$, $LS(n_j) \ \forall \ j=1,\ldots,n$ and the new critical paths cp. Go to step 1.

Note: The number of the critical paths is monotone non-decreasing. The cost-function $C := f(ES(n_n))$ is convex and piecewise linear.

222 Planning Networks

Example
Given the following CPM-planning network $\mathcal{N}$ and the following table, accelerate the project completion to $ES(n_n)=x=16$ t.u.

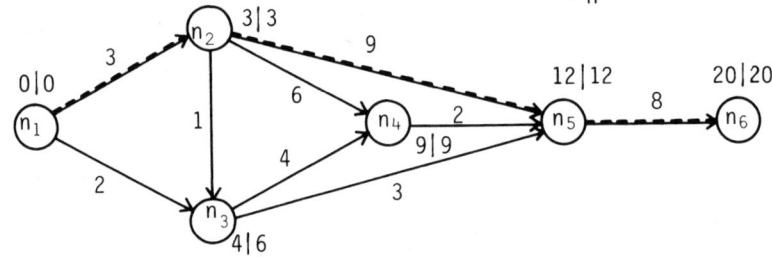

(n_i,n_j)	(n_1,n_2)	(n_1,n_3)	(n_2,n_3)	(n_2,n_4)	(n_2,n_5)	(n_3,n_4)	(n_3,n_5)
Δc_{ij}	7	5	9	3	2	1	6
λ_{ij}	3	1	0	2	5	2	0

(n_4,n_5)	(n_5,n_6)
3	8
1	4

$ES(n_6) \neq 16$; $C = 0$; $cp_1 = (n_1,n_2,n_5,n_6)$; $(n_r,n_s)^{(1)} = (n_2,n_5)$;
$\Delta c_{25}^{(1)} = 2$; $d_{25}^{(1)} = 9 - 1 = 8$; $C = 2$;

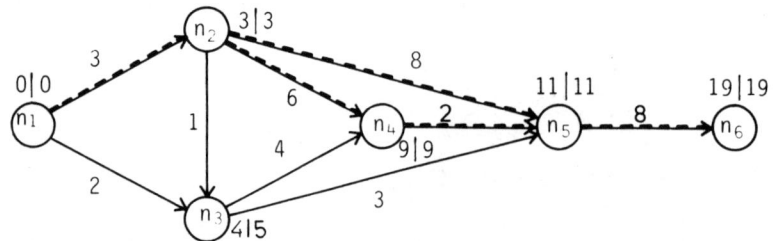

$ES(n_6) \neq 16$; $cp_1 = (n_1,n_2,n_5,n_6)$; $cp_2 = (n_1,n_2,n_4,n_5,n_6)$;
$(n_r,n_s)^{(1)} = (n_2,n_5)$; $(n_r,n_s)^{(2)} = (n_4,n_5)$; $\Delta c_{25}^{(1)} = 2$;

$\Delta c_{45}^{(2)} = 3$; $d_{25}^{(1)} = 8 - 1 = 7$; $d_{45}^{(2)} = 2 - 1 = 1$; $C = 2 + 2 + 3 = 7$

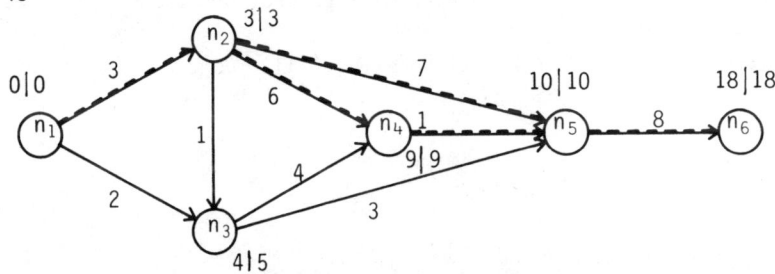

$ES(n_6) \neq 16$; $cp_1 = (n_1, n_2, n_5, n_6)$; $cp_2 = (n_1, n_2, n_4, n_5, n_6)$;
$(n_r, n_s)^{(1)} = (n_2, n_5)$; $(n_r, n_s)^{(2)} = (n_2, n_4)$; $\Delta c_{25}^{(1)} = 2$;
$\Delta c_{24}^{(2)} = 3$; $d_{25}^{(1)} = 7 - 1 = 6$; $d_{24}^{(2)} = 6 - 1 = 5$;
$C = 7 + 2 + 3 = 12$

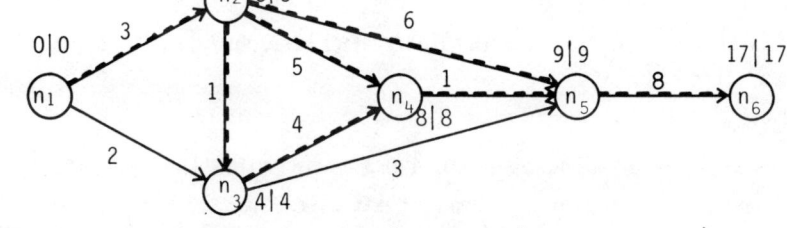

$ES(n_6) \neq 16$; $cp_1 = (n_1, n_2, n_5, n_6)$; $cp_2 = (n_1, n_2, n_4, n_5, n_6)$;
$cp_3 = (n_1, n_2, n_3, n_4, n_5, n_6)$; $(n_r, n_s)^{(1)} = (n_2, n_5)$;
$(n_r, n_s)^{(2)} = (n_2, n_4)$; $(n_r, n_s)^{(3)} = (n_3, n_4)$;
$\Delta c_{25}^{(1)} = 2$; $\Delta c_{24}^{(2)} = 3$; $\Delta c_{34}^{(3)} = 1$;
$d_{25}^{(1)} = 6 - 1 = 5$; $d_{24}^{(2)} = 5 - 1 = 4$; $d_{34}^{(3)} = 4 - 1 = 3$;
$C = 12 + 2 + 3 + 1 = 18$;

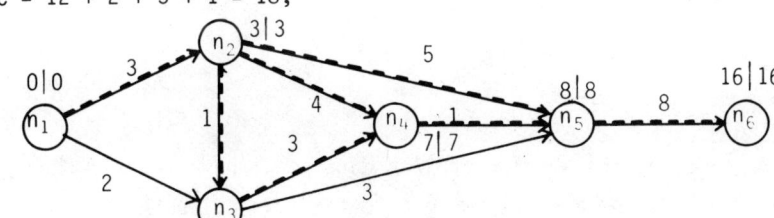

224 *Planning Networks*

$ES(n_6) \stackrel{!}{=} 16 \rightarrow$ Stop, the project is terminated after
$ES(n_6)=x=16$ t.u., the corresponding accelerations cost is C=18 m.u.

The corresponding cost-function $C=f(ES(n_6))$:

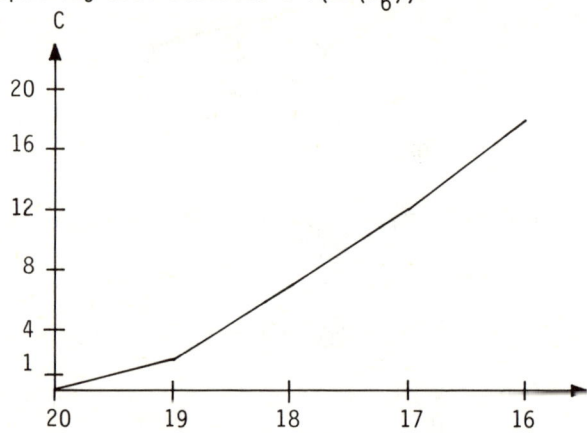

4.0.3 The Program Evaluation and Review Technique (PERT)

<u>Hypotheses</u>

Under the same hypotheses as in 4.0.1 , determine the probability for the termination of the project after x t.u.
Note: The difference between PERT and CPM is that in PERT the following three time-estimations are used:

k_{ij}: duration of the activity (n_i, n_j) under favorable conditions (optimistic estimation)

m_{ij}: most probable duration of the activity (n_i, n_j)

l_{ij}: duration of the activity (n_i, n_j) under unfavorable conditions (pessimistic estimation),

where $k_{ij} \leq m_{ij} \leq l_{ij}$ $\forall$ i,j .

<u>Principle</u>

After computing the average project duration, the method is the same as 4.0.1 . As a result we obtain the critical path with respect to time, not necessarily the actual critical path.

Description

Step 1: Compute the average duration

$$d_{ij} := \frac{1}{6}(k_{ij} + 4 \cdot m_{ij} + l_{ij}) \; \forall \; i,j$$

and the variance $\sigma_{ij}^2 := \frac{1}{36}(l_{ij} - k_{ij})^2 \; \forall \; i,j$.

Note: Step 1 is valid under the assumption that the variables have an Euler-Beta distribution.

Step 2: Using the d_{ij} from step 1 determine with 4.0.1 the unknown $ES(n_j); \; LS(n_j) \; \forall \; j=1,\ldots,n$ and the set of critical paths $\{cp\}$.

Step 3: Compute $\sigma_l^2 := \sum_{a_{ij} \in cp_l} \sigma_{ij}^2 \; \forall \; l=1,\ldots,q$

and determine $\sigma^2 := \max_l \{\sigma_l^2\}$.

Step 4: Determine $P(ES(n_n) \le x) := \Phi(\frac{x-ES(n_n)}{\sigma^2})$ from table 5 in the appendix.

Example

Given the following PERT-planning network $\mathcal{N} = (N,A,k,m,l)$, determine the probability for the termination of the project after $x = 25 \; (20)$ t.u.

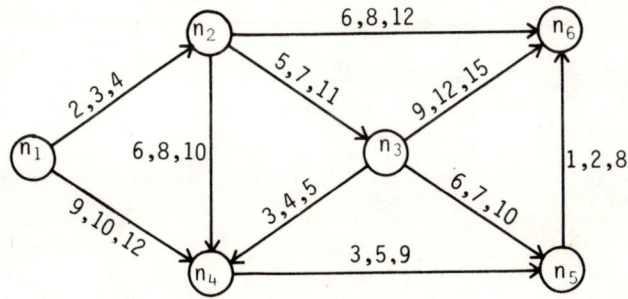

Step 1 yields the following planning network $\mathcal{N}' = (N, A, d, \sigma^2)$:

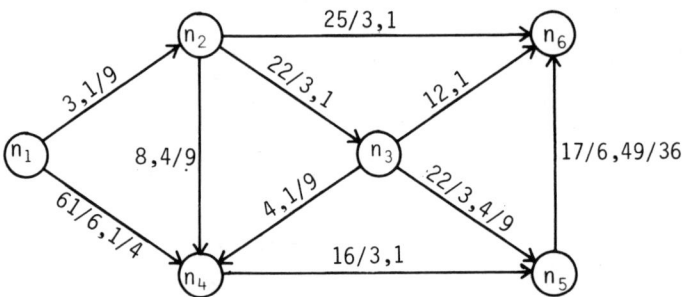

The critical path is represented by the dotted line.

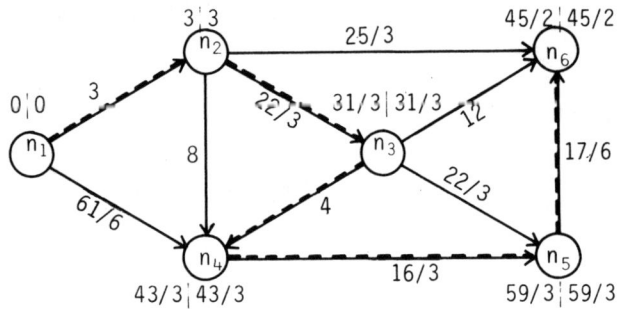

$$\sigma^2 = \sigma_1^2 = \sum_{a_{ij} \in cp_1} \sigma_{ij}^2 = 3.058\overline{3}$$

α) $P(ES(n_6) \leq 25) = \Phi(\dfrac{25 - 22.5}{3.058\overline{3}}) = \Phi(\dfrac{2.5}{3.058\overline{3}}) \approx 0.7939 \cong 79.39\ \%$

β) $P(ES(n_6) \leq 20) = \Phi(\dfrac{20 - 22.5}{3.058\overline{3}}) = \Phi(\dfrac{-2.5}{3.058\overline{3}}) \approx 0.2061 \cong 20.61\ \%$.

4.0.4 The Metra Potential Method (MPM)

Hypotheses

Given a node-oriented network $\mathcal{N}'=(N',A,l,d)$ (i.e. the (n-1) activities $n_j, j = 1,\ldots,n - 1$, are represented by the nodes), determine the minimal project duration. The arc-values l_{ij} represent the time relations between the activities n_i and n_j; the following graph should be interpreted as such:

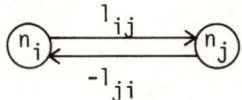

the earliest possible beginning for the activity n_j is l_{ij} t.u. after the beginning of the activity n_i; the latest acceptable beginning for the activity n_j is l_{ji} t.u. after the beginning of the activity n_i. Logically no cycles with strictly positive length may occur. The node-values $d_j \in \mathbb{N}_0$ represent the duration of the activities $n_j, j = 1,\ldots,n - 1$.

Define

$ES(n_j)$: earliest possible start time
$EF(n_j)$: earliest possible finish time
$LS(n_j)$: latest allowable start time ⟶ of activity n_j.
$LF(n_j)$: latest allowable finish time
$S_{tot}(n_j)$: total slack

Principle

In this method an artificial sink is included; then the longest paths from the source to this sink are determined as well as the longest paths from the sink to the source in the inverted graph.

Description

Step 1: Adjoin the node n_n and the arcs $(n_j, n_n) \forall j = 1,\ldots, n - 1$ to the given network. Let $l_{jn} := d_j \forall j = 1,\ldots, n - 1; d_n := 0$.

Result: A planning network $\mathcal{N} = (N,A,1)$. Construct the following table which will be completed during the course of the procedure:

n_j	d_j	$ES(n_j)$	$EF(n_j)$	$LS(n_j)$	$LF(n_j)$	$S_{tot}(n_j)$
n_1						
.						
.						
.						
n_n						

Step 2: Determine the longest paths from the source n_1 to all other nodes n_j, $j \neq 1$, with the FORD Algorithm II.
Result: The labels $M(n_j) \; \forall \; j=1,\ldots,n$.

Step 3: Define $ES(n_j) := M(n_j) \; \forall \; j=1,\ldots,n$
and compute $EF(n_j) := ES(n_j) + d_j \; \forall \; j=1,\ldots,n$.

Step 4: Construct the planning network $\tilde{\mathcal{N}} = (N,\tilde{A},\tilde{l})$, so that
$a_{ij} \in A \Leftrightarrow \tilde{a}_{ji} \in \tilde{A}; \; \tilde{l}_{ji} := l_{ij}$.

Step 5: Determine the longest paths from the sink n_n to all other nodes n_j, $j \neq n$, in $\tilde{\mathcal{N}}$ with the FORD Algorithm II.
Result: The labels $\tilde{M}(n_j) \; \forall \; j=1,\ldots,n$.

Step 6: Compute $LS(n_j) := EF(n_j) - \tilde{M}(n_j)$
$LF(n_j) := LS(n_j) + d_j$ $\Big\} \; \forall \; j=1,\ldots,n$.

Result: The minimal project duration is $LF(n_n)$ t.u.

Step 7: Determine the total slack of each activity:
$S_{tot}(n_j) := LS(n_j) - ES(n_j) \; \forall \; j=1,\ldots,n$.

Example

Given the following MPM-planning network, determine the minimal project duration.

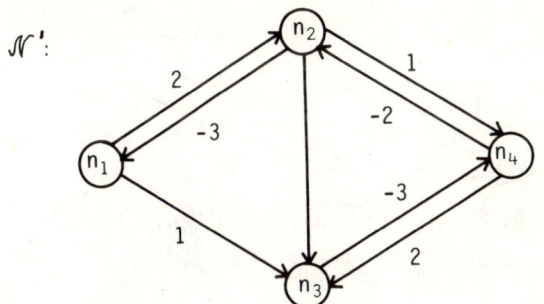

$D = (d_1; d_2; d_3; d_4) = (1; 7; 7; 3)$;
step 1 yields the following network $\mathcal{N}$:

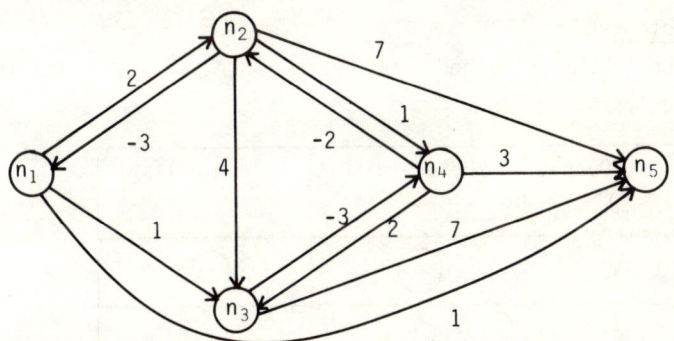

n_j	d_j	$ES(n_j)$	$EF(n_j)$	$LS(n_j)$	$LF(n_j)$	$S_{tot}(n_j)$
n_1	1	0	1	0	1	0
n_2	7	2	9	2	9	0
n_3	7	6	13	6	13	0
n_4	3	3	6	4	7	1
n_5	0	13	13	13	13	0

$M(n_1) = 0$; $M(n_2) = 2$; $M(n_3) = 6$; $M(n_4) = 3$; $M(n_5) = 13$;
step 4 yields the following network $\widetilde{\mathcal{N}}$:

$\mathcal{N}$:

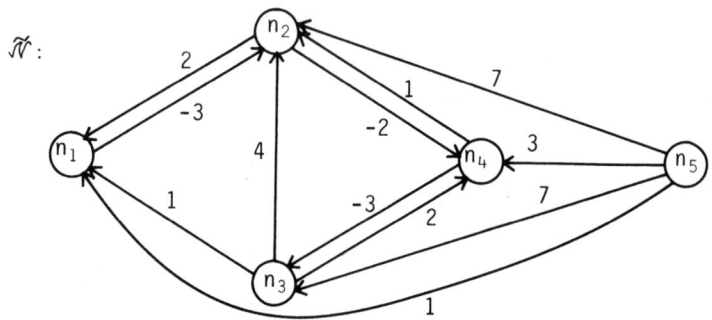

$\hat{M}(n_5) = 0$; $\hat{M}(n_4) = 9$; $\hat{M}(n_3) = 7$; $\hat{M}(n_2) = 11$; $\hat{M}(n_1) = 13$;
the minimal project duration is $LF(n_5) = 13$ t.u.

4.0.5 The Graphical Evaluation and Review Technique (GERT)

<u>Hypotheses</u>

Given a GERT decision-network $\mathcal{N} = (N,A,p,t)$, with the following node-types,

Input \ Output	Deterministic	Probabilistic
"AND"	◯	◠
inclusive "OR"	◇	◇
exclusive "OR"	⏢	⏢

the arc-values are defined as:

$P_{ij} \in (0;1]$: probability that the activity (n_i, n_j) will be realized.

Note: $\sum_j p_{ij} = 1 \; \forall$ probabilistic nodes n_i.

$t_{ij} \in \mathbb{R}_+$: (constant) duration of the activity (n_i, n_j),

Section 4.0.5   231

reduce the given network to only one arc. The value of that arc indicates, with what probability the total project will be realized and what time will be required.

Description
The reduction of the decision-network is done by the following reduction rules (the arc-value x means $x = (p_x; t_x)$):

(1) arcs in a series:

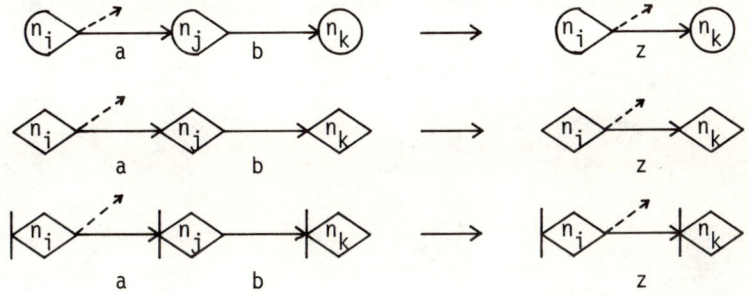

$p_z := p_a \cdot p_b; \quad t_z := t_a + t_b;$

(2) parallel arcs with "AND"-relation:

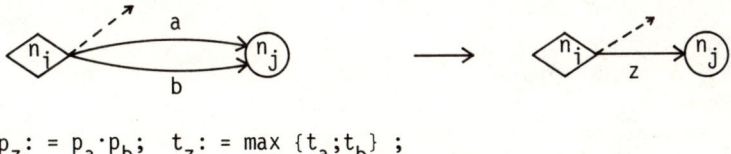

$p_z := p_a \cdot p_b; \quad t_z := \max \{t_a; t_b\};$

(3) parallel arcs with inclusive "OR"-relation:

$p_z := p_{a+b} := p_a + p_b - p_a \cdot p_b;$
$t_z := t_a \cdot p_a + t_b \cdot p_b + (\min \{t_a; t_b\} - t_a - t_b) \cdot p_a \cdot p_b;$

(4) parallel arcs with exclusive "OR"-relation:

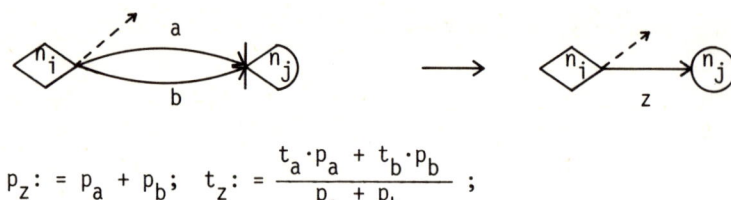

$$p_z := p_a + p_b; \quad t_z := \frac{t_a \cdot p_a + t_b \cdot p_b}{p_a + p_b} \; ;$$

(5) loops with unlimited repetition and inclusive "OR"-relation:

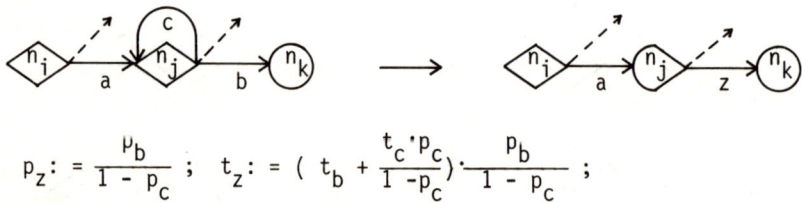

$$p_z := \frac{p_b}{1 - p_c} \; ; \quad t_z := \left(t_b + \frac{t_c \cdot p_c}{1 - p_c} \right) \cdot \frac{p_b}{1 - p_c} \; ;$$

(6) loops with unlimited repetition and exclusive "OR"-relation:

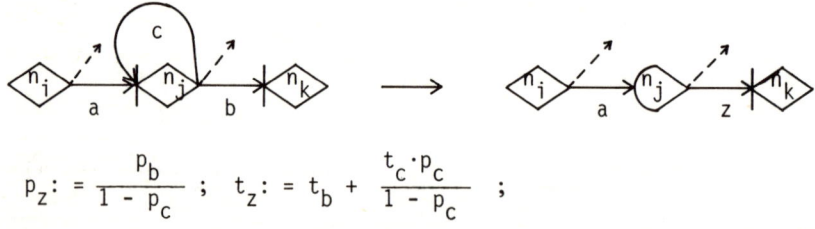

$$p_z := \frac{p_b}{1 - p_c} \; ; \quad t_z := t_b + \frac{t_c \cdot p_c}{1 - p_c} \; ;$$

(7) loops with limited repetition:
A loop, which can be repeated at most l times, may be represented by the following graph, where $x_i = (p_{x_i}; t_{x_i})$; $p_{b_i} = 1 - p_{c_i}$ $\forall \; i = 1, \ldots, l$:

Section 4.0.5   233

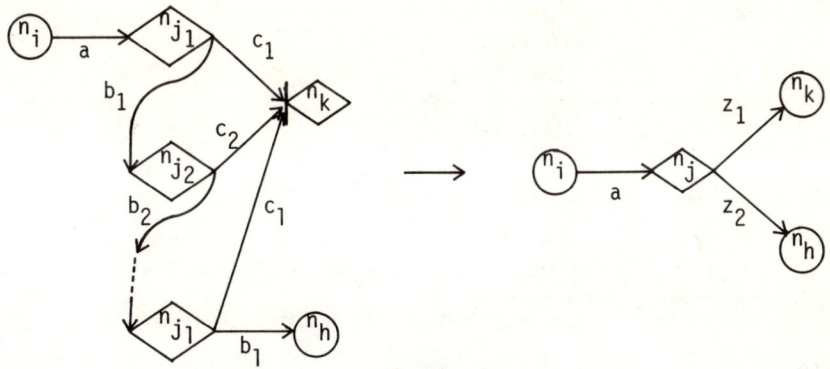

$$p_{z_1} := 1 - p_b^1; \quad p_{z_2} := p_b^1; \quad t_{z_2} := 1 \cdot t_b;$$

$$t_{z_1} := t_c \cdot (1 - p_b^{1+1}) + t_b \cdot p_b \cdot (1 - (1-1) \cdot p_b^{1-1} + (p_b - p_b^{1-1})/(1-p_b));$$

this loop may be interpreted as follows:
A product comes over arc a to the first test at node n_{j_1}. If accepted, it comes over arc c_1 to node n_k. If refused, it comes over arc b_1 to node n_{j_2}, where another test is conducted. After at most 1 tests, the product will be accepted or it will be rejected, i.e. will end in node n_k or node n_h respectively.

Example 1

Given the following decision-network $\mathcal{N}$, reduce it to only one arc.

Legend:

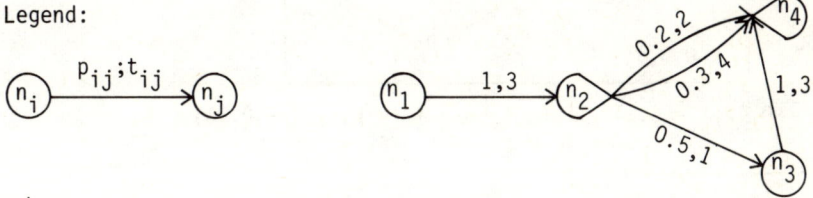

1st step:

$p_z = 0.2 + 0.3 - 0.2 \cdot 0.3 = 0.44;$

$t_z = \max \{2; 4\} = 4;$

234 Planning Networks

2nd step:

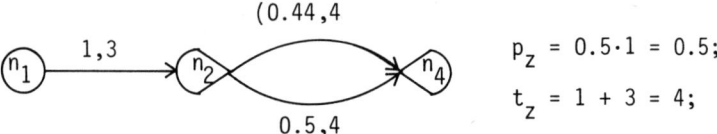

$p_z = 0.5 \cdot 1 = 0.5;$

$t_z = 1 + 3 = 4;$

3rd step:

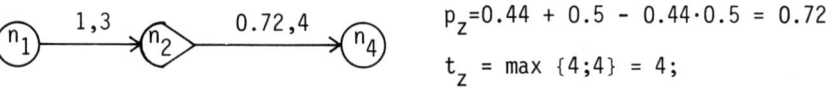

$p_z = 0.44 + 0.5 - 0.44 \cdot 0.5 = 0.72;$

$t_z = \max \{4;4\} = 4;$

4th step:

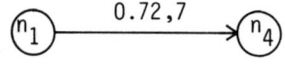

$p_z = 1 \cdot 0.72 = 0.72 ;$

$t_z = 3 + 4 = 7;$

Example 2

Given the following decision-network $\mathcal{N}$, reduce it to only one arc.

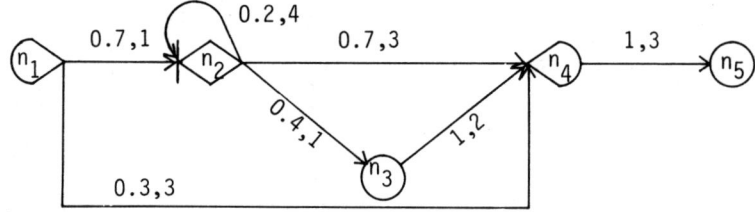

1st step:

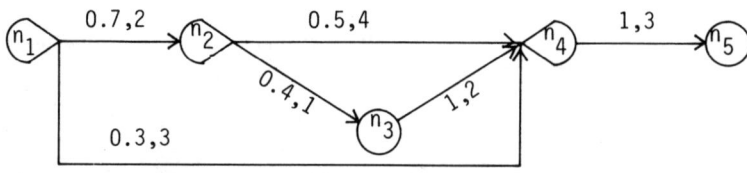

$p_z = 0.4/(1 - 0.2) = 0.5; \quad t_z = 3 + (4 \cdot 0.2/(1 - 0.2)) = 4;$

2nd step:

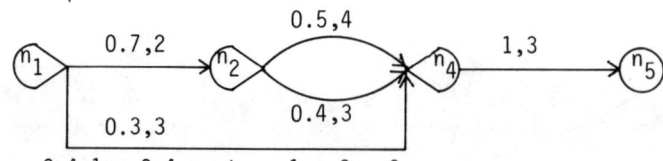

$p_z = 0.4 \cdot 1 = 0.4 ; \quad t_z = 1 + 2 = 3;$

3rd step:

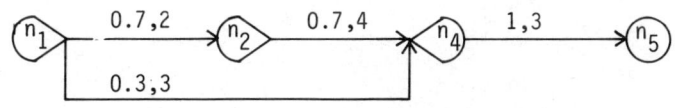

$p_z = 0.5 + 0.4 - 0.5 \cdot 0.4 = 0.7;$ $t_z = \max\{4;3\} = 4;$

4th step:

$p_z = 0.7 \cdot 0.7 = 0.49;$ $t_z = 2 + 4 = 6;$

5th step:

$p_z = 0.49 + 0.3 - 0.49 \cdot 0.3 = 0.643;$ $t_z = \max\{6;3\} = 6;$

6th step:

$p_z = 0.643 \cdot 1 = 0.643;$ $t_z = 6 + 3 = 9$.

Note: In decision-networks of the most general type, like GERT-planning networks, it is possible that the durations t of the activities are random variables with a certain density-function. Each random variable may be distributed differently. For the case when t is a variable, there are modified reduction rules which derive from "moment generating functions".

5. Game Theory

5.1 Non Matrix Games

5.1.1 The Normal Form

Hypotheses
Given two appliers of one product, let x_i be the quantity to be supplied with cost-function $C_i(x_i)$ and l_i the maximum marketable quantity of each supplier $i=1,2$.
Furthermore the price-function $p(x)$, corresponding to the supply $x = x_1 + x_2$ is known. So a duopoly model is defined. Let $i=1$, then $\bar{i}=2$ and reverse.

Principle
The optimal strategies of both players are computed taking into account different objectives.

Description
Step 1: Set up the payoff-function $A_i := p(x) \cdot x_i - C_i(x_i)$ for each player $i=1,2$.

Step 2: Determine the Bayes-strategy x_i^* for the player $i=1,2$, so that

$$\frac{\partial A_i}{\partial x_i} = 0 \quad \rightarrow \quad x_i^* := f(x_{\bar{i}}) ,$$

and the payoff A_i^* for the player $i=1,2$, playing the Bayes-strategy against $\bar{i}$, so that

$$A_i^*(x_i^*(x_{\bar{i}})) := p(x_i^*, x_{\bar{i}}) \cdot x_i^* - C_i(x_i^*).$$

Section 5.5.1   237

Step 3: Determine the maximum payoff for the player i=1,2 under the assumption that player i is monopolist (i.e. $x=x_i$; $x_{\bar{i}}=0$), so that

$$\frac{\partial A_i}{\partial x_i} = 0 \mid A_i = p(x_i) \cdot x_i - C_i(x_i) \rightarrow x_i(x_{\bar{i}}) = x_i(0).$$

It follows: $A_i(x_i(0)) := p(x) \cdot x_i(0) - C_i(x_i(0))$.

Step 4: Determine the equilibrium-strategies $\hat{x}_i, \forall\ i=1,2$, solving the system of equations, determined in step 2:
$x_1^* := f(x_2);\ x_2^* := f(x_1)$.

The resulting payoffs for both players in the equilibrium-point are: $A_i(\hat{x}_1;\hat{x}_2) := p(\hat{x}) \cdot \hat{x}_i - C_i(\hat{x}_i) \ \forall\ i=1,2$.

Step 5: Let inf $A_i(x_i;x_{\bar{i}}) := p(x_i;L_{\bar{i}}) \cdot x_i - C_i(x_i)$.
Determine the maximal guarantee for the player i=1,2, so that

$$\frac{\partial (\inf A_i(x_i;x_{\bar{i}}))}{\partial x_i} = 0 \rightarrow \tilde{x}_i(L_{\bar{i}}) .$$

The values $\tilde{x}_i$ represent the minimax-strategies of the two players i=1,2 .

Assume ruinous competition of the opponent $\bar{i}$, compute the guarantee for the player i=1,2, so that
$A_i(\tilde{x}_i;L_{\bar{i}}) := p(\tilde{x}_i;L_{\bar{i}}) \cdot \tilde{x}_i - C_i(\tilde{x}_i)$.

Example
Given two suppliers of one product, the cost-functions are
$C_1(x_1) = 5 x_1 + 8;\ L_1 = 18;\ C_2(x_2) = 6 x_2 + 2\ ;\ L_2 = 20$.
The price-function in given by

$$p(x) := \begin{cases} 20 - 3/4\ x, & \text{if } 0 \le x \le 80/3 \\ 0, & \text{if } x > 80/3 \end{cases}$$

$$A_1 = \begin{cases} [20 - 3/4(x_1 + x_2)] \cdot x_1 - 5 x_1 - 8 , & \text{if } x \leq 80/3 \\ -5 x_1 - 8 & , \text{if } x > 80/3 ; \end{cases}$$

$$A_2 = \begin{cases} [20 - 3/4(x_1 + x_2)] \cdot x_2 - 6 x_2 - 2, & \text{if } x \leq 80/3 \\ - 6 x_2 - 2 & , \text{if } x > 80/3 . \end{cases}$$

The Bayes-strategies are:

$$\frac{\partial A_1}{\partial x_1} = \frac{\partial (20 x_1 - 3/4 x_1^2 - 3/4 x_1 x_2 - 5 x_1 - 8)}{\partial x_1} = 0 \rightarrow$$

$$\rightarrow \boxed{x_1^* = 10 - 1/2 x_2} \; ;$$

$$\frac{\partial A_2}{\partial x_2} = \frac{\partial (20 x_2 - 3/4 x_1 x_2 - 3/4 x_2^2 - 6 x_2 - 2)}{\partial x_2} = 0 \rightarrow$$

$$\rightarrow \boxed{x_2^* = 28/3 - 1/2 x_1} \; .$$

$$A_1^*(x_1^*(x_2)) = [20 - 3/4(10 - 1/2 x_2 + x_2)] \cdot (10 - 1/2 x_2) - 5(10 - 1/2 x_2)$$
$$- 8 = 67 - 15/2 x_2 + 3/16 x_2^2 \; ;$$

is A_1^* maximal ?

$$\frac{\partial A_1^*}{\partial x_2} = -15/2 + 3/8 x_2 < 0 \rightarrow \text{for } x_2 < 20 \quad A_1^* \text{ is maximal.}$$

$$A_2^*(x_2^*(x_1)) = [20 - 3/4(x_1 + 28/3 - 1/2 x_1)] \cdot (28/3 - 1/2 x_1)$$
$$- 6(28/3 - 1/2 x_1) - 2 = 190/3 - 7 x_1 + 3/16 x_1^2 \; ;$$

is A_2^* maximal ?

$$\frac{\partial A_2^*}{\partial x_1} = -7 + 3/8 x_1 < 0 \rightarrow \text{for } x_1 < 56/3 \quad A_2^* \text{ is maximal.}$$

If player 1 is monopolist: $x = x_1$; $x_2 = 0$;

$A_1(x_1,0) = (20 - 3/4\ x_1) \cdot x_1 - 5\ x_1 - 8 = -3/4\ x_1^2 + 15\ x_1 - 8$

$\dfrac{\partial A_1}{\partial x_1} = -3/2\ x_1 + 15 = 0 \rightarrow \boxed{x_1(0) = 10}$;

$A_1(10;0) = [20 - 3/4(10 + 0)] \cdot 10 - 5 \cdot 10 - 8 = 67.$

If player 2 is monopolist: $x = x_2$; $x_1 = 0$;

$A_2(0;x_2) = (20 - 3/4\ x_2) \cdot x_2 - 6\ x_2 - 2 = -3/4\ x_2^2 + 14\ x_2 - 2$

$\dfrac{\partial A_2}{\partial x_2} = -3/2\ x_2 + 14 = 0 \rightarrow \boxed{x_2(0) = 28/3}$;

$A_2(0;28/3) = (20 - (3/4) \cdot (28/3)) \cdot 28/3 - 6 \cdot (28/3) - 2 = 190/3.$

The equilibrium-strategies and the equilibrium-points are:

$\tilde{x}_1 = 10 - 1/2\ \tilde{x}_2$; $\tilde{x}_2 = 28/3 - 1/2\ \tilde{x}_1 \rightarrow$

$\rightarrow \boxed{\tilde{x}_1 = 64/9}$; $\boxed{\tilde{x}_2 = 52/9}$;

the payoffs in the equilibrium-point are:

$A_1(64/9;52/9) = [20 - 3/4(64/9 + 52/9)] \cdot 64/9 - 5 \cdot (64/9) - 8 = 29.\overline{925}$;

$A_2(64/9;52/9) = [20 - 3/4(64/9 + 52/9)] \cdot 52/9 - 6 \cdot (52/9) - 2 = 23.\overline{037}$.

Guarantee for player 1:

$\inf A_1(x_1;x_2) = \inf \{[20 - 3/4 \cdot (x_1 + x_2)] \cdot x_1 - 5\ x_1 - 8\} =$
$= -3/4\ x_1^2 - 8;$

maximal guarantee: $\dfrac{\partial\ (\inf A_1(x_1;x_2))}{\partial\ x_1} = -3/2\ x_1 = 0 \rightarrow \boxed{\tilde{\tilde{x}}_1 = 0}$;

$\tilde{\tilde{x}}_1 = 0$; $x_2 = 20$;

$A_1(0;20) = [20 - 3/4 \cdot (20)] \cdot 0 - 5 \cdot 0 - 8 = -8$;

guarantee for player 2:

$$\inf A_2(x_1;x_2) = \inf \{[20 - 3/4(x_1 + x_2)]x_2 - 6 x_2 - 2\} =$$
$$= 1/2 x_2 - 3/4 x_2^2 - 2 \;;$$

maximal guarantee: $\dfrac{\partial (\inf A_2(x_1;x_2))}{\partial x_2} = 1/2 - 3/2 x_2 = 0 \rightarrow \boxed{\tilde{x}_2 = 1/3}$;

$x_1 = 18; \tilde{x}_2 = 1/3$;

$A_2(18;1/3) = (20 - (3/4) \cdot (55/3)) \cdot 1/3 - 2 - 2 = -23/12$.

The guarantee point of the game is defined by the maximum guaranteed payoffs A_1 und A_2: $(A_1;A_2) = (-8; -23/12)$.

For each pair of strategies $(x_1;x_2)$ there exists exactly one pair of payoffs $(A_1;A_2)$. These payoffs can be illustrated in a payoff-diagram:

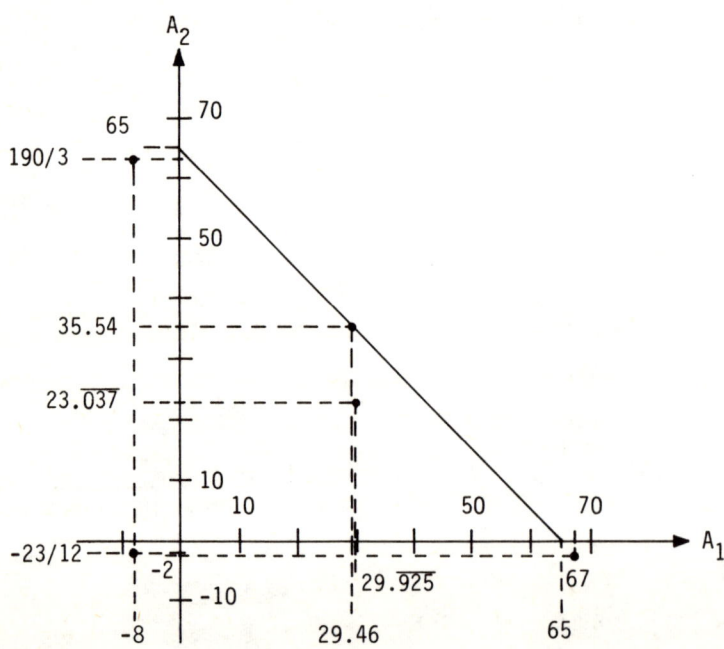

In the diagram above the payoff-points $A(A_1;A_2)$ represent:
$A(29.\overline{925} ; 23.\overline{037})$ the equilibrium-point;
$A(67 ; -2)$ the monopoly point for player 1;
$A(-8 ; 190/3)$ the monopoly point for player 2;
$A(-8 ; -23/12)$ the guarantee point;
$A(29.46; 35.54)$ Nash's solution of the bargaining problem (see 5.1.2).

5.1.2 NASH's Solution of the Bargaining Problem

Hypotheses
Given a duopoly model as in 5.1.1 , find a solution of a game, dividing the maximum total payoff $\bar{A}$ as evenly as possible.

Description
Step 1: Determine
$$\bar{A} := \bar{A}_1 + \bar{A}_2 := \max\{[A_2(0;x_2^*) + A_1(0;x_2^*)]; [A_1(x_1^*;0) + A_2(x_1^*;0)]\}$$
and set: $\bar{A}_1 := \bar{A} - \bar{A}_2$.

Step 2: Solve the optimization problem:
$$\max \pi = f(\bar{A}_2) = ((\bar{A} - \bar{A}_2 - \bar{A}_1(\tilde{x}_1; L_2)) \cdot (\bar{A}_2 - A_2(L_1; \tilde{x}_2)))$$ by differentiation, i.e.
$$\frac{df(\bar{A}_2)}{d\bar{A}_2} \stackrel{!}{=} 0 .$$

Result: The payoffs $\bar{A}_1^*$ and $\bar{A}_2^*$. This solution is of course always parento-optimal. Nash's solution of the bargaining problem may also be applied to bimatrix games, using the fixed-threat point instead of the guarantee point.

Example
Given the example of 5.1.1 .
$\bar{A} = \bar{A}_1 + \bar{A}_2 = \max \{[190/3 - 8]; [67 - 2]\} = 65 \rightarrow \bar{A}_1 = 65 - \bar{A}_2$;

242 Game - Theory

$$\max \pi = f(\bar{A}_2) = (65 - \bar{A}_2 - (-8)) \cdot (\bar{A}_2 - (-23/12)) = (73 - \bar{A}_2) \cdot (\bar{A}_2 + 23/12);$$

$$\frac{df(\bar{A}_2)}{d\bar{A}_2} = 73 - 2 \cdot \bar{A}_2 - 23/12 \stackrel{!}{=} 0 \rightarrow$$

$$\bar{A}_2^* = 853/24 \approx 35.54; \quad \bar{A}_1^* \approx 29.46.$$

5.1.3 The Extensive Form

Hypotheses

Given a game for a finite number of players $j=2,3,\ldots,n$.
Each player has at most $t_j < \infty$ moves, the game so consists of
$$k := \sum_{j=2}^{n} t_j$$ moves, which are taken in order one after the other. Before the i-th move, that player whose turn it is, may select from $s_i < \infty$ different strategies. After all moves have been taken, a certain progress of the game has been determined. For each possible progress of the game a corresponding payoff-function is given.

Procedure

Represent the game as a digraph $G=(N,A)$. The possible strategies are the nodes, connected with a directed arc from "their" source. Starting with an artificial source n_0, determine the nodes (strategies) of the first move and the arcs to each of these nodes.
This so defined digraph is a tree which is successively extended by considering the nodes of the i-th move as the sources for the nodes of the $(i + 1)$-th move, $\forall \; i = 1,\ldots, k - 1$.
After all moves are completed, there exists a tree with $N^{(k)} := \prod_{i=1}^{k} s_i$

terminal nodes, the complete tree, which is also called a game tree,
exists of $\tilde{N} := \sum_{x=1}^{k} \prod_{i=1}^{x} s_i$ nodes.

Section 5.1.3   243

These formulas only hold, if $|N^S(n_i)| = |N^S(n_j)|$ for all $u_i \neq u_j$ which can be reached after μ moves, $\forall\, \mu = 1,\ldots,k$; i.e. the game ends after exactly k moves; the resulting game tree is symmetric. Otherwise the number of nodes is $\tilde{N} := 1 + \bigcup_{i=1}^{k} N_i$, where N_i is the set of nodes, resulting from the i-th move. For each terminal node compute for each player according to the payoff-function.

Extension (solution of a game with total information)
If each player knows the current state of the game at any time (total information), then it is possible to determine one (or more) equilibrium-point(s) of the game.

Procedure
The player who takes the (k - 1)-th move regards each node $n_j \in N_{k-2}$; if there are alternative strategies at any node $n_j \in N_{k-2}$, then those strategy is selected that leads to the maximal payoff for himself (if several strategies lead to the same maximal payoff, then all these strategies are selected). All other strategies are eliminated from the current game tree and of course the successor nodes and arcs are eliminated, too. Then the player who takes the (k - 2)-th move regards the nodes $n_j \in N_{k-3}$ etc.
This procedure continuous until the first player has selected the optimal strategy of the first move.

Result: The remaining game tree determines a unique solution of the game (equilibrium-point) with a fixed payoff for all players. In the case of degeneration several solutions may exist, but with the same payoff.

Example 1 (game without total information)
Two players (i.e. n = 2) each have two cards (the seven and eight of hearts), lying covered in front of them. Both select a card at random and place it upside down on the table before them

(i.e. $s_1 = s_2 = 2$). The first player guesses how many hearts are lying on the table, then the second player does. Let be the set of all possible numbers of hearts; $M = \{14; 15; 16\}$, then $y_i \in M$ is that number, the i-th player guessed. Double-guesses are not permitted, which means that each player must guess a different number.

So we have: $y_1 \in \{14; 15; 16\}$; $s_3 = 3$

$y_2 \in \{\{14; 15; 16\} - y_1\}$; $s_4 = 2$.

The following payoff-function is given:

Whosever guess is correct, receives 1 m.u. from the pot.
Whosever guess is wrong, pays 1 m.u. to the pot.
If both players guess wrong, no one pays.
Note: This problem can be <u>solved</u> by determining the maximal expected value.

Section 5.1.3 245

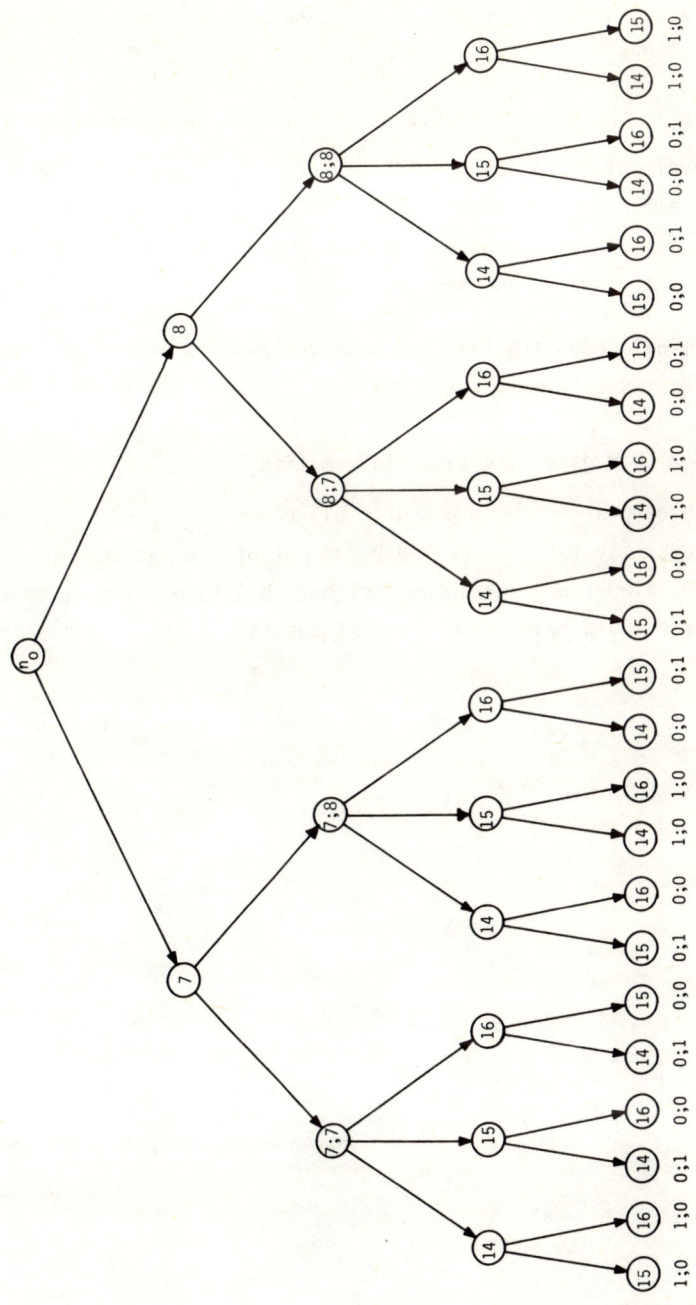

Note: To eliminate confusion the node-numbers $n_1,\ldots,n_{42}$ have not been included.

The data in the nodes indicate the strategies selected, i.e.

1-st level: player 1 selects "7" or "8"
2-nd level: player 2 selects "7" or "8"
3-rd level: player 1 guesses y_1
4-th level: player 2 guesses y_2.

The terminal nodes are labelled with the payoffs of player 1 and player 2, respectively.

Example 2 (NIM-game with total information)

Given three players and a pot with six beans.
Each player may take one or two beans out of the pot when it is his turn. The player who takes the last bean has to pay 1 m.u. to that player who would have taken the next bean(s) if the game had proceeded.

Section 5.1.3 247

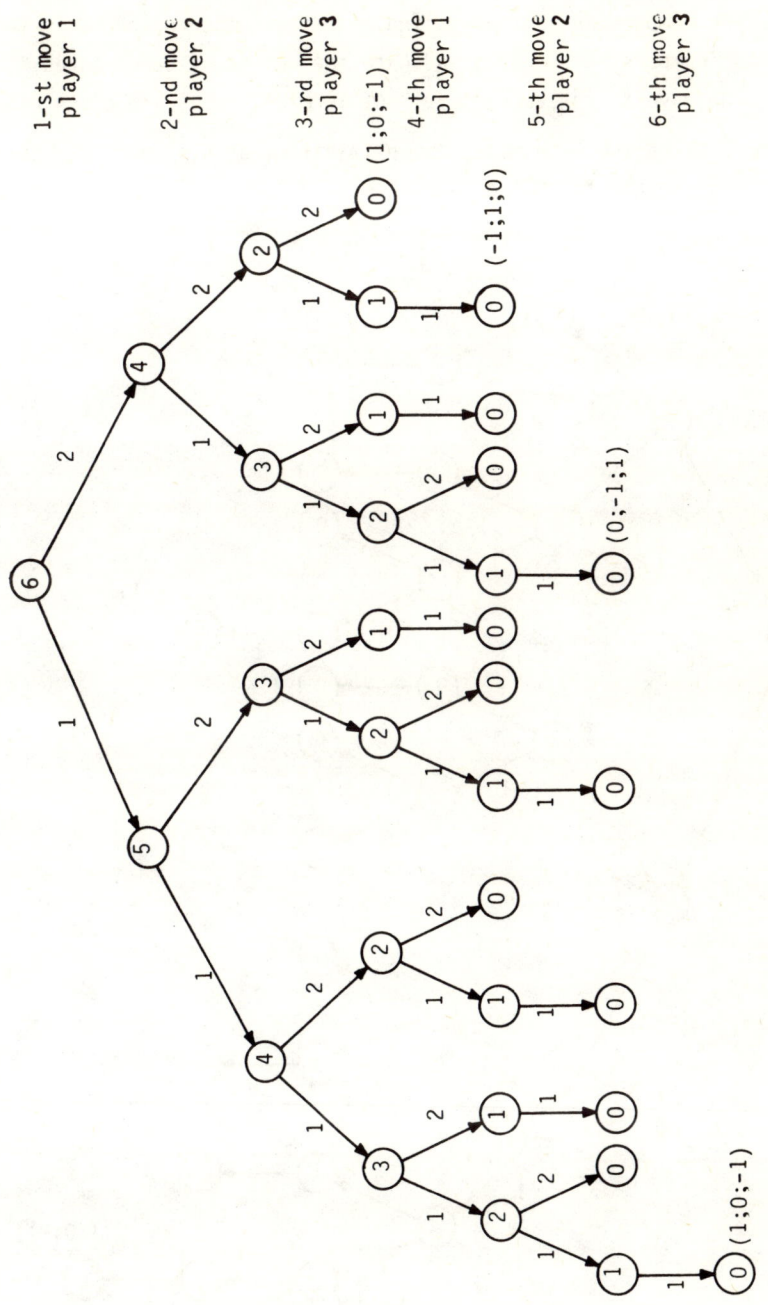

248 Game-Theory

Note: The numbers in the bounds indicate the number of beans in the pot, the numbers on the arcs indicate the number of beans taken away. The payoffs for each "level" of terminal nodes are given beside.

player 2 takes the 5-th move, the elimination of his "bad" strategies leads to the following game tree:

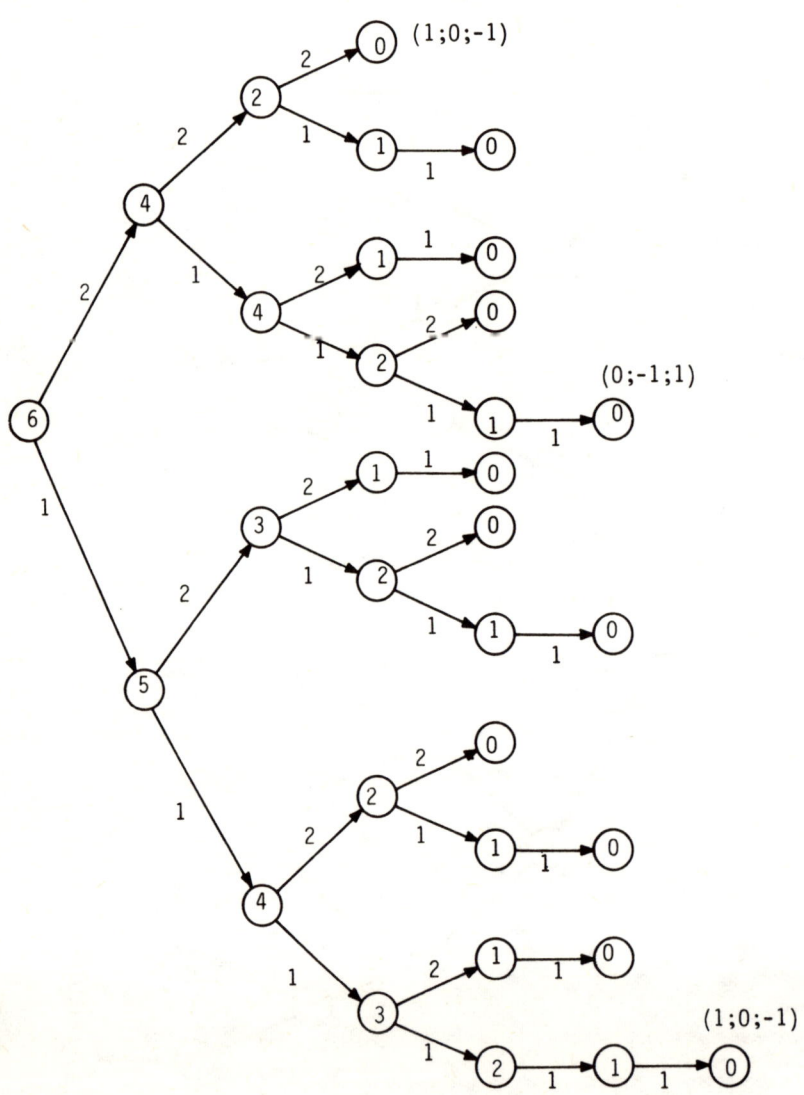

Section 5.1.3 249

player 1 takes the 4-th move, the elimination of his "bad" strategies leads to the following game tree:

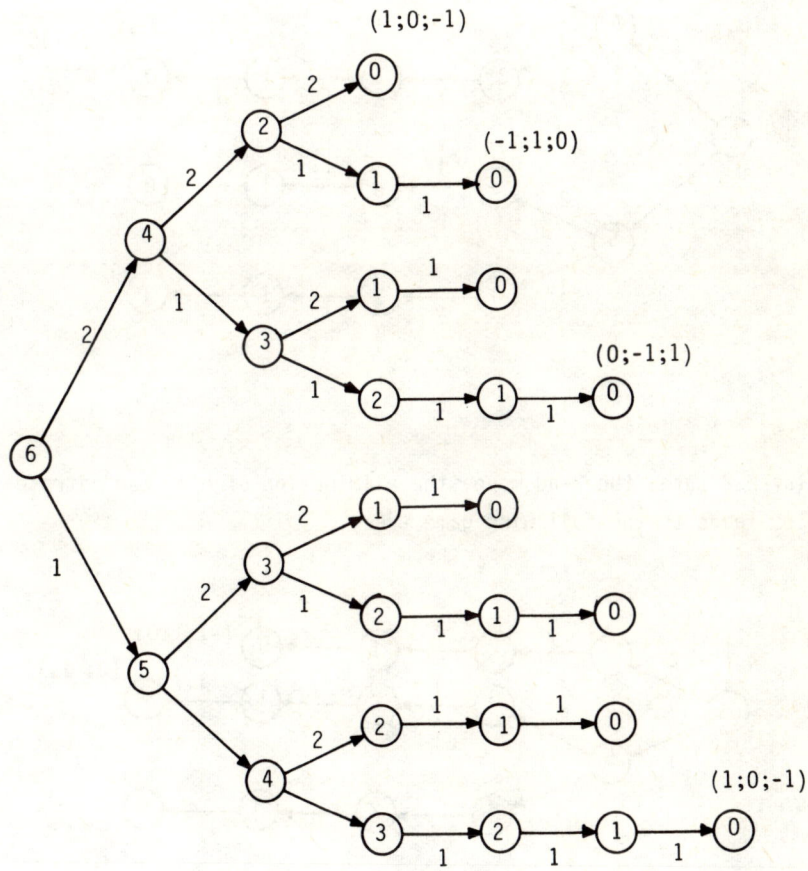

player 3 takes the 3-rd move, the elimination of his "bad" strategies leads to the following game tree:

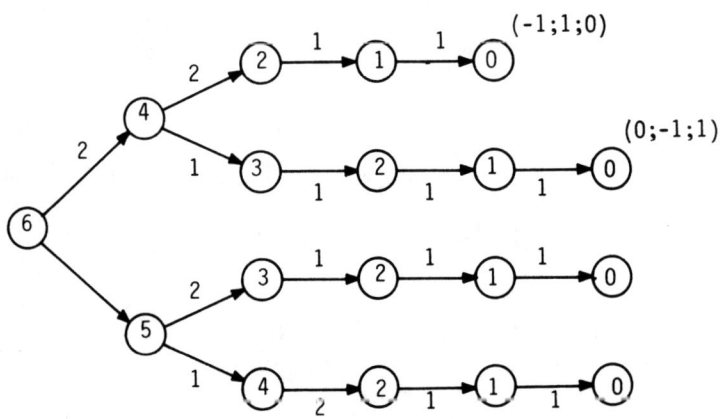

player 2 takes the 2-nd move, the elimination of his "bad" strategies leads to the following game tree:

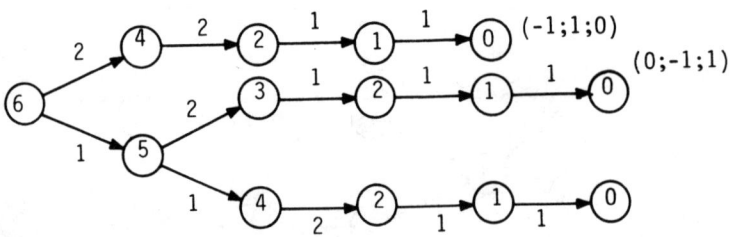

player 1 takes the 1-st move, the elimination of his "bad" strategies leads to the following game tree:

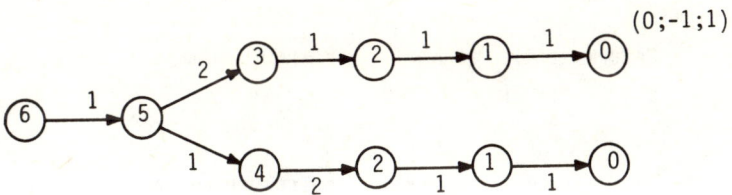

Stop, the solution of the game is fully determined.

5.2 Matrix Games

5.2.1 A Method for Determining Pure Strategy Pairs for Two-Person Zero-Sum Games

<u>Hypotheses</u>

Given a two-person game, player I has m strategies $s_i^{(I)}$, i=1,...,m and player II has n strategies $s_j^{(II)}$, j=1,...,n.

Let the profit (loss) of player I be equal to the loss (profit) of player II, so a zero-sum game is defined. The payoffs are given in the payoff-matrix A : $A_{[m \times n]}$, so that $a_{ij} \in \mathbb{R}$ defines the payoff for player I, playing his strategie $s_i^{(I)}$ against the strategy $s_j^{(II)}$ of his opponent, player II.

<u>Principle</u>

In this method the optimal strategy for each player is determined by attempting to maximize the minimal profit (minimize the maximal loss). The solution of the game has been found when the optimal payoffs have equal indices.

Description

Step 1: Determine the maximum of the row-minima:

$$a_{kl} := \max_i \{\min_j \{a_{ij}\}\}.$$

Step 2: Determine the minimum of the column-maxima:

$$a_{rs} := \min_j \{\max_i \{a_{ij}\}\}.$$

Step 3: Is $(k = r) \wedge (l = s)$?

If yes: Stop, player I plays his strategy $s_k^{(I)}$, player II plays his strategy $s_l^{(II)}$ as a pure strategy. The pair of strategies $(s_k^{(I)}; s_l^{(II)})$ is also called a saddlepoint or an equilibrium-point of the game.

If no : Stop, the game has no saddlepoint. The mixed strategies of the game can be determined with the Simplex-Algorithm or an approximation-method.

Example 1
Given the following payoff-matrix A :

$$A = \begin{pmatrix} 2 & 5 & 0 \\ 3 & -2 & -1 \end{pmatrix}$$

$a_{kl} = \max \{0; -2\} = 0 \rightarrow a_{kl} = a_{13}$; $a_{rs} = \min \{3; 5; 0\} = 0 \rightarrow a_{rs} = a_{13}$;
$(k \stackrel{!}{=} r) \wedge (l \stackrel{!}{=} s)$; Stop, $(s_1^{(I)}; s_3^{(II)})$ is the saddlepoint of the game.

Example 2
Given the following payoff-matrix A :

$$A = \begin{pmatrix} 1 & 4 & -3 \\ -1 & -2 & 3 \end{pmatrix}$$

$a_{kl} = \max \{-3; -2\} = -2 \rightarrow a_{kl} = a_{22}$; $a_{rs} = \min\{1; 4; 3\} = 1 \rightarrow a_{rs} = a_{11}$;
$(k \neq r) \wedge (l \neq s)$; Stop, the game has no saddlepoint.

5.2.2 A Method for Solving Two-Person Zero-Sum Games with the Simplex-Algorithm

Hypotheses
Given the payoff-matrix $\bar{A}$ of a two-person zero-sum game without an equilibrium-point for pure strategies.

Principle
First those strategies are eliminated which are worse than others in all cases. In the remaining game a formulation is given, which is always solvable with the Primal Simplex-Algorithm.

Description

Step 1: (elimination of strictly dominated strategies of player I)
Consider the rows i and k of the current matrix $\bar{A}$.
If $\bar{\bar{a}}_{ij} \geq \bar{\bar{a}}_{kj}\ \forall\ j$, eliminate row k. Continue this procedure until no other row may be eliminated.

Step 2: (elimination of strictly dominated strategies of player II)
Consider two columns j and l of the current matrix $\bar{A}$.
If $\bar{\bar{a}}_{il} \geq \bar{\bar{a}}_{ij}\ \forall\ i$, eliminate column l. Continue this procedure until no other column may be eliminated.

Step 3: Was at least one column in step 2 in this iteration eliminated?
If yes: Go to step 1.
If no : Go to step 4.

Step 4: Now the game has been reduced to the essential strategies. The corresponding payoff-matrix is $\bar{A} : \bar{A}_{[m\ \times\ n]}$, the remaining strategies are

$s_i^{(I)}$, $i = 1,\ldots, m$ for player I;

$s_j^{(II)}$, $j = 1,\ldots, n$ for player II.

Is $m + n \geq 3$?
If yes: Go to step 5.

If no : Stop, the remaining matrix-element $\bar{a}_{kl}$ determines a unique solution. Player I plays his strategy $s_k^{(I)}$, player II plays his strategy $s_l^{(II)}$.

Step 5: Is $\bar{a}_{ij} \geq 0 \;\forall\; i,j$?
If yes: Set $A := \bar{A}$. Go to step 7.
If no : Go to step 6.

Step 6: Determine $\bar{a}_{qt} := \min_{i,j}\{\bar{a}_{ij}\}$ and form the constant matrix $\tilde{A} : \tilde{A}_{[m \times n]}$, so that $\tilde{A} := (|\bar{a}_{qt}|)$. Compute $A := \bar{A} + \tilde{A}$.

Step 7: Solve the following problem P with the Primal Simplex-Algorithm:

$$P:\; \max \pi = (e;\theta)\cdot(x;y)^T$$

$$A\cdot x + y = e$$

$$x;\; y \geq \theta$$

Result: The vector $\bar{x} = (\bar{x}_1,\ldots,\bar{x}_n)$ of the optimal solution and the vector $\bar{v} = (\bar{v}_1,\ldots,\bar{v}_m)$, where $\bar{v}_i := \bar{c}_{n+i}$, i.e. the value under the slack variable y_i in the objective function.

Step 8: Compute $x_j^* := \dfrac{\bar{x}_j}{\sum_{j=1}^{n} \bar{x}_j} \;\forall\; j=1,\ldots,n$,

$$v_i^* := \dfrac{\bar{v}_i}{\sum_{i=1}^{m} \bar{v}_i} \;\forall\; i=1,\ldots,m .$$

Here x_j^* indicates the relative frequency for player II to play his strategy $s_j^{(II)}$, v_i^* indicates the relative frequency for player I to play his strategy $s_i^{(I)}$.

Section 5.2.2 255

If $\bar{\pi} = c \cdot \bar{x}$ is the value of the objective function in the optimal tableau, then $\pi^* = (\frac{1}{\bar{\pi}} + \bar{a}_{qt})$ is the average payoff for player I per game, when both players play their optimal strategies.

Example

Given the following payoff-matrix $\bar{A}$:

$$\bar{\bar{A}} = \begin{pmatrix} 5 & 3 & 4 \\ -4 & 4 & 8 \\ 1 & 3 & 2 \\ -6 & 4 & 3 \end{pmatrix} \quad s_1^{(I)} \text{ dominates } s_3^{(I)}; \ s_2^{(I)} \text{ dominates } s_4^{(I)};$$

$$\bar{\bar{A}} = \begin{pmatrix} 5 & 3 & 4 \\ -4 & 4 & 8 \end{pmatrix} \quad s_2^{(II)} \text{ dominates } s_3^{(II)}; \quad \bar{A} = \begin{pmatrix} 5 & 3 \\ -4 & 4 \end{pmatrix}$$

$$\tilde{A} = \begin{pmatrix} 4 & 4 \\ 4 & 4 \end{pmatrix} \quad A = \bar{A} + \tilde{A} = \begin{pmatrix} 9 & 7 \\ 0 & 8 \end{pmatrix}$$

P:

	x_1	x_2	y_1	y_2	1
	9	7	1	0	1
	0	8	0	1	1
	-1	-1	0	0	0

	x_1	x_2	y_1	y_2	1
	1	7/9	1/9	0	1/9
	0	8	0	1	1
	0	-2/9	1/9	0	1/9

x_1	x_2	y_1	y_2	1
1	0	1/9	-7/12	1/72
0	1	0	1/8	1/8
0	0	1/9	1/36	5/36

$x_1^* = \frac{1/72}{10/72} = 1/10; \quad x_2^* = \frac{1/8}{10/72} = 9/10;$

$v_1^* = \frac{1/9}{5/36} = 4/5; \quad v_2^* = \frac{1/36}{5/36} = 1/5;$

i.e. player I selects his strategy $s_1^{(I)}$ (of the reduced game) with the relative frequency 1/10 and his strategy $s_2^{(I)}$ with the relative frequency 9/10; player II selects his strategy $s_1^{(II)}$ (of the reduced game) with the relative frequency 4/5 and his strategy $s_2^{(II)}$ with the relative frequency 1/5. The payoff for player I per game is on the average $\pi^* = 36/5 - 4 = 16/5$ m.u.

5.2.3 An Approximization Method for Two-Person Zero-Sum Games ("learning method"; Gale; Brown)

Hypotheses
Given the payoff-matrix $\bar{A}$ of a two-person zero-sum game without equilibrium-point for pure strategies. Let $\bar{t} \in \mathbb{N}$ be the number of planned repetitions of the game. Set the running index $t:=1$ and define the sets $M_1 := M_2 := \emptyset$.

Principle
In this method each player begins with an arbitrary strategy. By succesive choices of strategies each player considers the post actions of his opponent.

Description

Step 1: Eliminate strictly dominated strategies $s_i^{(I)}$ and $s_j^{(II)}$ according to 5.2.2, step 1 to step 3.
Result: A payoff-matrix $A : A_{[m \times n]}$, which may be considered as an ordered set of row-vectors a_i, $i=1,\ldots,m$, or column-vectors $\bar{a}_j$, $j=1,\ldots,n$, respectively.

Step 2: Select an arbitrary row-vector $a_i^{(t)}$ and set $b^{(t)} := a_i^{(t)}$.
Determine $b_s^{(t)} := \min_j \{b_j^{(t)}\}$ and set $M_1 := M_1 \cup \{b_s^{(t)}\}$

Step 3: Select an arbitrary column-vector $\bar{a}_j^{(t)}$ and set $\bar{b}^{(t)} := \bar{a}_j^{(t)}$. Determine $\bar{b}_r^{(t)} := \max_i \{\bar{b}_i^{(t)}\}$ and set $M_2 := M_2 \cup \{\bar{b}_r^{(t)}\}$.

Step 4: Is $t = \bar{t}$?
If yes: Go to step 7.
If no : Set $t := t + 1$. Go to step 5.

Step 5: Select the row-vector $a_r^{(t)}$ and compute $b^{(t)} := b^{(t-1)} + a_r^{(t)}$.
Determine $b_k^{(t)} := \min_j \{b_j^{(t)}\}$ and set $M_1 := M_1 \cup \{b_k^{(t)}\}$.

Step 6: Select the column-vector $\bar{a}_s^{(t)}$ and compute
$\bar{b}^{(t)} := \bar{b}^{(t-1)} + \bar{a}_s^{(t)}$. Determine $\bar{b}_r^{(t)} := \max_i \{\bar{b}_i^{(t)}\}$
and set $M_2 := M_2 \cup \{\bar{b}_r^{(t)}\}$; $s := k$. Go to step 4.

Step 7: Let $\bar{h}_i$ be the number of elements $\bar{b}_i$ in the set M_2, $\forall$
$\forall i=1,\ldots,m$. Calculate $p_i^* := \bar{h}_i/\bar{t}$ $\forall i=1,\ldots,m$;

$$\bar{v} := \min_t \left\{ \frac{\bar{b}_r^{(t)}}{t} \right\}.$$

p_i^* is the relative frequency for player I to select his strategy $s_i^{(I)}$ (of the reduced game); $\bar{v}$ determines the upper bound for the value of the game.

Step 8: Let h_j be the number of elements b_j in the set M_1, $\forall$
$\forall j=1,\ldots,n$. Calculate $q_j^* := h_j/\bar{t}$ $\forall j=1,\ldots,n$;

$$\underline{v} := \max_t \left\{ \frac{b_k^{(t)}}{t} \right\}.$$

q_j^* is the relative frequency for player II to select his strategy $s_j^{(II)}$ (of the reduced game); $\underline{v}$ determines the lower bound for the value of the game. So the value of the game v lies in the interval $[\underline{v}; \bar{v}]$, i.e. $v \in [\underline{v}; \bar{v}]$.
The average payoff for player I is

$$\pi^* := \sum_{i=1}^{m} \sum_{j=1}^{n} a_{ij} \cdot p_i^* \cdot q_j^*.$$

Example

Given the following payoff-matrix $\bar{A}$:

$$\bar{A} = \begin{pmatrix} 4 & 3 & -2 \\ -1 & 2 & -3 \\ 5 & -2 & 1 \\ 3 & -3 & 2 \end{pmatrix}; \text{ the game will be played 8 times, i.e. } \bar{t}=8.$$

According to 5.2.2 the strictly dominated strategies $s_i^{(I)}$ and $s_j^{(II)}$ are eliminated. The resulting payoff-matrix A is

$$A = \begin{pmatrix} 3 & -2 \\ -2 & 1 \\ -3 & 2 \end{pmatrix}$$

In the following, for reasons of clearness the column-vectors $\bar{b}^{(t)}$ are noted one after another on the right side of the matrix A, the row-vectors $b^{(t)}$ one under another under the matrix A. The elements $\bar{b}_r^{(t)}$ and $b_k^{(t)}$ are especially labelled with "*".

$$A = \begin{pmatrix} 3 & -2 \\ -2 & 1 \\ -3 & 2 \end{pmatrix}$$

t =	1	2	3	4	5	6	7	8
	3^*	1^*	-1	-3	-5	-7	-4	-1
	-2	-1	0	1	2	3	1	1
	-3	-1	1^*	3^*	5^*	7^*	4^*	1^*

t =		
1	3	-2^*
2	6	-4^*
3	9	-6^*
4	6	-4^*
5	3	-2^*
6	0^*	0
7	-3^*	2
8	-6^*	4

$t = 1;$

$a_j^{(1)} = a_1^{(1)}; \quad b_s^{(1)} = b_2^{(1)}; \quad M_1 = \{b_2^{(1)}\};$

$\bar{a}_j^{(1)} = \bar{a}_1^{(1)}; \quad \bar{b}_r^{(1)} = \bar{b}_1^{(1)}; \quad M_2 = \{\bar{b}_1^{(1)}\};$

$t = 2;$

$a_r^{(2)} = a_1^{(2)}; \quad b_k^{(2)} = b_2^{(2)}; \quad M_1 = \{b_2^{(1)}, b_2^{(2)}\};$

$\bar{a}_s^{(2)} = \bar{a}_2^{(2)}; \quad \bar{b}_r^{(2)} = \bar{b}_1^{(2)}; \quad M_2 = \{\bar{b}_1^{(1)}, \bar{b}_1^{(2)}\};$

$s = 2; \quad t = 3;$

$a_r^{(3)} = a_1^{(3)}; \quad b_k^{(3)} = b_2^{(3)}; \quad M_1 = \{b_2^{(\tau)}, \tau = 1,2,3\};$

$\bar{a}_s^{(3)} = \bar{a}_2^{(3)}; \quad \bar{b}_r^{(3)} = \bar{b}_3^{(3)}; \quad M_2 = \{\bar{b}_1^{(1)}, \bar{b}_1^{(2)}, \bar{b}_3^{(3)}\};$

$s = 2; \quad t = 4;$

$a_r^{(4)} = a_3^{(4)}; \quad b_k^{(4)} = b_2^{(4)}; \quad M_1 = \{b_2^{(\tau)}, \tau = 1,\ldots,4\};$

$\bar{a}_s^{(4)} = \bar{a}_2^{(4)}; \quad \bar{b}_r^{(4)} = \bar{b}_3^{(4)}; \quad M_2 = \{\bar{b}_1^{(1)}, \bar{b}_1^{(2)}, \bar{b}_3^{(3)}, \bar{b}_3^{(4)}\};$

$s = 2; \quad t = 5;$

$a_r^{(5)} = a_3^{(5)}; \quad b_k^{(5)} = b_2^{(5)}; \quad M_1 = \{b_2^{(\tau)}, \tau = 1,\ldots,5\};$

$\bar{a}_s^{(5)} = \bar{a}_2^{(5)}; \quad \bar{b}_r^{(5)} = \bar{b}_3^{(5)}; \quad M_2 = \{\bar{b}_1^{(1)}, \bar{b}_1^{(2)}, \bar{b}_3^{(\tau)}, \tau = 3,4,5\};$

$s = 2; \quad t = 6;$

$a_r^{(6)} = a_3^{(6)}; \quad b_k^{(6)} = b_1^{(6)}; \quad M_1 = \{b_2^{(\tau)}, \tau = 1,\ldots,5, \ b_1^{(6)}\};$

$\bar{a}_s^{(6)} = \bar{a}_2^{(6)}; \quad \bar{b}_r^{(6)} = \bar{b}_3^{(6)}; \quad M_2 = \{\bar{b}_1^{(1)}, \bar{b}_1^{(2)}, \bar{b}_3^{(\tau)}, \tau = 3,\ldots,6\};$

$s = 1; \quad t = 7;$

$a_r^{(7)} = a_3^{(7)}; \quad b_k^{(7)} = b_1^{(7)}; \quad M_1 = \{b_2^{(\tau)}, \tau = 1,\ldots,5, b_1^{(6)}, b_1^{(7)}\};$

$\bar{a}_s^{(7)} = \bar{a}_1^{(7)}; \quad \bar{b}_r^{(7)} = \bar{b}_3^{(7)}; \quad M_2 = \{\bar{b}_1^{(1)}, \bar{b}_1^{(2)}, \bar{b}_3^{(\tau)}, \tau = 3,\ldots,7\};$

260 Game-Theory

$a_r^{(8)} = a_3^{(8)}$; $b_k^{(8)} = b_1^{(8)}$; $M_1 = \{b_2^{(\tau)}, \tau = 1,\ldots,5, b_1^{(6)}, b_1^{(7)}, b_1^{(8)}\}$;

$\bar{a}_s^{(8)} = \bar{a}_1^{(8)}$; $\bar{b}_r^{(8)} = \bar{b}_3^{(8)}$; $M_2 = \{\bar{b}_1^{(1)}, \bar{b}_1^{(2)}, \bar{b}_3^{(\tau)}, \tau = 3,\ldots,8\}$;

$t \stackrel{!}{=} \bar{t} = 8$; $\bar{h}_1 = 2$; $\bar{h}_2 = 0$; $\bar{h}_3 = 6$;

$p_1^* = 2/8 = 1/4$; $p_2^* = 0/8 = 0$; $p_3^* = 6/8 = 3/4$;

$\bar{v} = \min\{3/1; 1/2; 1/3; 3/4; 5/5; 7/6; 4/7; 1/8\} = 1/8$;

$h_1 = 3$; $h_2 = 5$; $q_1^* = 3/8$; $q_2^* = 5/8$;

$\underline{v} = \max\{-2/1; -4/2; -6/3; -4/4; -2/5; 0/6; -3/7; -6/8\} = 0$;

so we have $v \in [0; 1/8]$; the average payoff for player I
is $\pi^* = 1/16$ m.u.

5.2.4 The LEMKE–HOWSON Algorithm for the Solution of Bimatrix Games

Hypotheses
Given the matrices $A : A_{[m \times n]}$ and $B : B_{[m \times n]}$, where $a_{ij}(b_{ij})$ is the payoff for player I (player II), if player I plays his strategy $s_i^{(I)}$ and player II plays his strategy $s_j^{(II)}$.
Let $a_{ij}, b_{ij} \in \mathbb{R}\ \forall\ i,j$; $m,n < \infty$.

Principle
The algorithm determines a sequence of "nearly-complementary basic-solutions". It terminates in an equilibrium-point, when a complementary basic-solution is found.

Description

Step 1: Determine the column-maxima of the matrix A and the row-maxima of the matrix B and label them.

Step 2: Are there at least one pair of indices (k,l), so that
(a_{kl} is labelled)∧(b_{kl} is labelled) ?

Section 5.2.4   261

If yes: Stop, all pairs of strategies $(s_k^{(I)}; s_l^{(II)})$ are equilibrium-points of the game.
If no : Go to step 3.

Step 3: Determine $a_{rs} := \max_{i,j} \{a_{ij}\}$ and form the constant matrix $A' : A'_{[m \times n]}$, so that $A' := (a_{rs} + 1)$. Compute $\tilde{A} := A - A'$.

Step 4: Determine $b_{rs} := \max_{i,j} \{b_{ij}\}$ and form the constant matrix $B' : B'_{[m \times n]}$, so that $B' := (b_{rs} + 1)$. Compute $\tilde{B} := B - B'$.

Step 5: Set up the following tableau:

$r_1 \ldots r_m$	$s_1 \ldots s_n$	$p_1 \ldots p_m$	$q_1 \ldots q_n$	1
$I_{[m \times m]}$	$\Theta_{[m \times n]}$	$\Theta_{[m \times m]}$	$\tilde{A}_{[m \times n]}$	-e
$\Theta_{[n \times m]}$	$I_{[n \times n]}$	$\tilde{B}^T_{[n \times m]}$	$\Theta_{[n \times n]}$	-e

The pairs of variables $(r_i; p_i)$ and $(s_j; q_j)$ are called "pairs of complementary variables".

Step 6: Determine the pivot-element $\tilde{b}_{rp_1}$ in column p_1, so that $\tilde{b}_{rp_1} := \max_j \{\tilde{b}_{jp_1}\}$ and do one tableau-transformation (according to 1.1.1) .
Result: p_1 is basic variable (bv), s_r has become non-basic variable (nbv).

Step 7: Select as the pivot-element the maximal element in column q_r and do one tableau-transformation.
Result: $r_i \cdot p_i = 0 \; \forall \; i=2,\ldots,m$;
$s_j \cdot q_j = 0 \; \forall \; j=1,\ldots,n$.

Step 8: Select as the next pivot-column that column whose complementary variable became nbv in the last iteration;

262 Game-Theory

the pivot-row is determined as in 1.1.1 , step 5.
Do one tableau-transformation.

Step 9: Is $(p_1$ is nbv$) \vee (r_1$ is nbv$)$?
If yes: Go to step 10.
If no : Go to step 8.

Step 10: Let $\bar{p}_i$, $\bar{q}_j$ be the solution of the last tableau.

Compute $p_i^* := \dfrac{\bar{p}_i}{\sum_{i=1}^{m} \bar{p}_i}$ $\forall\ i=1,\ldots,m;$

$q_j^* := \dfrac{\bar{q}_j}{\sum_{j=1}^{n} \bar{q}_j}$ $\forall\ j=1,\ldots,n.$

Here p_i^* indicates the relative frequency for player I to play his strategy $s_i^{(I)}$, q_j^* indicates the relative frequency for player II to play his strategy $s_j^{(II)}$; the pair $(p^*;q^*)$ is an equilibrium-point of the bimatrix game.

The average payoff for player I is given by

$$\pi_I^* = \sum_{i=1}^{m} \sum_{j=1}^{n} a_{ij} \cdot p_i^* \cdot q_j^* ,$$

for player II by $\pi_{II}^* = \sum_{i=1}^{m} \sum_{j=1}^{n} b_{ij} \cdot p_i^* \cdot q_j^*$.

Example

Given the following payoff-matrices A and B :

$$A = \begin{pmatrix} 4^* & -4 & 7^* \\ 3 & 1^* & 6 \end{pmatrix} ; \quad B = \begin{pmatrix} 2 & 4^* & -1 \\ 6^* & -2 & 3 \end{pmatrix} ;$$

the column-maxima of A and the row-maxima of B are labelled with "*"; there is no equilibrium-point for pure strategies.

$$a_{rs} = 7 \rightarrow A' = \begin{pmatrix} 8 & 8 & 8 \\ 8 & 8 & 8 \end{pmatrix} ; \tilde{A}=A-A' = \begin{pmatrix} -4 & -12 & -1 \\ -5 & -7 & -2 \end{pmatrix} ;$$

$$b_{rs} = 6 \rightarrow B' = \begin{pmatrix} 7 & 7 & 7 \\ 7 & 7 & 7 \end{pmatrix} ; \tilde{B}=B-B' = \begin{pmatrix} -5 & -3 & -8 \\ -1 & -9 & -4 \end{pmatrix} ;$$

r_1	r_2	s_1	s_2	s_3	p_1	p_2	q_1	q_2	q_3	1
1	0	0	0	0	0	0	-4	-12	-1	-1
0	1	0	0	0	0	0	-5	-7	-2	-1
0	0	1	0	0	-5	-1	0	0	0	-1
0	0	0	1	0	-3	-9	0	0	0	-1
0	0	0	0	1	-8	-4	0	0	0	-1

r_1	r_2	s_1	s_2	s_3	p_1	p_2	q_1	q_2	q_3	1
1	0	0	0	0	0	0	-4	-12	-1	-1
0	1	0	0	0	0	0	-5	-7	-2	-1
0	0	1	-5/3	0	0	14	0	0	0	2/3
0	0	0	-1/3	0	1	3	0	0	0	1/3
0	0	0	-8/3	1	0	20	0	0	0	5/3

r_1	r_2	s_1	s_2	s_3	p_1	p_2	q_1	q_2	q_3	1
1	-12/7	0	0	0	0	0	32/7	0	17/7	5/7
0	-1/7	0	0	0	0	0	5/7	1	2/7	1/7
0	0	1	-5/3	0	0	14	0	0	0	2/3
0	0	0	-1/3	0	1	3	0	0	0	1/3
0	0	0	-8/3	1	0	20	0	0	0	5/3

r_1	r_2	s_1	s_2	s_3	p_1	p_2	q_1	q_2	q_3	1
1	-12/7	0	0	0	0	0	32/7	0	17/7	5/7
0	-1/7	0	0	0	0	0	5/7	1	2/7	1/7
0	0	1/14	-5/42	0	0	1	0	0	0	1/21
0	0	-3/14	-3/14	0	1	0	0	0	0	4/21
0	0	-10/7	-2/7	1	0	0	0	0	0	15/21

264 Game-Theory

r_1	r_2	s_1	s_2	s_3	p_1	p_2	q_1	q_2	q_3	1
7/32	-3/8	0	0	0	0	0	1	0	17/32	5/32
-5/32	1/8	0	0	0	0	0	0	1	-3/32	1/32
0	0	1/14	-5/42	0	0	1	0	0	0	1/21
0	0	-3/14	-3/14	0	1	0	0	0	0	4/21
0	0	-10/7	-2/7	1	0	0	0	0	0	15/21

r_1 is nbv !

$\bar{p}_1 = 4/21$; $\bar{p}_2 = 1/21$;
$\bar{q}_1 = 5/32$; $\bar{q}_2 = 1/32$; $\bar{q}_3 = 0$;

$p_1^* = \frac{4/21}{5/21} = \frac{4}{5}$; $p_2^* = \frac{1/21}{5/21} = \frac{1}{5}$; $q_1^* = \frac{5/32}{6/32} = \frac{5}{6}$; $q_2^* = \frac{1/32}{6/32} = \frac{1}{6}$;

$q_3^* = 0$;

the average payoff for player I is

$$\pi_I^* = \sum_{i=1}^{2} \sum_{j=1}^{3} a_{ij} \cdot p_i^* \cdot q_j^* = 8/3 \text{ m.u.},$$

the average payoff for player II is

$$\pi_{II}^* = \sum_{i=1}^{2} \sum_{j=1}^{3} b_{ij} \cdot p_i^* \cdot q_j^* = 14/5 \text{ m.u.}$$

5.3 Decisions under Uncertainty (games against nature)

Hypotheses

See 5.2.1. Additionally assume that player II (nature) is a naive player. He does not play to maximize his own profit or to minimize the profit of his opponent. His selection of strategies is given by a relative frequency distribution $q = (q_1, \ldots, q_n)$. Determine the optimal strategy $s_k^{(I)}$ for player I.

Principle

Player I who is optimistic or pessimistic determines his optimal solution, regarding a particular object which he has in view. This is in

5.3.1 : maximization of the minimal profit
5.3.2 : maximization of the weighted (optimism-parameter) sum of minimal and maximal profit
5.3.3 : maximization of the minimal divergence of the profit from the maximal profit
5.3.4 : maximization of the weighted (relative frequency distribution) average profit
5.3.5 : maximization of the average profit
5.3.6 : maximization of the weighted (optimism-parameter) sum of the weighted (relative frequency distribution) average profit and the minimal profit.

5.3.1 The Solution of WALD

Procedure
Determine $p_i := \min_j \{a_{ij}\} \ \forall i=1,\ldots,m$,

and $p_k := \max_i \{p_i\}$.

Strategy $s_k^{(I)}$ is optimal according to the solution of WALD.

Example
Given the following payoff-matrix A:

$$A = \begin{pmatrix} 6 & -3 & 2 & 5 \\ 5 & 1 & 0 & 3 \\ -1 & 4 & 7 & 0 \end{pmatrix} \ ; \ p = \begin{pmatrix} -3 \\ 0 \\ -1 \end{pmatrix} \leftarrow \max$$

$p_k = p_2 \to s_2^{(I)}$ is optimal according to the solution of WALD.

5.3.2 The Solution of HURWICZ

Hypotheses
Given an optimism-parameter $\lambda \in [0;1]$.

Procedure
Determine
$$p'_i := \max_j \{a_{ij}\}$$
$$p''_i := \min_j \{a_{ij}\}$$
$\forall i=1,\ldots,m$

and compute $\tilde{p}_i := \lambda \cdot p'_i + (1-\lambda) \cdot p''_i \quad \forall \ i=1,\ldots,m$.
Determine $p_k := \max_i \{\tilde{p}_i\}$.
The strategy $s_k^{(I)}$ is optimal according to the solution of HURWICZ.

Note: For $\lambda = 0$ the solution of HURWICZ is identical to the solution of WALD (total pessimism).

Example
Given the optimism-parameter $\lambda = 0.6$ and the following payoff-matrix A:

$$A = \begin{pmatrix} 6 & -3 & 2 & 5 \\ 5 & 1 & 0 & 3 \\ -1 & 4 & 7 & 0 \end{pmatrix} ; \ p' = \begin{pmatrix} 6 \\ 5 \\ 7 \end{pmatrix} ; \ p'' = \begin{pmatrix} -3 \\ 0 \\ -1 \end{pmatrix}$$

$$\tilde{p} = \begin{pmatrix} 0.6 \cdot 6 + 0.4 \cdot (-3) \\ 0.6 \cdot 5 + 0.4 \cdot 4 \\ 0.6 \cdot 7 + 0.4 \cdot (-1) \end{pmatrix} = \begin{pmatrix} 2.4 \\ 3.0 \\ 3.8 \end{pmatrix} \leftarrow \max$$

$p_k = \tilde{p}_3 \rightarrow s_3^{(I)}$ is optimal according to the solution of HURWICZ.

5.3.3 The Solution of SAVAGE and NIEHANS

Procedure
Determine $p_j := \max_i \{a_{ij}\} \ \forall j=1,\ldots,n$ and form the "frustration-matrix" $\tilde{A} : \tilde{A}_{[m \times n]}$, so that $\tilde{a}_{ij} := a_{ij} - p_j \ \forall \ i=1,\ldots,m; \ j=1,\ldots,n$.

Determine $p_i' := \min_j \{\tilde{a}_{ij}\} \ \forall \ i=1,\ldots,m;$

$p_k := \max_i \{p_i'\}.$

$s_k^{(I)}$ - the strategy of the "minimal regret" - is optimal according to the solution of SAVAGE and NIEHANS.

Example

Given the following payoff-matrix A:

$$A = \begin{pmatrix} 6 & -3 & 2 & 5 \\ 5 & 1 & 0 & 3 \\ -1 & 4 & 7 & 0 \end{pmatrix}$$

$$p = (\ 6 \quad 4 \quad 7 \quad 5\ )$$

$$\tilde{A} = \begin{pmatrix} 0 & -7 & -5 & 0 \\ -1 & -3 & -7 & -2 \\ -7 & 0 & 0 & -5 \end{pmatrix} \ ; \quad p' = \begin{pmatrix} -7 \\ -7 \\ -7 \end{pmatrix} \longleftrightarrow \max$$

$p_k = p_1' = p_2' = p_3' \rightarrow$ all strategies $s_i^{(I)}$, $i=1,2,3$, are optimal according to the solution of SAVAGE and NIEHANS.

5.3.4 The Solution of BAYES

Hypotheses

Given the relative frequency distribution $q = (q_1,\ldots,q_n)$ of player II, so that $\sum_{j=1}^{n} q_j = 1$.

Procedure

Compute $p_i := \sum_{j=1}^{n} a_{ij} \cdot q_j \ \forall \ i=1,\ldots,m$

and determine $p_k := \max_i \{p_i\}$.

The strategy $s_k^{(I)}$ is optimal according to the solution of BAYES.

268 Game-Theory

Example
Given the relative frequency distribution q=(0.2;0.4;0.3;0.1) for player II and the following payoff-matrix A:

$$A = \begin{pmatrix} 6 & -3 & 2 & 5 \\ 5 & 1 & 0 & 3 \\ -1 & 4 & 7 & 0 \end{pmatrix}$$

$$p = \begin{pmatrix} 0.2 \cdot 6 + 0.4 \cdot (-3) + 0.3 \cdot 2 + 0.1 \cdot 5 \\ 0.2 \cdot 5 + 0.4 \cdot 1 + 0.3 \cdot 0 + 0.1 \cdot 3 \\ 0.2 \cdot (-1) + 0.4 \cdot 4 + 0.3 \cdot 7 + 0.1 \cdot 0 \end{pmatrix} = \begin{pmatrix} 1.1 \\ 1.7 \\ 3.5 \end{pmatrix} \leftarrow \max$$

$p_k = p_3 \rightarrow s_3^{(I)}$ is optimal according to the solution of BAYES.

5.3.5 The Solution of LAPLACE

Procedure
Compute $p_i := 1/n \cdot \sum_{j=1}^{n} a_{ij}$ $\forall$ i=1,...,m and determine $p_k := \max_i \{p_i\}$.

The strategy $s_k^{(I)}$ is optimal according to the solution of LAPLACE.

Note: The solution of LAPLACE is a special case of the solution of BAYES. If $q_i = q_j$ $\forall$ i ≠ j in 5.3.4, both solutions are identical.

Example
Given the following payoff-matrix A:

$$A = \begin{pmatrix} 6 & -3 & 2 & 5 \\ 5 & 1 & 0 & 3 \\ -1 & 4 & 7 & 0 \end{pmatrix} \quad ; \quad p = \begin{pmatrix} 10/4 \\ 9/4 \\ 10/4 \end{pmatrix} \max$$

$p_k = p_1 = p_3 \rightarrow$ the strategies $s_1^{(I)}$ and $s_3^{(I)}$ are optimal according to the solution of LAPLACE.

5.3.6 The Solution of HODGES and LEHMANN

Hypotheses
Given an optimism-parameter $\lambda \in [0;1]$ and the relative frequency

distribution $q=(q_1,\ldots,q_n)$ of player II, so that $\sum_{j=1}^{n} q_j = 1$.

Procedure

Determine $\quad p_i' := \min_{j} \{a_{ij}\}$

and compute $\quad p_i'' := \sum_{j=1}^{n} a_{ij} \cdot q_j \qquad \forall\ i=1,\ldots,m;$

$\tilde{p}_i := \lambda \cdot p_i'' + (1-\lambda) \cdot p_i'$

determine $\quad p_k := \max_{i} \{\tilde{p}_i\}$.

The strategy $s_k^{(I)}$ is optimal according to the solution of HODGES and LEHMANN.

Note: For $\lambda=0$ the solution of HODGES and LEHMANN is identical to the solution of WALD (total pessimism); for $\lambda=1$ it is identical to the solution of BAYES.

Example

Given the relative frequency distribution $q=(0.2;0.4;0.3;0.1)$ for player II, the optimism-parameter $\lambda=0.6$, and the following payoff-matrix A:

$$A = \begin{pmatrix} 6 & -3 & 2 & 5 \\ 5 & 1 & 0 & 3 \\ -1 & 4 & 7 & 0 \end{pmatrix} ; \quad p' = \begin{pmatrix} -3 \\ 0 \\ -1 \end{pmatrix} ; \quad p'' = \begin{pmatrix} 1.1 \\ 1.7 \\ 3.5 \end{pmatrix}$$

$$\tilde{p} = \begin{pmatrix} 0.6 \cdot 1.1 + 0.4 \cdot (-3) \\ 0.6 \cdot 1.7 + 0.4 \cdot 0 \\ 0.6 \cdot 3.5 + 0.4 \cdot (-1) \end{pmatrix} = \begin{pmatrix} -0.54 \\ 1.02 \\ 1.70 \end{pmatrix} \longleftarrow \max$$

$p_k = \tilde{p}_3 \rightarrow s_3^{(I)}$ is optimal according to the solution of HODGES and LEHMANN.

6. Dynamic Programming

6.0.1 The n-Period Model

Hypotheses
Given the following problem:
The development of an enterprise is considered for $n < \infty$ periods. The enterprise works with only one product given in the beginning as well as at the end in a definite quantity. During the periods a change of the on-hand balance may occur though sale, purchase etc. Normally a finite storage capacity is given and a limit as to the quantity to be marketed. Under these assumptions a given (generally non-linear) objective function must be optimized.

Definitions
n: number of periods to be considered, $n < \infty$

X_j: set of feasible states ($\cong$ warehouse storage capacity) in the j-th period

Y_j: set of feasible actions ($\cong$ market capacity) in the j-th period

x_j: stored products at the end of the j-th period, $x_j \in X_j$.

y_j: the action taken in the j-th period, $y_j \in Y_j$

$v_j(x_{j-1}; y_j)$: transition function ($\cong$ quantitive change in the product from the (j-1)-th to the j-th period

$u_j(x_{j-1}; y_j)$: objective function.

Let x_n: = 0 .

Principle
In this algorithm first the optimal decisions (actions) are recursively determined by starting with a given final state x_n.

When this is accomplished the optimal decisions are executed beginning with the given initial state; thus the various interim states are determined.

Description

Step 1: Set $f^*_{n+1}(x_n) := x_n$; $j := n$.

Step 2: Compute BELLMAN's functional equation (BFE) f^* for the j-th period: $f^*_j(x_{j-1}) := \max\{u_j(x_{j-1}; y_j) + f^*_{j+1}(x_j)\}$.
Result: $f^*_j(x_{j-1})$ is dependent on x_j and y_j.

Step 3: Replace x_j by the transition function $v_j(x_{j-1}; y_j)$.
Result: $f^*_j(x_{j-1})$ is dependent on x_{j-1} and y_j.

Step 4: Determine the maximum value of BFE (if necessary by differentiating) and solve the equation for y_j.
Result: The optimal decision in the j-th period, y_j, has been determined, this optimal point is dependent on x_{j-1}.

Step 5: Replace y_j in the current BFE by the value of the optimal point.
Result: $f^*_j(x_{j-1})$ is dependent only on x_{j-1}.

Step 6: Is $j = 1$?
If yes: Go to step 7.
If no : Set $j := j-1$. Go to step 2.

Step 7: y_1 is now given as a function of x_0. Because x_0 is known, compute y_1. With the transition function compute x_1. With x_1 and the next optimal point, y_2 is uniquely determined and so on. In general: x_{j-1} is known. Put in into the according optimal point and y_j is determined. x_j is then determined by placing the values for x_{j-1} and y_j in the transition function. When all optimal decisions y_j are executed the value of the objective function is fully determined.

272 Dynamic Programming

Example 1
A hamster farmer must plan his breeding for the following four periods. His farm may handle a maximum of 350,000 hamsters. As he cannot buy any animals, he can at most sell as many animals as he owns. In the beginning he owns 3,000 hamsters and at the end of the four periods all the hamsters shall be sold. The hamsters will increase five-fold with births and deaths balancing out. The profit in the j-th period is $2 \cdot \sqrt{y_j}$ m.u. Under the neither humane nor realistic assumption of arbitrary divisibility of an animal, determine the optimal sales-policy and the total profit over the four periods.

$n = 4$; $X_j = [0; 350,000]$; $Y_j = [0; x_{j-1}]$;

$v_j(x_{j-1}; y_j) = 5 \cdot (x_{j-1} - y_j) = x_j$;

$u_j(x_{j-1}; y_j) = 2 \cdot \sqrt{y_j}$; $x_0 = 3,000$,

$f_5^*(x_4) = 0$;

$f_4^*(x_3) = \max \{u_4(x_3; y_4) + f_5^*(x_4)\} = 2 \cdot \sqrt{y_4}$;

$x_4 = 5 x_3 - 5 y_4$; $x_4 = 0 \rightarrow 5x_3 - 5y_4 = 0$;

$\rightarrow \boxed{y_4 = x_3}$;

$f_4^*(x_3) = 2 \cdot \sqrt{x_3}$;

$f_3^*(x_2) = \max \{u_3(x_2; y_3) + f_4^*(x_3)\} = \max \{2 \cdot \sqrt{y_3} + 2 \cdot \sqrt{x_3}\}$;

$x_3 = 5 x_2 - 5 y_3 \rightarrow f_3^*(x_2) = \max \{2 \cdot \sqrt{y_3} + 2 \cdot \sqrt{5x_2 - 5 y_3}\} =: \rho$;

$\dfrac{\partial \rho}{\partial y_3} = \dfrac{1}{\sqrt{y_3}} + \dfrac{-5}{\sqrt{5x_2 - 5y_3}} = 0 \rightarrow \boxed{y_3 = 1/6 \, x_2}$;

$f_3^*(x_2) = 2 \cdot \sqrt{1/6 \, x_2} + 2 \cdot \sqrt{5x_2 - (5/6)x_2} = 2 \cdot \sqrt{6x_2}$;

$f_2^*(x_1) = \max\{u_2(x_1; y_2) + 2 \cdot \sqrt{6x_2}\} = \max \{2 \cdot \sqrt{y_2} + 2 \cdot \sqrt{6x_2}\} =$
$\max \{2 \cdot \sqrt{y_2} + 2 \cdot \sqrt{6(5x_1 - 5y_2)}\} = \max \{2 \cdot \sqrt{y_2} + 2 \cdot \sqrt{30x_1 - 30y_2}\} =: \rho$;

$$\frac{\partial \rho}{\partial y_2} = \frac{1}{\sqrt{y_2}} + \frac{-30}{\sqrt{30x_1-30y_2}} = 0 \rightarrow \boxed{y_2 = 1/31\ x_1} \quad ;$$

$$f_2^*(x_1) = 2\cdot\sqrt{(1/31)\cdot x_1} + 2\cdot\sqrt{30x_1 - 30\cdot(1/31)\cdot x_1} = 2\cdot\sqrt{31x_1} \quad ;$$

$$f_1^*(x_0) = \max\{u_1(x_0;y_1) + 62\cdot\sqrt{1/31\ x_1}\} = \max\{2\cdot\sqrt{y_1} + 62\cdot\sqrt{(1/31)\cdot x_1}\} =$$
$$\max\{2\cdot\sqrt{y_1} + 62\cdot\sqrt{(5/31)\cdot x_0 - (5/31)\cdot y_1}\} =: \rho \quad ;$$

$$\frac{\partial \rho}{\partial y_1} = \frac{1}{\sqrt{y_1}} + \frac{62(-5/31)}{2\cdot\sqrt{5/31\ x_0 - 5/31\ y_1}} = 0 \rightarrow \boxed{y_1 = 1/156\ x_0} \quad ;$$

$$f_1^*(x_0) = 2\cdot\sqrt{(1/156)\cdot x_0} + 62\cdot\sqrt{(5/31)\cdot x_0 - (5/31)\cdot(1/156)\cdot x_0} = \sqrt{624\ x_0};$$

determination of the sales-policy:

$x_0 = 3{,}000$;

$y_1 = \frac{3{,}000}{156} \approx 19{,}23$;

$x_1 = 5(x_0 - y_1) = 5(3{,}000 - \frac{3{,}000}{156}) \approx 14{,}903.85$;

$y_2 = 1/31\ x_1 \approx \frac{14{,}904}{31} \approx 480.77$;

$x_2 = 5(x_1 - y_2) \approx 5(14{,}904 - 481) \approx 72{,}115.4$;

$y_3 \approx (1/6)\cdot 72{,}115.4 \approx 12{,}019.23$;

$x_3 = 5(x_2 - y_3) \approx 300{,}480.85$;

$y_4 \approx 300{,}480.85$;

$x_4 = 0$

total profit: $\pi = 2\cdot\sqrt{y_1} + 2\cdot\sqrt{y_2} + 2\cdot\sqrt{y_3} + 2\cdot\sqrt{y_4} \approx 1{,}368.21$ m.u.,
this is equivalent to
$\pi = f_1^*(x_0) = \sqrt{624\ x_0} \approx 1{,}368.21$ m.u.

Example 2

A speculator wants to invest 200 m.u. in such a way that his long term profit is maximized. The following three investments are

available:
(1) participation in a sky-scraper construction plan. Upon investing z_1 m.u. a long term profit of $\pi_1 = 4 \cdot z_1 - 2$ m.u. will be realized.
(2) financiation of a very promising film-project. When z_2 m.u. are invested, a long term profit $\pi_2 = 1/16 \cdot z_2^2 - 1/5 \cdot z_2 + 2$ m.u. will be realized.
(3) buying properties in the suburban area. Investments of z_3 m.u. bring long term profits of $\pi_3 = 3 \cdot z_3 + 1$ m.u.

In order to distribute the risk let:

$z_1 \leq b_1 = 95$ m.u.

$z_2 \leq b_2 = 50$ m.u.

$z_3 \leq b_3 = 100$ m.u.

The problem now reads:

$$\max \; \pi = \pi_1 + \pi_2 + \pi_3 =$$

$$= 4 \cdot z_1 + 1/16 \cdot z_2^2 - 1/5 \cdot z_2 + 3 \cdot z_3 + 1$$

$$z_1 + z_2 + z_3 = 200$$

$$z_1 \leq 95$$

$$z_2 \leq 50$$

$$z_3 \leq 100$$

$$z_i \geq 0 \quad \forall \; i=1,2,3$$

This problem may be formulated as one in dynamic programming as follows:

$n = 3$; $X_j = [0; \infty]$; $Y_j = [0; \min\{x_{j-1}; b_j\}]$;

$v_j(x_{j-1}; y_j) = (x_{j-1} - y_j) = x_j$; $x_0 = 200$; $x_3 = 0$;

$u_1(x_0; y_1) = 4 \cdot y_1 - 2$;

$u_2(x_1; y_2) = 1/16 \cdot y_2^2 - 1/5 \cdot y_2 + 2$;

$u_3(x_2; y_3) = 3 \cdot y_3 + 1$;

$f_4^*(x_3) = 0;$

$f_3^*(x_2) = \max \{3 \cdot y_3 + 1 + 0\} ; x_3 = x_2 - y_3 = 0;$
$\rightarrow \boxed{y_3 = (x_2 \mid y_3 \leq 100)}$;
$f_3^*(x_2) = (3 \cdot x_2 + 1);$

$f_2^*(x_1) = \max \{1/16 \cdot y_2^2 - 1/5 \cdot y_2 + 2 + 3 \cdot x_2 + 1\}; x_2 = x_1 - y_2 \rightarrow$
$f_2^*(x_1) = \max \{1/16 \cdot y_2^2 - 16/5 \cdot y_2 + 3 \cdot x_1 + 3\} =: \rho ;$

$\frac{\partial \rho}{\partial y_2} = 1/8 \cdot y_2 - 16/5 = 0 \rightarrow \boxed{y_2 = 128/5 = 25.6}$;

$f_2^*(x_1) = (40.96 - 81.92 + 3 \cdot x_1 + 3) = (3 \cdot x_1 - 37.96) ;$

$f_1^*(x_0) = \max \{4 \cdot y_1 - 2 + 3 \cdot x_1 - 37.96\} ; x_1 = x_0 - y_1 \rightarrow$
$f_1^*(x_0) = \max \{y_1 + 3 \cdot x_0 - 39.96\} \rightarrow y_1 = (x_0 \mid y_1 \leq 95)$
$\rightarrow \boxed{y_1 = 95}$;
$f_1^*(x_0) = (3 \cdot x_0 + 55.04);$

determination of the optimal policy:

$x_0 = 200 ;\qquad y_1 = 95 ;$
$x_1 = 105 ;\qquad y_2 = 25.6 ;$
$x_2 = 79.4 ;\qquad y_3 = 79.4 ;$
$x_3 = 0 ;$

total profit: $\pi = (4 \cdot 95 - 2) + (1/16 \cdot 655.36 - 1/5 \cdot 25.6 + 2) + (3 \cdot 79.4 + 1) =$
$= 655.04$ m.u., equivalent to
$\pi = f_1^*(x_0) = (3 \cdot 200 + 55.04) = 655.04$ m.u.

The speculator will invest
$y_1 = z_1 = 95$ m.u. in the sky-scraper-project
$y_2 = z_2 = 25.6$ m.u. in the film-project
$y_3 = z_3 = 79.4$ m.u. in the properties,
his long term profit is $\pi = 655.04$ m.u.

6.0.2 The Infinite-Period Model (policy iteration routine)

Hypotheses
See 6.0.1 , with $n \to \infty$. Furthermore assume that the transition function is stationary. That is, if the process is in a state x , then the action y defines a transition to the state x' independent of the period. For this reason the period-index for the set of feasible states, the set of feasible actions, the transition function and the objective function is unnecessary.

Definitions
X : set of feasible states, $|X| < \infty$
Y : set of feasible actions, $|Y| < \infty$
x : stored products at the end of a period, $x \in X$
y : taken action, $y \in Y$
α : discount rate
v(x,y): transition function
u(x,y): objective function

Principle
In this algorithm an initial decision (action) is improved until further improvement is impossible to obtain.

Description
Step 1: Determine a starting policy $n_1(x) \in Y$, so that
$n_1(x) := (y \mid u(x;y) \to \max) \; \forall \; x \in X$,
i.e. determine the optimal action y for the one-stage process. Set the running index t: = 1.
Note: The algorithm also starts with any feasible policy.

Step 2: Set up the functional equations
$g_t(x) := \left[u(x;n_t(x)) + \alpha \cdot g_t(v(x;n_t(x))) \right] \; \forall \; x \in X$
and solve them.
Result: Functional values $g_t(x) \forall \; x \in X$.
Set t: = t + 1.

Step 3: Determine the improved policy $n_t(x) \in Y$, so that
$$n_t(x) := (y|\{u(x;y) + \alpha \cdot g_{t-1}(v(x;y))\} \to \max) \; \forall \; x \in X \; .$$

Step 4: Is $n_t(x) = n_{t-1}(x) \; \forall \; x \in X$?
If yes: Stop, $n_t(x)$ is the optimal policy for the infinite period model.
If no : Go to step 2.

Note: Another method for the solution of infinite-period problems is the Value Determination Operation VDO. Generally however, the PIR seems to work better, because the VDO often does not converge to the optimum even after a large number of iterations.

Example

A catering service must hire additional kitchen help from time to time to insure high quality food preparation. Help may be obtained on a day to day basis through an employment agency. The number $\tilde{x}$ of meals which can be prepared with y helpers is given in the following table:

y	3	4	5	6
$\tilde{x}$	5,160	5,300	5,450	5,600

The daily demand x, according to past experience is 5,200; 5,340 or 5,600, depends in part upon the quality of the food from the previous day. The quality of the food decreases noticably if insufficient help had been hired. The exact situation is given in the following table (transition function):

v(x;y):	3	4	5	6
5,200	5,340	5,600	-	-
5,340	-	5,340	5,600	-
5,600	-	-	-	5,200

The caterer may fall 1 % short of the daily demand without losing costumers. Each helper is paid 90 m.u. per day.

278 *Dynamic Programming*

Each meal is produced at a cost of 2 m.u. and sold for 4 m.u. With a discount rate $\alpha = 0.8$ determine the number of helpers to be hired daily which will maximize the long term profit of the catering service.

$X = \{5,200; 5,340; 5,600\}; \quad Y = \{3;4;5;6\};$
$u(x;y) = (4-2) \cdot \min\{\hat{x}(y);x\} - 90 \cdot y \ ;$

$n_1(5,200) = (y \mid \max \{(2 \cdot 5,160 - 270); (2 \cdot 5,200 - 360)\}) =$
$\qquad (y \mid \max \{10,050; 10,040\}) = 3;$
$n_1(5,340) = (y \mid \max \{(2 \cdot 5,300 - 360); (2 \cdot 5,340 - 450)\}) =$
$\qquad (y \mid \max \{10,240; 10,230\}) = 4;$
$n_1(5,600) = (y \mid \max \{2 \cdot 5,600 - 540\}) = 6;$

$t = 1;$
$g_1(5,200) = [2 \cdot 5,160 \quad 270 + 0.8 \cdot g_1(5,340)];$
$g_1(5,200) = 10,050 + 0.8 \cdot g_1(5,340) =: d_1;$

$g_1(5,340) = [2 \cdot 5,300 - 360 + 0.8 \cdot g_1(5,600)];$
$g_1(5,340) = 10,240 + 0.8 \cdot g_1(5,600) =: d_2;$

$g_1(5,600) = [2 \cdot 5,600 - 540 + 0.8 \cdot g_1(5,200)];$
$g_1(5,600) = 10,660 + 0.8 \cdot g_1(5,200) =: d_3;$

$d_1 = 10,050 + 0.8 \cdot (10,240 + 0.8 \cdot d_3) \to d_1 = 18,242 + 0.64 \, d_3;$
$d_1 = 51,361.475 = g_1(5,200);$

$d_3 = 10,660 + 0.8 \cdot (18,242 + 0.64 \cdot d_3) \to d_3 = 51,749.18 = g_1(5,600);$

$d_2 = 51,639.344 = g_1(5,340);$

$t = 2;$
$n_2(5,200) = (y \mid \max \{(2 \cdot 5,160 - 270 + 0.8 \cdot 51,639.344);$
$\qquad\qquad (2 \cdot 5,200 - 360 + 0.8 \cdot 51,749.18)\}) =$
$\qquad (y \mid \max \{51,361.475; 51,439.344\}) = 4;$

$n_2(5,340) = (y \mid \max \{(2 \cdot 5,300 - 360 + 0.8 \cdot 51,639.344);$
$\qquad\qquad\qquad (2 \cdot 5,340 - 450 + 0.8 \cdot 51,749.18)\}) =$
$\qquad\qquad (y \mid \max \{51,551.475; 51,629.344\}) = 5;$
$n_2(5,600) = (y \mid \max \{2 \cdot 5,600 - 540 + 0.8 \cdot 51,361.475\}) = 6;$

$g_2(5,200) = 2 \cdot 5,200 - 360 + 0.8 \cdot g_2(5,600);$
$g_2(5,200) = 10,040 + 0.8 \cdot g_2(5,600) =: d_1;$

$g_2(5,340) = 2 \cdot 5,340 - 450 + 0.8 \cdot g_2(5,600);$
$g_2(5,340) = 10,230 + 0.8 \cdot g_2(5,600) =: d_2;$

$g_2(5,600) = 2 \cdot 5,600 - 540 + 0.8 \cdot g_2(5,200);$
$g_2(5,600) = 10,660 + 0.8 \cdot g_2(5,200) =: d_3;$

$d_1 = 10,040 + 0.8 \cdot (10,660 + 0.8 \cdot d_1) \rightarrow d_1 = 51,577.77 = g_2(5,200);$

$d_3 = 10,660 + 0.8 \cdot 51,577.77 \rightarrow d_3 = 51,922.221 = g_2(5,600);$

$d_2 = 51,767.776 = g_2(5,340);$

$t = 3;$

$n_3(5,200) = (y \mid \max \{(2 \cdot 5,160 - 270 + 0.8 \cdot 51,767.776);$
$\qquad\qquad\qquad (2 \cdot 5,200 - 360 + 0.8 \cdot 51,922.221)\}) =$
$\qquad\qquad (y \mid \max \{51,464.22; 51,577.776\}) = 4;$
$n_3(5,340) = (y \mid \max \{(2 \cdot 5,300 - 360 + 0.8 \cdot 51,767.776);$
$\qquad\qquad\qquad (2 \cdot 5,340 - 450 + 0.8 \cdot 51,922.221)\}) =$
$\qquad\qquad (y \mid \max \{51,654.22; 51,767.776\}) = 5;$
$n_3(5,600) = (y \mid \max \{2 \cdot 5,600 - 540 + 0.8 \cdot 51,577.77\}) = 6;$

$n_3(x) \stackrel{!}{=} n_2(x) \;\forall x \in X \rightarrow$ Stop, $n_3(x)$ is the optimal policy,

i.e. for a demand of

$x = 5,200$ meals $\rightarrow y = 4$ helpers;
$x = 5,340$ meals $\rightarrow y = 5$ helpers;
$x = 5,600$ meals $\rightarrow y = 6$ helpers are hired.

An average profit per day of $\pi = 10,310$ m.u. results.

7. Queueing Models

Hypotheses

Given the following situation:
There exists a system with k parallel serving stations, having the same function and capabilities (channels). At certain intervals customers enter the system, are served at the first available station (waiting if necessary), and depart the system. If this service is accomplished in several stages, then the customer may depart the system only after each of the stages in series has been visited. In general the system may be visualized as below:

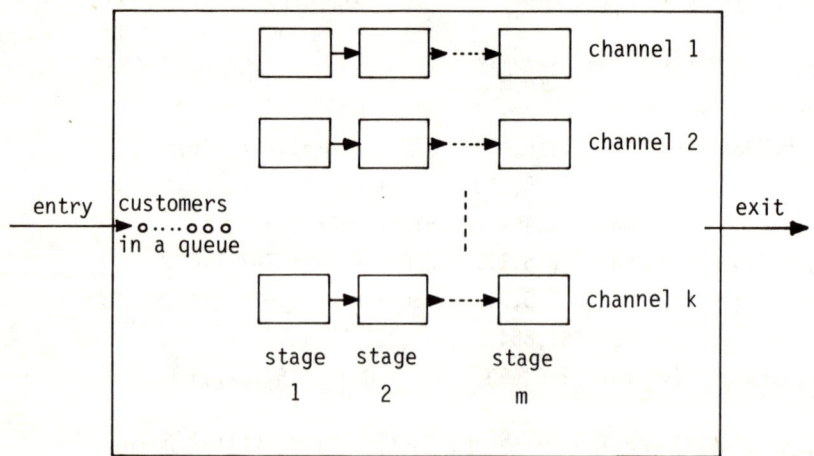

Suppose the arrival of the customers is poisson-distributed (no peak loads). The following rule pertains to all customers: First come; first served. Assume the departure of the customers is exponential-distributed.

We distinguish the following models:

(1) 1 - channel, 1 - stage model
(2) 1 - channel, r - phase model; $r \in \mathbb{N}$
(3) k - channel, 1 - stage model
(4) k - channel, m - stage model; $k \geq 1$; $m > 1$.

Case (4) will not be considered here, because the mathematical formulas and the required calculations are beyond the scope of this book. Case (4) models are generally solved with simulation.

A system has r phases iff the service is accomplished in more than one step, and the next customer may be served only after the preceding customer has completely departed the system.

Special case: If the serving times are not exponential-distributed random variables but <u>exactly</u> determined, then the r-phase model with $r = \infty$ is applicable.

Definitions

λ: average number of customers, entering the system per time unit (arrival rate)

μ: average number of customers, departing the system per time unit (service rate)

$\bar{t}_a = \frac{1}{\lambda}$: average time-interval between the arrival of two customers

$\bar{t}_b = \frac{1}{\mu}$: average service time

$\rho = \frac{\lambda}{\mu}$: capacity utilization of one service station

$\kappa = \frac{\rho}{k}$: capacity utilization of the system

For 1 - channel models $\rho < 1$ must hold, for k - channel models only $\kappa < 1$ must hold.

P_n : probability that the length of the queue is equal to n (persons), n = 0,1,...

P(w): probability that a customer must wait

$\bar{n}$: average length of the queue <u>including</u> the customer that is being served

$\bar{n}_s$: average length of the queue <u>excluding</u> the customer that is being served

$\bar{t}_w$: average waiting time for a customer

$\bar{t}_{tot}$: average waiting and service time for a customer

$P(t_w \leq t)$: probability that a customer must wait at most t time units

$P(t_w > t)$: probability that a customer must wait at least t time units

$P(t_{tot} \leq t)$: probability that a customer remains in the system at most t time units

$P(t_{tot} > t)$: probability that a customer remains in the system at least t time units

$\bar{n} \mid n > 0$: average length of the queue including the customer that is being served under the assumption that a queue with more than 0 elements (customers) exists

$\bar{t}_w \mid n > 0$: average waiting time for a customer under the assumption that a queue with more than 0 elements (customers) exists.

Principle
Formulas are given, with special states in a queueing system can be determined.

7.0.1 The I-Channel, I-Stage Model

Formulas
$P_n = \rho^n \cdot (1 - \rho)$ $\qquad P(t_w \leq t) = 1 - \rho \cdot e^{-(\mu - \lambda) \cdot t}$

$P_0 = \rho^0 \cdot (1 - \rho) = (1 - \rho)$ $\qquad P(t_w > t) = \rho \cdot e^{-(\mu - \lambda) \cdot t}$

$P(w) = \rho$

$\bar{n} = \frac{\rho}{1-\rho} = \frac{\lambda}{\mu-\lambda}$

$\bar{n}_s = \frac{\rho^2}{1-\rho} = \bar{n}\cdot\rho$

$\bar{t}_w = \frac{\rho}{\mu-\lambda}$

$\bar{t}_{tot} = \frac{1}{\mu-\lambda}$

$P(t_{tot} \leq t) = 1 - e^{-(\mu-\lambda)\cdot t}$

$P(t_{tot} > t) = e^{-(\mu-\lambda)\cdot t}$

$\bar{n}\,|n>0 = \frac{1}{1-\rho} = \frac{\mu}{\mu-\lambda}$

$\bar{t}_w|n>0 = \frac{1}{\mu-\lambda}$

Example

8 customers per hour enter the shop of a retailer in a little village. The arrivals are poisson-distributed. Let the service time be exponential-distributed at 5 minutes.

$\lambda = 8/\text{hr.}; \quad \mu = 12/\text{hr.}; \quad \rho = 2/3;$

(a) Determine the probability that a customer must wait.

$P(w) = \rho = 2/3 \cong 66.\bar{6}\ \%$.

(b) What is the average length of the queue including the customer that is being served?

$\bar{n} = \frac{\rho}{1-\rho} = \frac{2/3}{1-2/3} = 2$.

(c) What is the average total time that a customer must spend in the shop?

$\bar{t}_{tot} = \frac{1}{\mu-\lambda} = \frac{1}{12-8} \cong \frac{1}{4}\ \text{hr.} = 15\ \text{min.}$

(d) If the retailer determines that a customer must wait on the average for more than 15 minutes, then he will hire a helper to ensure that customers will not leave his shop without having bought anything. How many <u>more</u> customers must arrive before the retailer hires help ?

$$\bar{t}_w = \frac{\rho}{\mu - \lambda} = \frac{2/3}{12-8} \cong \frac{1}{6} \text{ hr.} = 10 \text{ min};$$

$$\bar{t}'_w \cong \frac{1}{4} \text{ hr.}; \quad \lambda' = \lambda + \Delta\lambda; \quad \mu' = \mu;$$

$$\bar{t}'_w = \frac{\frac{\lambda'}{\mu'}}{\mu' - \lambda'} = \frac{\frac{\lambda'}{12}}{12 - \lambda'} \stackrel{!}{=} \frac{1}{4} \rightarrow \frac{\lambda'}{12} = \frac{12 - \lambda'}{4};$$

$\lambda' = 9/\text{hr.} \rightarrow \Delta\lambda = 1/\text{hr.}$, i.e. the number of customers must increase by 1 customer per hour.

(e) What is the probability that a customer remains longer than 15 minutes in the shop ?

$$P(t_{tot} > 1/4) = e^{-(12-8) \cdot 1/4} = e^{-1} = 0.3679 \cong 36.79 \text{ \%}.$$

(f) What is the average waiting time for a customer under the assumption that he must wait ?

$$\bar{t}_w \mid n > 0 = \frac{1}{\mu - \lambda} = \frac{1}{12 - 8} \cong \frac{1}{4} \text{ hr.} = 15 \text{ min}.$$

7.0.2 The l-Channel, r-Phase Model

Formulas

$$\bar{n} = \frac{r+1}{2 \cdot r} \cdot \frac{\lambda^2}{\mu \cdot (\mu - \lambda)} + \frac{\lambda}{\mu} \qquad \bar{t}_w = \frac{r+1}{2 \cdot r} \cdot \frac{\lambda}{\mu \cdot (\mu - \lambda)}$$

$$\bar{n}_s = \frac{r+1}{2 \cdot r} \cdot \frac{\lambda^2}{\mu \cdot (\mu - \lambda)} \qquad \bar{t}_{tot} = \frac{r+1}{2 \cdot r} \cdot \frac{\lambda}{\mu \cdot (\mu - \lambda)} + \frac{1}{\mu}$$

Note: Let the durations of the r phases be $\mu_1, \ldots, \mu_r$,
then $\mu := \sum_{j=1}^{r} \mu_j$.

Example 1
A small motor-vehicle repair shop receives on the average 3 cars per day. The foreman, who has no other help, must write up the defects in these cars. Because he has devised a system for this diagnosis, his job may be devided into 8 parts per car, each lasting 15 minutes, and accomplished one after the other before starting another car. The foreman works 8 hours per day.

$r = 8$; $\lambda = 3$/day; $\mu' = 4$/hr. $\cong 32$/day; $\mu = \dfrac{\mu'}{r} = 4$/day.

(a) How many cars (except for the one being inspected) are on the average in the repair shop?

$$\bar{n}_s = \dfrac{8+1}{2 \cdot 8} \cdot \dfrac{3^2}{4(4-3)} = \dfrac{81}{64} \approx 1.2656 .$$

(b) What is the average time until a customer receives his car again?

$$\bar{t}_{tot} = \dfrac{8+1}{2 \cdot 8} \cdot \dfrac{3}{4(4-3)} + \dfrac{1}{4} \cong \dfrac{43}{64} \text{ days} = 322.5 \text{ min}.$$

Example 2
A modern car wash receives on the average of one car every 5 minutes for servicing. The service time at this station is exactly 3 minutes. What is the average waiting time for a customer?

$\bar{t}_a \cong \dfrac{1}{12}$ hr.; $\bar{t}_b \cong \dfrac{1}{20}$ hr.; $\lambda = 12$/hr.; $\mu = 20$/hr.;

$r = \infty \to \dfrac{r+1}{2 \cdot r} = \dfrac{1}{2}$; $\bar{t}_w = \dfrac{1}{2} \cdot \dfrac{12}{20(20-12)} \cong \dfrac{3}{80}$ hr. $= 2.25$ min.

7.0.3 The k-Channel, l-Stage Model

Formulas

$$P_o = \dfrac{1}{\left[\displaystyle\sum_{n=0}^{k-1} \dfrac{1}{n!} \cdot \left(\dfrac{\lambda}{\mu}\right)^n\right] + \dfrac{1}{k!} \cdot \left(\dfrac{\lambda}{\mu}\right)^k \cdot \dfrac{k \cdot \mu}{k \cdot \mu - \lambda}}$$

286 *Queueing Models*

$$P(w) = \frac{\left(\frac{\lambda}{\mu}\right)^k}{(k-1)! \, (k \cdot \mu - \lambda)} \cdot P_o$$

$$\bar{n} = \frac{\lambda \cdot \mu \cdot \left(\frac{\lambda}{\mu}\right)^k}{(k-1)! \cdot (k \cdot \mu - \lambda)^2} \cdot P_o + \frac{\lambda}{\mu}$$

$$\bar{n}_s = \frac{\lambda \cdot \mu \cdot \left(\frac{\lambda}{\mu}\right)^k}{(k-1)! \cdot (k \cdot \mu - \lambda)^2} \cdot P_o$$

$$\bar{t}_w = \frac{\mu \cdot \left(\frac{\lambda}{\mu}\right)^k}{(k-1)! \, (k \cdot \mu - \lambda)^2} \cdot P_o$$

$$\bar{t}_{tot} = \bar{t}_w + \frac{1}{\mu} = \frac{\mu \cdot \left(\frac{\lambda}{\mu}\right)^k}{(k-1)! \, (k \cdot \mu - \lambda)^2} \cdot P_o + \frac{1}{\mu}$$

Example

A post-office, having 3 windows providing the same services, receives on the average of 30 customers per hour. (Arrivals are poisson-distributed.) The service time for a customer is exponential-distributed at 5 minutes.

$k = 3;$ $\lambda = 30/\text{hr.};$ $\mu = 12/\text{hr.};$

$$P_o = \frac{1}{\frac{1}{0!} \cdot \left(\frac{30}{12}\right)^0 + \frac{1}{1!} \cdot \left(\frac{30}{12}\right)^1 + \frac{1}{2!} \left(\frac{30}{12}\right)^2 + \frac{1}{3!} \left(\frac{30}{12}\right)^3 \cdot \frac{3 \cdot 12}{3 \cdot 12 - 30}} = \frac{4}{89}.$$

(a) What is the probability that a customer must wait?

$$P(w) = \frac{12 \cdot \left(\frac{30}{12}\right)^3}{(3-1)! \cdot (3 \cdot 12 - 30)} \cdot \frac{4}{89} = \frac{125}{178} \approx 70.22\,\%$$

(b) What is the average length of the queue including the customer being served ?

$$\bar{n} = \frac{30 \cdot 12 \cdot \left(\frac{30}{12}\right)^3}{(3-1)! \cdot (3 \cdot 12 - 30)^2} \cdot \frac{4}{89} + \frac{30}{12} = \frac{535}{89} \approx 6.0112$$

(c) What is the average total time that a customer must spend in the post-office ?

$$\bar{t}_{tot} = \frac{12 \cdot \left(\frac{30}{12}\right)^3}{(3-1)! \cdot (3 \cdot 12 - 30)^2} \cdot \frac{4}{89} + \frac{1}{12} = \frac{107}{534} \approx 0.2004 \text{ hr.}$$

8. Nonlinear Programming

8.1 Theorems and Special Methods

8.1.1 The Theorem of KUHN and TUCKER

Hypotheses
Given a convex minimization problem P:

P: min $\pi = f(x)$
$g_i(x) \leq 0 \quad \forall \; i=1,\ldots,m$
$x_j \geq 0 \quad \forall \; j=1,\ldots,n.$

The partial derivatives exist for both the (nonlinear) objective function $f(x)$ and the restrictions $g_i(x)$. The Lagrange-function $L(x,\lambda)$ is defined as:

$$L(x,\lambda) := f(x) + \sum_{i=1}^{m} \lambda_i \cdot g_i(x) .$$

A point $(\bar{x},\bar{\lambda})$ is called a saddlepoint of $L(x,\lambda)$, if $L(\bar{x},\lambda) \leq L(\bar{x},\bar{\lambda}) \leq L(x,\bar{\lambda})$ or, equivalently if the following six Kuhn-Tucker conditions are fulfilled:

(1) $\dfrac{L(x,\lambda)}{x_j} \geq 0 \quad \forall \; j$ 	(4) $\dfrac{L(x,\lambda)}{\lambda_i} \leq 0 \quad \forall \; i$

(2) $\dfrac{L(x,\lambda)}{x_j} \cdot x_j \stackrel{!}{=} 0 \quad \forall \; j$ 	(5) $\dfrac{L(x,\lambda)}{\lambda_i} \cdot \lambda_i \stackrel{!}{=} 0 \quad \forall \; i$

(3) $x_j \geq 0 \quad \forall \; j = 1,\ldots,n$ 	(6) $\lambda_i \geq 0 \quad \forall \; i = 1,\ldots,m$

Theorem 1
If $(\bar{x},\bar{\lambda})$ is a saddlepoint of $L(x,\lambda)$, then $\bar{x}$ is an optimal solution of P.

Theorem 2
Let $\bar{x}$ be an optimal solution of P. If there exists at least one feasible solution $\tilde{x}$ of P, so that $g_i(\tilde{x}) < 0 \quad \forall \quad i=1,\ldots,m$ (SLATER's condition) and if there exists an $\bar{\lambda} \geq 0$, then $(\bar{x},\bar{\lambda})$ is a saddlepoint of $L(x,\lambda)$.

8.1.2 The Method of LAGRANGE

Hypotheses
Given the following problem P:

$$P: \left.\begin{array}{c}\min\\\max\end{array}\right\} \pi = f(x)$$
$$A \cdot x = b \quad ,$$

where $rk(A) < n$, i.e. the system of equations must not be fully determined.

Principle
In this method the Lagrange-function is formulated as the sum of objective function and the weighted restrictions. Each partial derivative of the function is set equal to zero resulting in a unique system of equations.

Description
Step 1: Formulate the Lagrange-function

$$L(x,\lambda) := f(x) + \sum_{i=1}^{m} \lambda_i \cdot (\sum_{j=1}^{n} a_{ij} \cdot x_j - b_i) \quad .$$

290 Nonlinear Programming

Step 2: Determine the gradient of the Lagrange-function and set it equal to zero:

$$\text{grad } L(x,\lambda) := \frac{\partial L(x_j,\lambda_i)}{\partial (x_j,\lambda_i)} \stackrel{!}{=} 0 \quad \forall \; i,j \;.$$

Step 3: Solve the so-determined system of equations (with one of the well-known methods (algorithm of Gauss; method of Gauss-Jordan etc.)).

Example
Given the following problem P:

$$\max \; \pi = 3 \cdot x_1^2 + 2 \cdot x_2^2 + 5 \cdot x_1 \cdot x_2 + 3 \cdot x_1 + 4 \cdot x_3 - x_1 \cdot x_3$$

$$2 \cdot x_1 + 3 \cdot x_2 = 10$$

$$x_2 + x_3 = 8$$

The Lagrange-function is

$$L(x, \lambda) = 3 \cdot x_1^2 + 2 \cdot x_2^2 + 5 \cdot x_1 \cdot x_2 + 3 \cdot x_1 + 4 \cdot x_3 - x_1 \cdot x_3 +$$

$$\lambda_1 \cdot (2 \cdot x_1 + 3 \cdot x_2 - 10) +$$

$$\lambda_2 \cdot (x_2 + x_3 - 8) \;;$$

$$\text{grad } L(x,\lambda) = \begin{pmatrix} 6 \cdot x_1 + 5 \cdot x_2 + 3 - x_3 + 2 \cdot \lambda_1 \\ 4 \cdot x_2 + 5 \cdot x_1 + 3 \cdot \lambda_1 + \lambda_2 \\ 4 - x_1 + \lambda_2 \\ 2 \cdot x_1 + 3 \cdot x_2 - 10 \\ x_2 + x_3 - 8 \end{pmatrix} \stackrel{!}{=} \Theta$$

Solution of the system of equations with the method of Gauss-Jordan:

Note: Tableau-transformations are executed, until the matrix of coefficients A is the identity matrix I.

x_1	x_2	x_3	λ_1	λ_2	1
6	5	-1	2	0	-3
5	4	0	3	1	0
-1	0	0	0	1	-4
2	3	0	0	0	10
0	1	1	0	0	8
1	5/6	-1/6	1/3	0	-1/2
0	-1/6	5/6	4/3	1	5/2
0	5/6	-1/6	1/3	1	-9/2
0	4/3	1/3	-2/3	0	11
0	1	1	0	0	8
1	0	4	7	5	12
0	1	-5	-8	-6	-15
0	0	4	7	6	8
0	0	7	10	8	31
0	0	6	8	6	23
1	0	0	0	-1	4
0	1	0	3/4	3/2	-5
0	0	1	7/4	3/2	2
0	0	0	-9/4	-5/2	17
0	0	0	-5/2	-3	11
1	0	0	0	-1	4
0	1	0	0	2/3	2/3
0	0	1	0	-4/9	15 2/9
0	0	0	1	10/9	-7 5/9
0	0	0	0	-2/9	-7 8/9

292 Nonlinear Programming

x_1	x_2	x_3	λ_1	λ_2	1
1	0	0	0	0	79/2
0	1	0	0	0	-23
0	0	1	0	0	31
0	0	0	1	0	-47
0	0	0	0	1	71/2

The optimal solution is:
$\bar{x}=(79/2;-23;31)$;
$\bar{\lambda}=(-47; 71/2)$;
the value of the objective function is
$\bar{\pi} = 214.25$.

8.1.3 A Method for the Optimization of Nonlinear Separable Objective Functions under Linear Constraints

Hypotheses

Given the following problem:

P: $\left.\begin{array}{c}\min\\\max\end{array}\right\}$ $f(x) = c \cdot x^\beta$

$A \cdot x \gtrless b \geq \theta$

$x \geq \theta$

$\beta \in \mathbb{R}$,

the object is to linearize the nonlinear objective function and solve the corresponding problem with one of the simplex-methods. For maximization problems (minimization problems) the objective function must be concave (convex).

Principle

In this method the nonlinear objective function is replaced by linear parts, i.e. each variable is replaced by a set of new variables, which are valid in a specified interval. The coefficients of the new objective function are determined from the gradient of the chord of the nonlinear function in each interval. The solution of the problem is obtained with one of the simplex-methods.

Description

Step 1: Determine in which interval the variable x_j may assume feasible values, ∀ j=1,...,n (e.g. using the restrictions).

Let $x_j \in [0; u_j]$, where u_j is the upper bound of the feasible region of x_j.

Step 2: Determine appropriate (equidistant or non-equidistant) intervals of "length" $\alpha_r^{(j)}$ and split up the variable x_j as follows:

$$x_j := \sum_{r=1}^{k_j} \tilde{x}_{jr} \quad \forall \ j=1,\ldots,n.$$

(The value k_j for the upper summation-index is determined by u_j and the selected size of the interval. For equidistant intervals of length $\alpha^{(j)}$ we have $k_j = u_j/\alpha^{(j)}$.)

Let u_{jr} be the upper bound of the r-th interval of the variable x_j, so that $u_{jo} = 0$, then

$$u_{jr} = u_{jr-1} + \alpha_r^{(j)} \quad \forall \ r=1,\ldots,k_j.$$

For the variables $\tilde{x}_{jr}$ holds:

$$\tilde{x}_{jr} \in [0; u_{jr} - u_{jr-1}] \quad \forall \ j=1,\ldots,n; \ r=1,\ldots,k_j.$$

Step 3: Determine the coefficients $\tilde{c}_{jr}$ of the objective function for the variable $\tilde{x}_{jr}$, so that

$$\tilde{c}_{jr} := \frac{f(x_j = u_{jr}) - f(x_j = u_{jr-1})}{(u_{jr} - u_{jr-1})} \quad \forall \ j=1,\ldots,n; \ r=1,\ldots,k_j.$$

Step 4: Formulate the problem P:

$$P: \ \begin{matrix} \min \\ \max \end{matrix} \Bigg\} \ f(\tilde{x}) = \sum_{j=1}^{n} \sum_{r=1}^{k_j} \tilde{c}_{jr} \cdot \tilde{x}_{jr}$$

$$\sum_{j=1}^{n} \sum_{r=1}^{k_j} a_{ij} \cdot \tilde{x}_{jr} \gtreqless b_i \quad \forall \ i=1,\ldots,m$$

$$\left. \begin{matrix} \tilde{x}_{jr} \leq \alpha_r^{(j)} \\ \tilde{x}_{jr} \geq 0 \end{matrix} \right\} \ \forall \ j=1,\ldots,n; \ r=1,\ldots,k_j.$$

Step 5: Solve the problem $\tilde{P}$ with one of the simplex-methods.
Result: The solution vector $\tilde{x}'$ (if a feasible solution exists).

Is $\sum_{r=1}^{k_j} \tilde{x}'_{jr} < \sum_{r=1}^{k_j} \alpha_r^{(j)}$ $\forall\, j=1,\ldots,n$?

If yes: Go to step 7.
If no : Go to step 6.

Step 6: Suppose the above inequality is not fulfilled for the variables $\tilde{x}_{1r}$. Set $k_1 := k_1 + 1$, i.e. add another interval of length $\alpha_r^{(1)}$ to the variables $\tilde{x}_{1r}$. The new upper bound is now $u_1 := u_1 + \alpha_r^{(1)}$.
Go to step 3.

Step 7: Determine the optimal solution for the problem P, so that

$$\bar{x}_j := \sum_{r=1}^{k_j} \tilde{x}'_{jr} \quad \forall\, j=1,\ldots,n, \quad f(\bar{x}) = c \cdot \bar{x}^\beta.$$

Example

Given the following problem:

P: max $\pi = \sqrt{4 \cdot x_1} + \sqrt{3 \cdot x_2}$
(1) $2 \cdot x_1 + 4 \cdot x_2 \leq 8$
(2) $3 \cdot x_1 + 2 \cdot x_2 \leq 6$
$x_1, x_2 \geq 0$

For the value $f(x) = 5$ of the objective function the "iso-profit-line" results, as indicated on the graph. A translation of the curve in the direction of the origin corresponds to a smaller profit. A conjecture over the optimal solution is $\bar{x} = (1; 3/2);\ f(\bar{x}) < 5$.

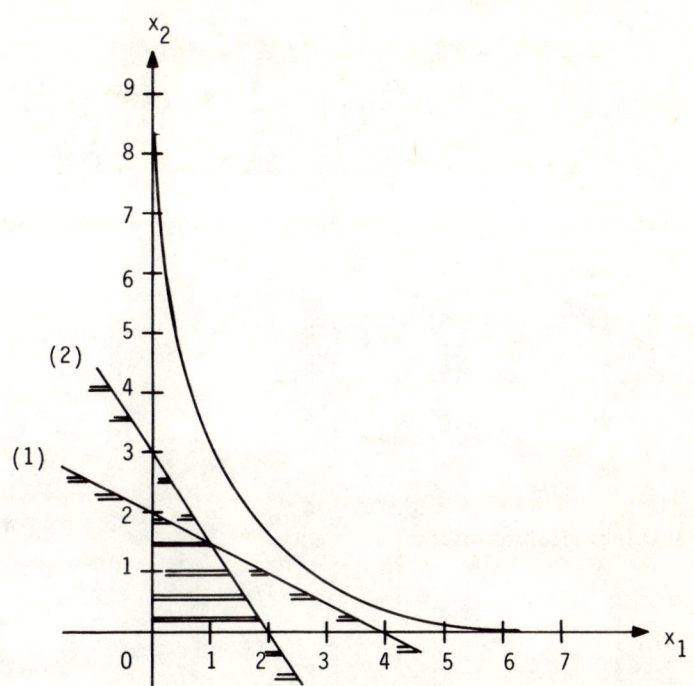

For x_1 holds: $x_1 \in [0\,;\,2]$, for x_2 holds: $x_2 \in [0\,;\,2]$.
For both variables equidistant intervals of length $\alpha = 1/2$ are chosen, thus:

$$x_1 = \sum_{r=1}^{4} \tilde{x}_{1r};\quad 0 \leq \tilde{x}_{1r} \leq 1/2 \quad \forall\, r=1,\ldots,4;$$

$$x_2 = \sum_{r=1}^{4} \tilde{x}_{2r};\quad 0 \leq \tilde{x}_{2r} \leq 1/2 \quad \forall\, r=1,\ldots,4.$$

The coefficients $\tilde{c}_{jr}$ of the objective function are:

$\tilde{c}_{11} = 2.82;\ \tilde{c}_{12} = 1.17;\ \tilde{c}_{13} = 0.9;\ \tilde{c}_{14} = 0.76;$

$\tilde{c}_{21} = 2.45;\ \tilde{c}_{22} = 1.01;\ \tilde{c}_{23} = 0.78;\ \tilde{c}_{24} = 0.66;$

thus the problem $\tilde{P}$ reads:

$\tilde{P}$: max $\pi = (2.82 \cdot \tilde{x}_{11} + 1.17 \cdot \tilde{x}_{12} + 0.9 \cdot \tilde{x}_{13} + 0.76 \cdot \tilde{x}_{14} +$
$\qquad\qquad 2.45 \cdot \tilde{x}_{21} + 1.01 \cdot \tilde{x}_{22} + 0.78 \cdot \tilde{x}_{23} + 0.66 \cdot \tilde{x}_{24})$

(1) $2 \cdot \tilde{x}_{11} + 2 \cdot \tilde{x}_{12} + 2 \cdot \tilde{x}_{13} + 2 \cdot \tilde{x}_{14} + 4 \cdot \tilde{x}_{21} + 4 \cdot \tilde{x}_{22} + 4 \cdot \tilde{x}_{23} + 4 \cdot \tilde{x}_{24} \leq 8$
(2) $3 \cdot \tilde{x}_{11} + 3 \cdot \tilde{x}_{12} + 3 \cdot \tilde{x}_{13} + 3 \cdot \tilde{x}_{14} + 2 \cdot \tilde{x}_{21} + 2 \cdot \tilde{x}_{22} + 2 \cdot \tilde{x}_{23} + 2 \cdot \tilde{x}_{24} \leq 6$
(3) $\tilde{x}_{11} \leq 1/2$;   (7) $\tilde{x}_{21} \leq 1/2$;
(4) $\tilde{x}_{12} \leq 1/2$;   (8) $\tilde{x}_{22} \leq 1/2$;
(5) $\tilde{x}_{13} \leq 1/2$;   (9) $\tilde{x}_{23} \leq 1/2$;
(6) $\tilde{x}_{14} \leq 1/2$;   (10) $\tilde{x}_{24} \leq 1/2$;

$\tilde{x}_{jr} \geq 0 \quad \forall\ j=1,2;\ r=1,\ldots,4$.

The resulting solution of $\tilde{P}$ is:

$\begin{aligned}\tilde{x}'_{11} &= 1/2; \\ \tilde{x}'_{12} &= 1/2; \\ \tilde{x}'_{13} &= 0; \\ \tilde{x}'_{14} &= 0;\end{aligned}\Bigg\} \rightarrow \bar{x}_1 = 1;$
$\begin{aligned}\tilde{x}'_{21} &= 1/2\ ; \\ \tilde{x}'_{22} &= 1/2\ ; \\ \tilde{x}'_{23} &= 1/2\ ; \\ \tilde{x}'_{24} &= 0\ ;\end{aligned}\Bigg\} \rightarrow \bar{x}_2 = 3/2\ ;$

$\bar{\pi} = f(\bar{x}) = \sqrt{4 \cdot 1} + \sqrt{3 \cdot 3/2} \approx 4.1213$.

8.2 General Methods

8.2.1 The Method of WOLFE (short form)

<u>Hypotheses</u>
Given the following problem:
P: min $\pi = x^T \cdot C \cdot x + d^T \cdot x$
$\qquad A \cdot x \gtreqless b$
$\qquad\ \ x \geq \Theta$,

where $C : C_{[n \times n]}$; $d : d_{[n \times 1]}$; A, x, b as defined. C is a diagonal matrix, i.e. $c_{ij} = 0 \; \forall \; i \neq j$; C must be positiv definite (or positive semidefinite, if $d = \Theta$).

Principle
In this method a special simplex-tableau is used which derives from the formulation of the nonlinear problem as a general Lagrange function. Simplex-iterations are used to minimize the degree of infeasibility or the deviation from the optimum.

Description
Step 1: Construct the following simplex tableau:

x	y	v	u	z	w	1
A	± I	Θ	Θ	Θ	I	b
2·C			A^T			-d
		- I		I		
Θ			I			Θ

where $y, v, u, z, w \geq \Theta$ holds.
Note: The artificial variables w are only used, if type-II inequalities appear.

Step 2: (1. phase)
Minimize the sum of the artificial variables w with the first phase of the Two-Phase-Method.
Result: A feasible solution of P (if one exists).

Step 3: (2. phase)
Minimize the sum of the artificial variables z with the first phase of the Two-Phase-Method.
Result: An optimal solution of the problem. The value of the objective function can be determined by substituting the optimal values $\bar{x}$ for x in the objective function.

298 Nonlinear Programming

Note: It is important to remember that no two variables x_j and v_j are basic variables with current values greater zero.

Example
Given the following problem:

P: min $\pi = 1/2 \cdot x_1^2 + 1/2 \cdot x_2^2 - 10 \cdot x_1 - 5 \cdot x_2$

(1) $x_1 + 2 \cdot x_2 \leq 12$

(2) $5 \cdot x_1 + x_2 \leq 15$

$x_1, x_2 \geq 0$

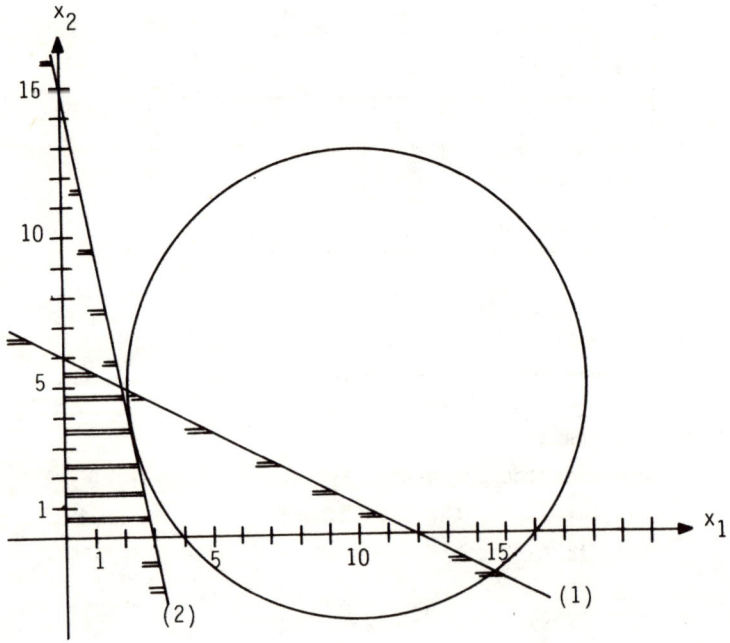

$$C = \begin{pmatrix} 1/2 & 0 \\ 0 & 1/2 \end{pmatrix} ; \quad A = \begin{pmatrix} 1 & 2 \\ 5 & 1 \end{pmatrix} ; \quad b = \begin{pmatrix} 12 \\ 15 \end{pmatrix} ; \quad d = \begin{pmatrix} -10 \\ -5 \end{pmatrix}$$

Section 8.2.1   299

$T^{(1)}$:

x_1	x_2	y_1	y_2	v_1	v_2	v_3	v_4	u_1	u_2	z_1	z_2	z_3	z_4	w_1	w_2	1
1	2	1	0	0	0	0	0	0	0	0	0	0	0	1	0	12
5	1	0	1	0	0	0	0	0	0	0	0	0	0	0	1	15
1	0	0	0	-1	0	0	0	1	5	1	0	0	0	0	0	10
0	1	0	0	0	-1	0	0	2	1	0	1	0	0	0	0	5
0	0	0	0	0	0	-1	0	1	0	0	0	1	0	0	0	0
0	0	0	0	0	0	0	-1	0	1	0	0	0	1	0	0	0
-6	-3	-1	-1	0	0	0	0	0	0	0	0	0	0	0	0	-27

$T^{(2)}$:

x_1	x_2	y_1	y_2	v_1	v_2	v_3	v_4	u_1	u_2	z_1	z_2	z_3	z_4	w_1	w_2	1
0	9/5	1	-1/5	0	0	0	0	0	0	0	0	0	0	1	-1/5	9
1	1/5	0	1/5	0	0	0	0	0	0	0	0	0	0	0	1/5	3
0	-1/5	0	-1/5	-1	0	0	0	1	5	1	0	0	0	0	-1/5	7
0	1	0	0	0	-1	0	0	2	1	0	1	0	0	0	0	5
0	0	0	0	0	0	-1	0	1	0	0	0	1	0	0	0	0
0	0	0	0	0	0	0	-1	0	1	0	0	0	1	0	0	0
0	-9/5	-1	1/5	0	0	0	0	0	0	0	0	0	0	0	6/5	-9

$T^{(3)}$:

x_1	x_2	y_1	y_2	v_1	v_2	v_3	v_4	u_1	u_2	z_1	z_2	z_3	z_4	w_1	w_2	1
0	1	5/9	-1/9	0	0	0	0	0	0	0	0	0	0	5/9	-1/9	5
1	0	-1/9	2/9	0	0	0	0	0	0	0	0	0	0	-1/9	2/9	2
0	0	1/9	-2/9	-1	0	0	0	1	5	1	0	0	0	1/9	-2/9	8
0	0	-5/9	1/9	0	-1	0	0	2	1	0	1	0	0	-5/9	1/9	0
0	0	0	0	0	0	-1	0	1	0	0	0	1	0	0	0	0
0	0	0	0	0	0	0	-1	0	1	0	0	0	1	0	0	0
0	0	0	0	0	0	0	0	0	0	0	0	0	0	1	1	0

Here the first phase ends. The artificial variables w can be eliminated from the tableau, as they are both nbv.

300 *Nonlinear Programming*

$T^{(4)}$:

x_1	x_2	y_1	y_2	v_1	v_2	v_3	v_4	u_1	u_2	z_1	z_2	z_3	z_4	1
0	1	5/9	-1/9	0	0	0	0	0	0	0	0	0	0	5
1	0	-1/9	2/9	0	0	0	0	0	0	0	0	0	0	2
0	0	1/9	-2/9	-1	0	0	0	1	5	1	0	0	0	8
0	0	-5/9	1/9	0	-1	0	0	2	1	0	1	0	0	0
0	0	0	0	0	0	-1	0	1	0	0	0	1	0	0
0	0	0	0	0	0	0	-1	0	1	0	0	0	1	0
0	0	4/9	1/9	1	1	1	1	-4	-7	0	0	0	0	-8

$T^{(5)}$:

x_1	x_2	y_1	y_2	v_1	v_2	v_3	v_4	u_1	u_2	z_1	z_2	z_3	z_4	1
0	1	5/9	-1/9	0	0	0	0	0	0	0	0	0	0	5
1	0	-1/9	2/9	0	0	0	0	0	0	0	0	0	0	2
0	0	1/9	-2/9	-1	0	0	5	1	0	1	0	0	-5	8
0	0	-5/9	1/9	0	-1	0	1	2	0	0	1	0	-1	0
0	0	0	0	0	0	-1	0	1	0	0	0	1	0	0
0	0	0	0	0	0	0	-1	0	1	0	0	0	1	0
0	0	4/9	1/9	1	1	1	-6	-4	0	0	0	0	7	-8

$T^{(6)}$:

x_1	x_2	y_1	y_2	v_1	v_2	v_3	v_4	u_1	u_2	z_1	z_2	z_3	z_4	1
0	1	5/9	-1/9	0	0	0	0	0	0	0	0	0	0	5
1	0	-1/9	2/9	0	0	0	0	0	0	0	0	0	0	2
0	0	26/9	-7/9	-1	5	0	0	-9	0	1	-5	0	0	8
0	0	-5/9	1/9	0	-1	0	1	2	0	0	1	0	-1	0
0	0	0	0	0	0	-1	0	1	0	0	0	1	0	0
0	0	-5/9	1/9	0	-1	0	0	2	1	0	1	0	0	0
0	0	-26/9	7/9	1	-5	1	0	8	0	0	6	0	1	-8

$T^{(7)}$:

x_1	x_2	y_1	y_2	v_1	v_2	v_3	v_4	u_1	u_2
0	1	0	-1/26	5/26	-25/26	0	0	45/26	0
1	0	0	5/26	-1/26	5/26	0	0	-9/26	0
0	0	1	-7/26	-9/26	45/26	0	0	-81/26	0
0	0	0	-1/26	-5/26	-1/26	0	1	7/26	0
0	0	0	0	0	0	-1	0	1	0
0	0	0	-1/26	-5/26	-1/26	0	0	7/26	1
0	0	0	0	0	0	1	0	-1	0

z_1	z_2	z_3	z_4	1
-5/26	25/26	0	0	45/13
1/26	-5/26	0	0	30/13
9/26	-45/26	0	0	36/13
5/26	1/26	0	-1	20/13
0	0	1	0	0
5/26	1/26	0	0	20/13
1	1	0	1	0

$T^{(8)}$:

x_1	x_2	y_1	y_2	v_1	v_2	v_3	v_4	u_1	u_2
0	1	0	-1/26	5/26	-25/26	45/26	0	0	0
1	0	0	5/26	-1/26	5/26	-9/26	0	0	0
0	0	1	-7/26	-9/26	45/26	-81/26	0	0	0
0	0	0	-1/26	-5/26	-1/26	7/26	1	0	0
0	0	0	0	0	0	-1	0	1	0
0	0	0	-1/26	-5/26	-1/26	7/26	0	0	1
0	0	0	0	0	0	0	0	0	0

$T^{(8)}$:

z_1	z_2	z_3	z_4	1
-5/26	25/26	-45/26	0	45/13
1/26	-5/26	9/26	0	30/13
9/26	-45/26	81/26	0	36/13
5/26	1/26	-7/26	-1	20/13
0	0	1	0	0
5/26	1/26	-7/26	0	20/13
1	1	1	1	0

The optimal solution is:

$\bar{x}^T = (30/13; 45/13)$, with:

$\bar{\pi} = \bar{x}^T \cdot C \cdot \bar{x} + d^T \cdot \bar{x} \approx -31.7$.

8.2.2 The Method of FRANK and WOLFE

Hypotheses See 8.2.1

Note: If type-II inequalities are present, start with a basic feasible solution.

Principle See 8.2.1

Description

Step 1: Construct the initial tableau Z :

	x	y	v	u	w	t
Z:	A	I	Θ	Θ	Θ	b
	2·C	Θ	-I	A^T	I	-d

where $t : t_{[(m+n) \times 1]}$, $t^T = (b; -d)$.

Step 2: (1.phase)

Minimize the sum of the artificial variables w with the first phase of the Two-Phase Method.

Step 3: Define the vector $p : p_{[(2m+2n) \times 1]}$, so that

$p^T = (x; y; v; u)$, then the vector $\tilde{p}^T = (v; u; x; y)$ is adjunctive to p.
Is $p^T \cdot \tilde{p} = 0$?
If yes: Stop, the optimal solution has been obtained.
If no : Go to step 4.

Step 4: Let $p_k \in p$ be the basic variable in the i-th row of the current tableau, then $\lambda_i := \tilde{p}_k$. Determine the values $\lambda_i \ \forall \ i=1,\ldots,m+n$.

Step 5: Let the new objective function be given by:

$$\tilde{c}_j := -\sum_{i=1}^{m+n} z_{ij} \cdot \lambda_i \quad \forall \ j : nbv$$

$$\tilde{\pi} := -\sum_{i=1}^{m+n} t_i \cdot \lambda_i \ ,$$

then the initial tableau of the next phase is given by the last tableau of the preceding phase with the new objective function.

Step 6: (next phase)
Execute one tableau-transformation according to the Primal Simplex-Algorithm.
Result: A solution $x^*; y^*; v^*; u^*$ with the value π^* of the objective function.

Step 7: Define the vector $p^* : p^*[(2m + 2n) \times 1]$, so that $p^{*T} = (x^*; y^*; v^*; u^*)$, then the vector $\tilde{p}^{*T} = (v^*; u^*; x^*; y^*)$ is adjunctive to p^*.
Is $p^{*T} \cdot \tilde{p}^* = 0$?
If yes: Stop, the optimal solution has been obtained.
If no : Go to step 8.

Step 8: Is $\pi^* \leq \tilde{\pi}/2$?
If yes: Go to step 9.

304 Nonlinear Programming

If no : Go to step 6.

Step 9: Compute

$$\mu := \min \left\{ 1; \frac{(p-p^*)^T \cdot \tilde{p}}{(p^*-p)^T \cdot (p^*-p)} \right\};$$

$$\hat{p} := p + \mu \cdot (p^*-p).$$

Step 10: Is $\mu = 1$?

If yes: Set $p := \hat{p}$; $\tilde{\pi} := \pi^*/2$. Go to step 6 with the current solution.

If no : Set $p := \hat{p}$ and determine the vector $\tilde{p}$ which is adjunctive to p. Go to step 4 with the current solution.

Note: If in step 6 more than one column may be selected as the pivot-column, then select that one, which assures the following:
$x_j \cdot v_j = 0 \ \forall \ j=1,\ldots,n$; $y_i \cdot u_i = 0 \ \forall \ i=1,\ldots,m$.

Example

Given the following problem:

P: min $\pi = 1/2 \cdot x_1^2 + 1/2 \cdot x_2^2 - 10 \cdot x_1 - 5 \cdot x_2$

$\quad x_1 + 2 \cdot x_2 \leq 12$

$\quad 5 \cdot x_1 + x_2 \leq 15$

$\quad x_1, x_2 \geq 0$

(the same as in 8.2.1)

$Z^{(1)}$:

x_1	x_2	y_1	y_2	v_1	v_2	u_1	u_2	w_1	w_2	t
1	2	1	0	0	0	0	0	0	0	12
5	1	0	1	0	0	0	0	0	0	15
1	0	0	0	-1	0	1	5	1	0	10
0	1	0	0	0	-1	2	1	0	1	5
-1	-1	0	0	1	1	-3	-6	0	0	-15

$Z^{(2)}$:

x_1	x_2	y_1	y_2	v_1	v_2	u_1	u_2	w_1	w_2	t
1	2	1	0	0	0	0	0	0	0	12
5	1	0	1	0	0	0	0	0	0	15
1/5	0	0	0	-1/5	0	1/5	1	1/5	0	2
-1/5	1	0	0	1/5	-1	9/5	0	-1/5	1	3
1/5	-1	0	0	-1/5	1	-9/5	0	6/5	0	-3

$Z^{(3)}$:

x_1	x_2	y_1	y_2	v_1	v_2	u_1	u_2	w_1	w_2	t
1	2	1	0	0	0	0	0	0	0	12
5	1	0	1	0	0	0	0	0	0	15
0	1	0	0	0	-1	2	1	0	1	5
-1	5	0	0	1	-5	9	0	-1	5	15
0	0	0	0	0	0	0	0	1	1	0

Here the first phase ends, the artificial variables w can be eliminated from the following tableaus.

$p^T = (0; 0; 12; 15; 15; 0; 0; 5)$; $\tilde{p}^T = (15; 0; 0; 5; 0; 0; 12; 15)$;

$p^T \cdot \tilde{p} \neq 0$; $\lambda_1 = 0$; $\lambda_2 = 5$; $\lambda_3 = 15$; $\lambda_4 = 0$; $\tilde{\pi} = -150$;

$Z^{(4)}$:

x_1	x_2	y_1	y_2	v_1	v_2	u_1	u_2	t
1	2	1	0	0	0	0	0	12
5	1	0	1	0	0	0	0	15
0	1	0	0	0	-1	2	1	5
-1	5	0	0	1	-5	9	0	15
-25	-20	0	0	0	15	-30	0	-150

$Z^{(5)}$:

x_1	x_2	y_1	y_2	v_1	v_2	u_1	u_2	t
0	9/5	1	-1/5	0	0	0	0	9
1	1/5	0	1/5	0	0	0	0	3
0	1	0	0	0	-1	2	1	5
0	26/5	0	1/5	1	-5	9	0	18
0	-15	0	5	0	15	-30	0	-75

$p^{*T} = (3; 0; 9; 0; 18; 0; 0; 5);\quad \tilde{p}^{*T} = (18; 0; 0; 5; 3; 0; 9; 0);$
$p^{*T} \cdot \tilde{p}^* \neq 0;\quad \pi^* = -75 \stackrel{!}{=} \tilde{\pi}/2;$

$$\mu = \min\left\{1;\ \frac{(-3;0;3;15;-3;0;0;0)\cdot(15;0;0;5;0;0;12;15)^T}{(3;0;-3;-15;3;0;0;0)\cdot(3;0;0;0;3;0;-3;15)^T}\right\} =$$

$= \min\ \{1;\ 5/3\}\ =\ 1;$

$\hat{p}^T = (3;0;9;0;18;0;0;5);\quad \mu \stackrel{!}{=} 1 \rightarrow$
$p^T = (3;0;9;0;18;0;0;5);\quad \tilde{\pi} = \pi^*/2 = -75/2;$

$Z^{(6)}$:

x_1	x_2	y_1	y_2	v_1	v_2	u_1	u_2	t
0	0	1	-7/26	-9/26	45/26	-81/26	0	36/13
1	0	0	5/26	-1/26	5/26	-9/26	0	30/13
0	0	0	-1/26	-5/26	-1/26	7/26	1	20/13
0	1	0	1/26	5/26	-25/26	45/26	0	45/13
0	0	0	145/26	75/26	15/26	-105/26	0	-300/13

$p^{*T} = (30/13;\ 45/13;\ 36/13;\ 0;\ 0;\ 0;\ 0;\ 20/13);$

$\tilde{p}^{*T} = (0;\ 0;\ 0;\ 20/13;\ 30/13;\ 45/13;\ 36/13;\ 0);$

$p^{*T} \cdot p^* \stackrel{!}{=} 0 \rightarrow$ Stop, $\bar{x}^T = (30/13;\ 45/13)$ is an optimal solution with the value of the objective function $\bar{\pi} \approx -31.7$.

Note: In this example the pivot-elements have been selected in such a way that steps 4 - 10 were necessary. By selecting other pivots, an optimal solution for this problem could be obtained after step 3.

8.2.3 The Method of BEALE

Hypotheses
Given the following problem:

$$P: \min \pi = x^T \cdot C \cdot x + d^T \cdot x$$
$$A \cdot x \leq b$$
$$x \geq \Theta ,$$

where $C : C_{[n \times n]}$; $d : d_{[n \times 1]}$; A, x, b as defined.
The matrix C must be positive semidefinite.

Principle
Starting with a basic feasible solution, an alternating sequence of intermediate tableaus and end tableaus is constructed by simplex-iterations. Whereas the coefficients of the "simplex"-portion of an end tableau are already computed for the next basic solution in the intermediate tableau, those of the "objective"-portion are only partially determined. They must undergo another pivoting.

Description
Step 1: Formulate the matrix $\tilde{C} : \tilde{C}_{[(n + 1) \times (n + 1)]}$, so that

$$\tilde{C} = \begin{array}{|c|c|} \hline C & d/2 \\ \hline d/2 & \pi \\ \hline \end{array}$$

Step 2: Add slack variables y to the system of inequalities and solve the system for y.

308 Nonlinear Programming

Step 3: Construct the initial tableau Z :

	x_1 ... x_n	1
x_1 $\vdots$ x_n	I	Θ
y_1 $\vdots$ y_m	-A	b
x_1 $\vdots$ x_n	$\tilde{c}$	
1		

simplex-tableau

objective-tableau

objective - row

Step 4: Does an element $\tilde{c}_{n+1,s} \neq 0$ exist in the objective-row in a column s, corresponding to a "free" variable u ?
 If yes: Select this column s for the pivot-column.
  Go to step 6.
 If no : Go to step 5.

Step 5: Is $\tilde{c}_{n+1,j} \geq 0 \; \forall \; j |$ column j corresponds to a variable x or y ?
 If yes: Stop, an optimal solution has been obtained.
 If no : Select one of these columns for the pivot-column (let this be column s). Go to step 6.

Step 6: Determine the row r, so that

$$q_r := \min_i \{\frac{b_i}{|a_{is}|} ; \frac{|\tilde{c}_{n+1,s}|}{\tilde{c}_{ss}} (|\text{sign } a_{is} = \text{sign } \tilde{c}_{n+1,s}) \wedge (\tilde{c}_{ss} > 0)\}.$$

Section 8.2.3

Herewith is the pivot-element z_{rs} given.

Step 7: Is $z_{rs} \in A$?
If yes: Go to step 8.
If no : Go to step 9.

Step 8: Replace the variable at the top of column s of the tableau Z by the variable, corresponding to row r. Go to step 10.

Step 9: Replace the variable at the top of the column s of the tableau Z by a "free" variable u. (If necessary, these variables must be indexed.)

Step 10: Construct the intermediate tableau Z^*, so that

$$z^*_{ij} := z_{ij} - \frac{z_{is} \cdot z_{rj}}{z_{rs}} \quad \forall\ i \neq r;\ j \neq s$$

$$z^*_{is} := \frac{z_{is}}{z_{rs}} \quad \forall\ i;\quad z^*_{rj} := 0 \ \forall\ j \neq s\ .$$

Step 11: Consider row s in the "objective"-portion of tableau Z^*. (It is always determined by the intersection of column s with the main diagonal in the objective-tableau.) Replace the variable in row s at the side of tableau Z^* by the variable, corresponding to column s.

Step 12: Construct the tableau Z^{**}, so that

$$z^{**}_{ij} := z_{ij} \ \forall\ z^*_{ij} \in \{\text{simplex-tableau}\}$$

$$z^{**}_{sj} := \frac{z^*_{sj}}{z_{rs}} \quad \forall\ j \text{ in row s in the objective-tableau}$$

$$z^{**}_{ij} := z^*_{ij} - z_{ri} \cdot z^{**}_{sj} \quad \text{otherwise with } i \neq s\ .$$

Set $Z := Z^{**}$; go to step 4.

Note: For a problem with restrictions $A \cdot x \gtreqless b$ start with a basic feasible solution.

Example
Given the following problem:

P: min $\pi = 1/2 \cdot x_1^2 + 1/2 \cdot x_2^2 - 10 \cdot x_1 - 5 \cdot x_2$

(1) $x_1 + 2 \cdot x_2 \leq 12$
(2) $5 \cdot x_1 + x_2 \leq 15$

$x_1, x_2 \geq 0$

(the same as in 8.2.1)

$z^1 = $

	x_1	x_2	1
x_1	1	0	0
x_2	0	1	0
y_1	-1	-2	12
y_2	-5	-1	15
x_1	1/2	0	-5
x_2	0	1/2	-5/2
1	-5	-5/2	0

$z^{1*} = $

	y_2	x_2	1
x_1	-1/5	-1/5	3
x_2	0	1	0
y_1	1/5	-9/5	9
y_2	1	0	0
x_1	-1/10	-1/10	-7/2
x_2	0	1/2	-5/2
1	1	-3/2	-15

(pivot-column s = column 1;

$q_r = \min \{ (\frac{12}{|-1|} ; \frac{15}{|-5|}); \frac{|-5|}{1/2} \} = 3 \to$ pivot-row r = row 4;

this leads to the tableau z^{1*})

Section 8.2.3   311

$$Z^{1**} = Z^2 =$$

	y_2	x_2	1
x_1	-1/5	-1/5	3
x_2	0	1	0
y_1	1/5	-9/5	9
y_2	1	0	0
y_2	1/50	1/50	7/10
x_2	1/50	13/25	-9/5
1	7/10	-9/5	-51/2

$$Z^{2*} =$$

	y_2	u_1	1
x_1	-5/26	-5/13	30/13
x_2	-1/26	25/13	45/13
y_1	7/26	45/13	36/13
y_2	1	0	0
y_2	1/52	1/26	10/13
x_2	0	1	0
1	10/13	-45/13	-825/26

(pivot-column s = column 2; q_r = min $\{(\frac{3}{|-1/5|}; \frac{9}{|-9/5|}); \frac{|-9/5|}{13/25}\} = \frac{45}{13}$
$\rightarrow$ pivot-row r = row 6; this leads to the tableau Z^{2*})

$$Z^{2**} = Z^3 =$$

	y_2	u_1	1
x_1	-5/26	-5/13	30/13
x_2	-1/26	25/13	45/13
y_1	7/26	45/13	36/13
y_2	1	0	0
y_2	1/52	0	10/13
u_1	0	25/13	0
1	10/13	0	-825/26

No pivot-column can be selected → Stop, an optimal solution has been obtained. $\bar{x}$ = (30/13; 45/13); the value of the objective function is $\bar{\pi}$ = -825/26 ≈ -31.7 .

312 *Nonlinear Programming*

8.2.4 An Algorithm for the Solution of Linear Complementarity Problems (Lemke)

Hypotheses
Given the following problem:

P': min $\pi = x^T \cdot C \cdot x + d^T \cdot x$

$A \cdot x \geq b$

$x \geq \Theta$,

where $C : C_{[n \times n]}$; $d : d_{[n \times 1]}$. C must be positive semidefinite.

Principle
In this algorithm a special simplex-tableau is used which derives from the formulation of the nonlinear problem as a linear complementary problem. An iterative process determines a complementary basic solution.

Description
Step 1: Formulate the linear complementary problem P from P':

P: $y = A \cdot x - b + e_m \cdot z_0$

$v = d + 2 \cdot C \cdot x - u \cdot A + e_n \cdot z_0$,

or in matrix notation:

P: $\begin{pmatrix} y \\ v \end{pmatrix} = \begin{pmatrix} -b \\ d \end{pmatrix} + \begin{pmatrix} \Theta & A \\ -A^T & 2 \cdot C \end{pmatrix} \cdot \begin{pmatrix} u \\ x \end{pmatrix} + \begin{pmatrix} e_m \\ e_n \end{pmatrix} \cdot z_0$,

or in tableau notation:

	y	v	u	x	z_0	t
P:	I	Θ	Θ	-A	$-e_m$	-b
	Θ	I	A^T	$-2 \cdot C$	$-e_n$	d

where:

- x : problem variables ⎫ of P'; $x,y \geq 0$
- y : slack variables ⎭
- u : dual variables ⎫ of P'; $u,v \geq 0$
- v : dual slack variables ⎭
- z_0 : artificial variable, $z_0 \geq 0$
- e_k : k-dimensional summing vector
- t : $t_{[(m+n) \times 1]}$; $t^T := (-b; d)$;

the variables (y_i, u_i) and (v_j, x_j) are called "pairs of complementary variables".

Step 2: Is $t_i \geq 0 \;\; \forall \; i=1,\ldots,m+n$?

If yes: Stop, the current solution $\bar{x}$ is optimal, the value of the objective function is $\bar{\pi} = f(\bar{x})$.

If no : Go to step 3.

Step 3: Determine $t_r := \min_i \{t_i\}$.

The pivot-element is given by row r and column z_0. Execute one tableau transformation according to the Primal Simplex-Algorithm.

Result: z_0 has become a bv, one variable y_i or v_j has become nonbasic. Now the following holds:

$(z_0 \geq 0) \wedge (y_i \cdot u_i = 0 \;\; \forall \; i=1,\ldots,m) \wedge (v_j \cdot x_j = 0 \;\; \forall \; j=1,\ldots,n)$,

i.e. an "almost" complementary basic solution has been obtained.

Step 4: Is $z_0 = 0$?

If yes: Stop, the current solution $\bar{x}$ is a complementary basic solution, the value of the objective function is $\bar{\pi} = f(\bar{x})$.

If no : Go to step 5.

Step 5: Select that column for pivot-column, whose complementary variable became nonbasic in the last iteration. The

314 Nonlinear Programming

pivot-row is determined according to the Primal Simplex-Algorithm.

Step 6: Does a feasible pivot-element still exist?
If yes: Execute one tableau transformation. Go to step 4.
If no : Stop, P' has an unbounded solution.

Example

Given the following problem P':

P': min $\pi = 1/2 \cdot x_1^2 + 1/2 \cdot x_2^2 - 10 \cdot x_1 - 5 \cdot x_2$

$\quad - x_1 - 2 \cdot x_2 \geq -12$

$\quad -5 \cdot x_1 - x_2 \geq -15$

$\quad x_1, x_2 \geq 0$

(the same as in 8.2.1)

$$A = \begin{pmatrix} -1 & -2 \\ -5 & -1 \end{pmatrix} ; \quad b = \begin{pmatrix} -12 \\ -15 \end{pmatrix} ; \quad 2 \cdot C = \begin{pmatrix} 1 & 0 \\ 0 & 1 \end{pmatrix} ; \quad d = \begin{pmatrix} -10 \\ -5 \end{pmatrix} ;$$

problem P reads:

y_1	y_2	v_1	v_2	u_1	u_2	x_1	x_2	z_0	1
1	0	0	0	0	0	1	2	-1	12
0	1	0	0	0	0	5	1	-1	15
0	0	1	0	-1	-5	-1	0	-1	-10
0	0	0	1	-2	-1	0	-1	-1	-5

y_1	y_2	v_1	v_2	u_1	u_2	x_1	x_2	z_0	1
1	0	-1	0	1	5	2	2	0	22
0	1	-1	0	1	5	6	1	0	25
0	0	-1	0	1	5	1	0	1	10
0	0	-1	1	-1	4	1	-1	0	5

y_1	y_2	v_1	v_2	u_1	u_2	x_1	x_2	z_0	1
1	-1/3	-4/3	0	2/3	10/3	0	5/3	0	41/3
0	1/6	-1/6	0	1/6	5/6	1	1/6	0	25/6
0	-1/6	-5/6	0	5/6	25/6	0	-1/6	1	35/6
0	-1/6	-5/6	1	-7/6	19/6	0	-7/6	0	5/6

y_1	y_2	v_1	v_2	u_1	u_2	x_1	x_2	z_0	1
1	-3/19	-26/57	-20/19	36/19	0	0	55/19	0	243/19
0	4/19	1/19	-5/19	9/19	0	1	9/19	0	75/19
0	1/19	5/19	-25/19	45/19	0	0	26/19	1	90/19
0	-1/19	-5/19	6/19	-7/19	1	0	-7/19	0	5/19

y_1	y_2	v_1	v_2	u_1	u_2	x_1	x_2	z_0	1
1	-7/26	-79/78	45/26	-81/26	0	0	0	-55/26	36/13
0	5/26	-1/26	5/26	-9/26	0	1	0	-9/26	30/13
0	1/26	5/26	-25/26	45/26	0	0	1	19/26	45/13
0	-1/26	-5/26	-1/26	7/26	1	0	0	7/26	20/13

$z_0 \stackrel{!}{=} 0 \rightarrow$ Stop, $\bar{x} = (30/13; 45/13)$ is an optimal solution, the value of the objective function is $\bar{\pi} \approx -31.7$.

8.2.5 The Gradient Projection Method (Rosen)

Hypotheses

Given the following problem :

P: min $\pi = f(x)$
   $A \cdot x \leq b$,

where $A : A_{[(m+n) \times n]}$; $b : b_{[(m+n) \times 1]}$.

$f(x)$ is a nonlinear convex objective function. In the system of restrictions the non-negativity constraints are included as: $-x \leq 0$. If the matrix of coefficients A is considered as an ordered set of row-vectors a_i, $i=1,\ldots,m+n$, then the restrictions are :

316 Nonlinear Programming

$a_i \cdot x \leq b_i \; \forall \; i=1,\ldots,m+n$. In the following this notation is used.

Principle

Starting with a basic feasible solution the method determines the projection of the gradient on the egde of the convex set (determined by the constraints), so that the direction and the length of the next move (given by the next corner of the convex set) is determined. After each move a test is performed to ascertain if the optimum has been bypassed in that move.

Description

Step 1: Determine the gradient of the objective function grad f(x). Set the running-index k: = 0 and determine an initial basic feasible solution $x^{(k)}$. (If necessary, this feasible solution can be determined by the 1. phase of the Two-Phase Method.)

Step 2: Does a solution $x^{(k)}$ exist, so that $A \cdot x^{(k)} \leq b$?
If yes: Go to step 3.
If no : Stop, P has no feasible solution.

Step 3: Let $\tilde{A}_k := (a_i \mid a_i \cdot x^{(k)} = b_i)$ be the matrix of coefficients of those restrictions, fulfilled as equations in the current solution $x^{(k)}$.
Determine $M_k := (\tilde{A}_k \cdot \tilde{A}_k^T)^{-1} \cdot \tilde{A}_k$;

$\tilde{P}_k := I - \tilde{A}_k^T \cdot M_k$;

$U_k := M_k \cdot \text{grad } f(x^{(k)})$.

Step 4: Is grad $f(x^{(k)}) = \Theta$?
If yes: Stop, $\bar{x} := x^{(k)}$ is an optimal solution, the value of the objective function is $\bar{\pi} = f(\bar{x})$.
If no : Go to step 5.

Step 5: Is $(\tilde{P}_k \cdot \text{grad } f(x^{(k)}) = \Theta) \wedge (U_k \geq \Theta)$?

If yes: Stop, $\bar{x} := x^{(k)}$ is an optimal solution, the value
of the objective function is $\bar{\pi} = f(\bar{x})$.
If no : Go to step 6.

Step 6: Determine $u_r := \max \{u_i \mid u_i \in U_k\}$.
Let u_r correspond to the restriction $a_s \cdot x^{(k)} = b_s$.
Define $\hat{A}_k := (a_i \mid (a_i \in \tilde{A}_k) \wedge (i \neq s))$
and compute $\hat{P}_k := I - \hat{A}_k^T \cdot (\hat{A}_k \cdot \hat{A}_k^T)^{-1} \cdot \hat{A}_k$;
$s_k := -\hat{P}_k \cdot \text{grad } f(x^{(k)})$.

Step 7: Determine $\lambda_i := \dfrac{b_i - a_i \cdot x^{(k)}}{a_i \cdot s_k} \; \forall \; i \mid a_i \notin \tilde{A}_k$;

$$\lambda := \min \{\lambda_i \mid \lambda_i > 0\}.$$

Step 8: Determine the provisional new basic solution
$x*^{(k)} := x^{(k)} + \lambda \cdot s_k$.

Step 9: Is $(s_k^T \cdot \text{grad } f(x*^{(k)})) \leq 0$?
If yes: Set $x^{(k+1)} := x*^{(k)}$; $k := k + 1$.
Go to step 3.
If no : Go to step 10.

Step 10: The optimum lies between $x^{(k)}$ and $x*^{(k)}$, compute
$\bar{x} := \mu \cdot x*^{(k)} + (1 - \mu) \cdot x^{(k)}$, where

$$\mu := \frac{s_k^T \cdot \text{grad } f(x^{(k)})}{s_k^T \cdot \text{grad } f(x^{(k)}) - s_k^T \cdot \text{grad } f(x*^{(k)})},$$

and $\bar{\pi} := f(\bar{x})$.

Example
Given the following problem:
P: min $\pi = 1/2 \cdot x_1^2 + 1/2 \cdot x_2^2 - 10 \cdot x_1 - 5 \cdot x_2$

318 Nonlinear Programming

$$x_1 + 2 \cdot x_2 \leq 12$$
$$5 \cdot x_1 + x_2 \leq 15$$
$$-x_1 \leq 0$$
$$-x_2 \leq 0$$

(the same as in 8.2.1)

$$A = \begin{pmatrix} 1 & 2 \\ 5 & 1 \\ -1 & 0 \\ 0 & -1 \end{pmatrix} ; \quad b = \begin{pmatrix} 12 \\ 15 \\ 0 \\ 0 \end{pmatrix} ; \quad \text{grad } f(x) = \begin{pmatrix} x_1 - 10 \\ x_2 - 5 \end{pmatrix} ;$$

$$k = 0; \quad x^{(0)} = \begin{pmatrix} 0 \\ 0 \end{pmatrix} ; \quad \text{grad } f(x^{(0)}) = \begin{pmatrix} -10 \\ -5 \end{pmatrix} ;$$

$$\tilde{A}_o = \begin{pmatrix} a_3 \\ a_4 \end{pmatrix} = \begin{pmatrix} -1 & 0 \\ 0 & -1 \end{pmatrix} ;$$

$$M_o = \left[\begin{pmatrix} -1 & 0 \\ 0 & -1 \end{pmatrix} \cdot \begin{pmatrix} -1 & 0 \\ 0 & -1 \end{pmatrix} \right]^{-1} \cdot \begin{pmatrix} -1 & 0 \\ 0 & -1 \end{pmatrix} = \begin{pmatrix} -1 & 0 \\ 0 & -1 \end{pmatrix} ;$$

$$\tilde{P}_o = \begin{pmatrix} 1 & 0 \\ 0 & 1 \end{pmatrix} - \begin{pmatrix} -1 & 0 \\ 0 & -1 \end{pmatrix} \cdot \begin{pmatrix} -1 & 0 \\ 0 & -1 \end{pmatrix} = \begin{pmatrix} 0 & 0 \\ 0 & 0 \end{pmatrix} ;$$

$$U_o = \begin{pmatrix} -1 & 0 \\ 0 & -1 \end{pmatrix} \cdot \begin{pmatrix} -10 \\ -5 \end{pmatrix} = \begin{pmatrix} 10 \\ 5 \end{pmatrix} ;$$

$u_r = 10; \quad r = 1; \quad s = 3;$

$\hat{A}_o = (a_4) = (0; -1);$

$$\hat{P}_o = \begin{pmatrix} 1 & 0 \\ 0 & 1 \end{pmatrix} - \begin{pmatrix} 0 \\ -1 \end{pmatrix} \cdot \left[(0;-1) \cdot \begin{pmatrix} 0 \\ -1 \end{pmatrix} \right]^{-1} \cdot (0;-1) = \begin{pmatrix} 1 & 0 \\ 0 & 1 \end{pmatrix} - \begin{pmatrix} 0 & 0 \\ 0 & 1 \end{pmatrix} = \begin{pmatrix} 1 & 0 \\ 0 & 0 \end{pmatrix} ;$$

$$s_o = - \begin{pmatrix} 1 & 0 \\ 0 & 0 \end{pmatrix} \cdot \begin{pmatrix} -10 \\ -5 \end{pmatrix} = \begin{pmatrix} 10 \\ 0 \end{pmatrix} ;$$

$$\lambda_1 = \frac{12-(1;2)\cdot\begin{pmatrix}0\\0\end{pmatrix}}{(1;2)\cdot\begin{pmatrix}10\\0\end{pmatrix}} = 6/5; \quad \lambda_2 = \frac{15-(5;1)\cdot\begin{pmatrix}0\\0\end{pmatrix}}{(5;1)\cdot\begin{pmatrix}10\\0\end{pmatrix}} = 3/10;$$

$\lambda = \min\{6/5; 3/10\} = 3/10;$

$$x*^{(0)} = \begin{pmatrix}0\\0\end{pmatrix} + 3/10 \cdot \begin{pmatrix}10\\0\end{pmatrix} = \begin{pmatrix}3\\0\end{pmatrix}; \quad \text{grad } f(x*^{(0)}) = \begin{pmatrix}-7\\-5\end{pmatrix};$$

$$s_0^T \cdot \text{grad } f(x*^{(0)}) = -70 \rightarrow x^{(1)} = x*^{(0)} = \begin{pmatrix}3\\0\end{pmatrix};$$

$$k = 1; \text{ grad } f(x^{(1)}) = \begin{pmatrix}-7\\-5\end{pmatrix};$$

$$\tilde{A}_1 = \begin{pmatrix}a_2\\a_4\end{pmatrix} = \begin{pmatrix}5 & 1\\0 & -1\end{pmatrix};$$

$$M_1 = \left[\begin{pmatrix}5 & 1\\0 & -1\end{pmatrix}\cdot\begin{pmatrix}5 & 0\\1 & -1\end{pmatrix}\right]^{-1}\cdot\begin{pmatrix}5 & 1\\0 & -1\end{pmatrix} = \begin{pmatrix}1/25 & 1/25\\1/25 & 26/25\end{pmatrix}\cdot\begin{pmatrix}5 & 1\\0 & -1\end{pmatrix} = \begin{pmatrix}1/5 & 0\\1/5 & -1\end{pmatrix};$$

$$\tilde{P}_1 = \begin{pmatrix}1 & 0\\0 & 1\end{pmatrix} - \begin{pmatrix}5 & 0\\1 & -1\end{pmatrix}\cdot\begin{pmatrix}1/5 & 0\\1/5 & -1\end{pmatrix} = \begin{pmatrix}0 & 0\\0 & 0\end{pmatrix};$$

$$U_1 = \begin{pmatrix}1/5 & 0\\1/5 & -1\end{pmatrix}\cdot\begin{pmatrix}-7\\-5\end{pmatrix} = \begin{pmatrix}-7/5\\18/5\end{pmatrix};$$

$u_r = 18/5; r = 2; s = 4;$

$$\hat{A}_1 = (a_2) = (5;1);$$

$$\hat{P}_1 = \begin{pmatrix}1 & 0\\0 & 1\end{pmatrix} - \begin{pmatrix}5\\1\end{pmatrix}\cdot\left[(5;1)\cdot\begin{pmatrix}5\\1\end{pmatrix}\right]^{-1}\cdot(5;1) = \begin{pmatrix}1/26 & -5/26\\-5/26 & 25/26\end{pmatrix};$$

$$s_1 = \begin{pmatrix}1/26 & -5/26\\-5/26 & 25/26\end{pmatrix}\cdot\begin{pmatrix}-7\\-5\end{pmatrix} = \begin{pmatrix}-9/13\\45/13\end{pmatrix};$$

$$\lambda_1 = \frac{12-(1;2)\cdot\begin{pmatrix}3\\0\end{pmatrix}}{(1;2)\cdot\begin{pmatrix}-9/13\\45/13\end{pmatrix}} = 13/9; \quad \lambda_3 = \frac{0-(-1;0)\cdot\begin{pmatrix}3\\0\end{pmatrix}}{(-1;0)\cdot\begin{pmatrix}-9/13\\45/13\end{pmatrix}} = 13/3;$$

$\lambda = \min \{13/9;\ 13/3\} = 13/9\ ;$

$$x*^{(1)} = \begin{pmatrix}3\\0\end{pmatrix} + 13/9\cdot\begin{pmatrix}-9/13\\45/13\end{pmatrix} = \begin{pmatrix}2\\5\end{pmatrix};\ \text{grad}\ f(x*^{(1)}) = \begin{pmatrix}-8\\0\end{pmatrix};$$

$$s_1^T\cdot\text{grad}\ f(x*^{(1)}) = (-9/13;\ 45/13)\cdot\begin{pmatrix}-8\\0\end{pmatrix} = 72/13 \rightarrow$$
the optimum lies between $x^{(1)}$ and $x*^{(1)}$;

$$\mu = \frac{(-9/13;45/13)\cdot\begin{pmatrix}-7\\-5\end{pmatrix}}{(-9/13;\ 45/13)\cdot\begin{pmatrix}-7\\-5\end{pmatrix} - (-9/13;45/13)\cdot\begin{pmatrix}-8\\0\end{pmatrix}} = 9/13;$$

$\bar{x}^T = 9/13\cdot(2;\ 5) + (1 - 9/13)\cdot(3;\ 0) = (18/13;\ 45/13) + (12/13;\ 0);$

$\bar{x}^T = (30/13;\ 45/13);\quad \bar{\pi} = f(\bar{x}) \approx -31.7\ .$

9. Generation of Random Numbers (Simulation)

Note

An important aid in simulation is the generation of random numbers. We distinguish:

(1) random numbers (numbers, whose sequence is not repeated after a finite number of steps

(2) pseudo random numbers (numbers, whose sequence is repeated after a finite number of steps (finite cycle length)).

(Pseudo)random numbers may be generated in different ways.

9.0.1 The AWF-Cubes (Graf)

Description
The surfaces of the cubes are insribed as follows:

1-st cube:	0;	1;	2;	3;	4;	5;
2-nd cube:	0;	6;	12;	18;	24;	30;
3-rd cube:	0;	36;	72;	108;	144;	180;
4-th cube:	0;	216;	432;	648;	864;	1080;

⋮

k-th cube: n_1; n_2; n_3; n_4; n_5; n_6

where $n_i := 6^{k-1} \cdot (i - 1) \quad \forall \; i=1,\ldots,6$.

If the cubes are not loaded, then all numbers between 0 and 6^k-1 have the same probability to appear, i.e. are equal-distributed. Consequently random numbers are generated.

9.0.2 The Midsquare Method (J. v. Neumann)

Procedure

Take any number with $2 \cdot n$ places, $n \in \mathbb{N}$; square it; take the middle $2 \cdot n$ places from the result (if necessary add zeros to the left) and square it; etc...
The numbers, consisting of the middle $2 \cdot n$ places, selected each time, are the (pseudo) random numbers.
This method requires the following tests to be performed:

(a) a test for equal-distribution of the numerals
(b) a test for the dependence of the generated numbers
(c) a test for degeneration

Example 1

$$(1234)^2 = 01\ 5227\ 56$$
$$(5227)^2 = 27\ 3215\ 29$$
$$(3215)^2 = 10\ 3362\ 25$$
$$(3362)^2 = 11\ 3030\ 44$$
$$(3030)^2 = 09\ 1809\ 00$$
$$(1809)^2 = 03\ 2724\ 81$$
$$(2724)^2 = 07\ 4201\ 76$$
$$(4201)^2 = 17\ 6484\ 01$$
$$(6484)^2 = 42\ 0422\ 56$$
$$(0422)^2 = 00\ 1780\ 84$$

test (a):

numeral	0	1	2	3	4	5	6	7	8	9
relative frequency	3/20	1/10	9/40	1/8	1/8	1/20	1/20	3/40	3/40	1/40

the numerals are approximately equal-distributed; test (b),(c): positive.

Example 2

$$(50)^2 = 2\ 50\ 0$$
$$(50)^2 = \ldots\ldots\ \text{all tests negative !}$$

Example 3

$(36)^2 = 1\ 29\ 6$
$(29)^2 = 0\ 84\ 1$
$(84)^2 = 7\ 05\ 6$
$(05)^2 = 0\ 02\ 5$
$(02)^2 = 0\ 00\ 4$
$(00)^2 = 0\ 00\ 0$
$(00)^2 = \ldots\ldots$ all tests negative !

9.0.3 A Mixed Congruence Method

Procedure
Take any constants a, b, c, $x_0 \in \mathbb{N}$ and generate the (pseudo) random numbers x_i as follows:

$$x_i := (a \cdot x_{i-1} + b) \bmod c \quad \forall\ i=1,2,\ldots$$

If the constants are appropriately selected, the distribution of the numbers $x_i \in [0;c)$ approximates the equal-distribution.
Note: Select prime numbers for constants.

Random numbers s_i in the unit-interval [0;1) are easily computed by:

$$s_i := x_i/c \quad \forall\ i=1,2,\ldots$$

Example
Let b = 3; c = 9; x_0 = 11; for various constants a:

a	x_0	x_1	x_2	x_3	x_4	x_5	x_6	x_7	x_8	...	cycle length
1	11	5	8	2	5	8	....				3
2	11	7	8	1	5	4	2	7	8	...	6
5	11	4	5	1	8	7	2	4	5	...	6
7	11	8	3	6	0	3	6	....			3

For a = 2; b = 3; c = 9; x_0 = 11 the following random numbers s are generated: s = (7/9; 8/9; 1/9; 5/9; 4/9; 2/9; 7/9; ...) .

9.0.4 A Multiplicative Congruence Method

Procedure

Take any constants a, c, $x_0 \in \mathbb{N}$ and generate the (pseudo) random numbers x_i as follows:

$$x_i := (a \cdot x_{i-1}) \bmod c \quad \forall \; i=1,2,\ldots$$

Note 1: The Multiplicative Congruence Method is a special case of the Mixed Congruence Method for b = 0.

Note 2: For use with a computer, it is suitable to select $c = 2^q$, $q \in \mathbb{N}$.

Note 3: For random numbers s_i in the unity-interval [0;1] see 9.0.3.

Example

Let $c = 2^4 = 16$; $x_0 = 11$; for various constants a:

a	x_0	x_1	x_2	x_3	x_4	x_5	x_6 ...	cycle length
1	11	11	11	....				1
2	11	6	12	8	0	0	....	1
5	11	7	3	15	11	7	....	4
7	11	13	11	13	....			2

10. Replacement Models

10.1 Replacement Models with Respect to Gradually Increasing Maintenance Costs

10.1.1 A Model Disregarding the Rate of Interest

Hypotheses

Given the following information:

C_p : purchase cost of the (considered) product

C_{o_t} : operating cost of the product in the t-th period

C_t^* : trade-in value of the product in the t-th period.

Assume the attrition of the product is approximately constant in all periods.

Principle

The costs in each period are enumerated for the replacement (total replacement) when it is accopmlished after t=1,2,... periods. The replacement policy with the minimal cost, is optimal.

Procedure

Set up the following table:

t	C_{o_t}	C_t^*	$S := \sum_{\tau=1}^{t} C_{o_\tau}$	$C_{tot}(t) := S + C_p + C_t^*$	$\tilde{C}_t := \frac{1}{t} \cdot C_{tot}(t)$
1					
2					
.					
.					
.					

326 Replacement Models

and determine $\tilde{C}_r := \min_t \{\tilde{C}_t\}$.

Result: The optimal replacement time is after the r-th period. The average cost per period is $\tilde{C}_r$ m.u.

Example
Assume the purchase cost of a car is $10,000, the trade-in value C_t^* in the following six periods is $C^* = (5,000; 4,000; 2,900; 1,700; 1,000; 500)$ \$. The operating cost C_{o_t} is $C_o = (3,000; 3,500; 4,000; 4,600; 5,200; 6,000)$ \$, the number of miles driven per period is approximately equal. Determine the optimal replacement time and the average cost per period.

t	C_{o_t}	C_t^*	S	$C_{tot}(t)$	$\tilde{C}_t$	
1	3,000	5,000	3,000	8,000	8,000	
2	3,500	4,000	6,500	12,500	6,250	
3	4,000	2,900	10,500	17,600	$5,866.\bar{6}$	
4	4,600	1,700	15,100	23,400	5,850	← min
5	5,200	1,000	20,300	29,300	5,860	
6	6,000	500	26,300	35,800	$5,966.\bar{6}$	

$\tilde{C}_r = \tilde{C}_4$ → the optimal replacement time is after 4 periods with an average cost per period of \$ 5,850 .

10.1.2 A Model Regarding the Rate of Interest

Hypotheses
Given the following information:
C_p : purchase cost of the (considered) product
C_{o_t} : operating cost of the product in the t-th period
C_t^* : trade-in value of the product in the t-th period
C_{d_t} : decrease in the value of the product in the t-th period relative to the (t - 1)-th period; i.e.

$$C_{d_t} := C^*_{t-1} - C^*_t; \quad C^*_0 := C_p;$$

i : rate of interest.

Let $C_t := C_{o_t} + C_{d_t}; \quad q := (1 + i)$.

Assume the attrition of the product is approximately constant in all periods.

Principle   See 10.1.1

Procedure

Set up the following table:

t	C_{o_t}	C^*_t	C_{d_t}	C_t	q^{-t+1}	$C_t \cdot q^{-t+1}$	$C_{tot}(t)$	v_t	$\tilde{C}_t$
1									
2									
.									

where $C_{tot}(t) := \sum_{\tau=1}^{t} C_\tau \cdot q^{-\tau+1}$

$$v_t := \sum_{\tau=1}^{t} q^{\tau-1}$$

$$\tilde{C}_t := \frac{1}{v_t} \cdot C_{tot}(t) .$$

Determine $\tilde{C}_r := \min_t \{\tilde{C}_t\}$.

Result: The optimal replacement time is after the r-th period. The average discounted cost per period is $\tilde{C}_r$ m.u.

Example

Assume the purchase cost of a car is $ 20,000, the trade-in value

C_t^* in the following six periods is C*=(15,000; 11,000; 8,000; 5,500; 3,500; 2,000) \$, the operating cost C_{o_t} is C_o=(5,000; 5,500; 6,200; 7,000; 8,000; 9,300) \$. The number of miles driven per period is approximately equal; the rate of interest is 10%, i.e. i=0.1 . Determine the optimal replacement time and the average cost per period with respect to the rate of interest.

t	C_{o_t}	C_t^*	C_{d_t}	C_t	q^{-t+1}	$C_t \cdot q^{-t+1}$
1	5,000	15,000	5,000	10,000	1.0000	10,000.00
2	5,500	11,000	4,000	9,500	0.9091	8,636.45
3	6,200	8,000	3,000	9,200	0.8264	7,602.88
4	7,000	5,500	2,500	9,500	0.7513	7,137.35
5	8,000	3,500	2,000	10,000	0.6830	6,830.00
6	9,300	2,000	1,500	10,800	0.6209	6,705.72

$C_{tot}(t)$	v_t	$\tilde{C}_t$	
10,000.00	1.0000	10,000.00	
18,636.45	1.9091	9,761.90	
26,239.33	2.7355	9,592.15	
33,376.68	3.4868	9,572.30	← min
40,206.68	4.1698	9,642.35	
46,912.40	4.7907	9,792.39	

$\tilde{C}_r = \tilde{C}_4 \rightarrow$ the optimal replacement time is after 4 periods with an average cost per period of \$ 9,572.30 .

10.2 Replacement Models with Respect to Sudden Failure

10.2.1 A Model Disregarding the Rate of Interest

Hypotheses
Given the following information:

$P(t)$: probability that an element breaks in the t-th period
$\bar{t}$: maximum workingtime of an element (in periods)
N : sum total of all elements
C_1 : cost of replacing one element, when all elements are replaced
C_2 : cost of replacing one element, when only certain elements are replaced.

Assume: $C_1 < C_2$; $\sum_{t=1}^{\bar{t}} P(t) = 1$.

Principle See 10.1.1

Procedure
Set up the following table:

t	P(t)	W(t)	$C_{tot}(t)$	$\tilde{C}_t$
1				
.				
.				
$\bar{t}$				

where $W(t) := \sum_{\tau=1}^{t} p(\tau) \cdot W(t - \tau)$

$W(0) := N$

$C_{tot}(t) := C_1 \cdot W(0) + C_2 \cdot \sum_{\tau=1}^{t} W(\tau)$

$\tilde{C}_t := \frac{1}{t} \cdot C_{tot}(t)$.

Determine $\tilde{C}_r := \min_{t} \{\tilde{C}_t\}$.

Result: After the r-th period <u>all</u> elements should be replaced; individual defective elements are replaced at once. The average cost per period is $\tilde{C}_r$ m.u.

330 *Replacement Models*

Example

In a computer of the "first generation" there are 2,000 tubes. In the t-th month the tubes become defective with the following probabilities:

t	1	2	3	4	5
P(t)	0.1	0.15	0.2	0.3	0.25

i.e. none of the built-in tubes has a workingtime longer than 5 months. If a single tube must be replaced, it costs $ 3 per tube; if all tubes are replaced, it costs $ 2,000 . After how many months should all the tubes be replaced ?

$W(0) = N = 2,000$; $\bar{t} = 5$; $C_1 = 3$ \$/tube; $C_2 = \frac{2,000}{2,000} = 1$ \$/tube;

t	P(t)	W(t)	$C_{tot}(t)$	$\tilde{C}_t$
1	0.1	0.1·2,000 = 200	1·2,000+3·200=2,600	2,600
2	0.15	0.1·200+0.15·2,000 = 320	1·2,000+3·(200+320) = 3,560	1,780
3	0.2	0.1·320+0.15·200 + 0.2·2,000= 462	1·2,000+3·(200+320 + 462) = 4,946	1,648.$\bar{6}$ ←min
4	0.3	0.1·462+0.15·320 +0.2·200+0.3· 2,000 = 734.2	1·2,000+3·(200+320 +462+734.2)=7,148.6	1,787.15
5	0.25	0.1·734.2+0.15·462 +0.2·320+0.3·200 +0.25·2,000=766.72	1·2,000+3·(200+320 +462+734.2+766.72) =9,448.8	1,889.75

$\tilde{C}_r = \tilde{C}_3$ → after 3 months all tubes are replaced, individual defective tubes are replaced at once. The average monthly cost is $ 1,648.$\bar{6}$.

10.2.2 A Model Regarding the Rate of Interest

Hypotheses
See 10.2.1 , furthermore let i be the rate of interest, and
q: = (1 + i).
Note: It is assumed that the cost result at the end of a period.

Principle See 10.1.1

Procedure
Set up the following table:

t	P(t)	W(t)	q^{-t}	$W(t) \cdot q^{-t}$	$\sum_{\tau=1}^{t} W(\tau) \cdot q^{-\tau}$	$C_{tot}(t)$	$\tilde{C}_t$
1							
.							
.							
.							
$\bar{t}$							

where $W(t) := \sum_{\tau=1}^{t} P(\tau) \cdot W(t - \tau)$

$W(0) := N$

$C_{tot}(t) := C_1 \cdot W(0) \cdot q^{-t} + C_2 \cdot \sum_{\tau=1}^{t} W(\tau) \cdot q^{-\tau}$

$\tilde{C}_t := \frac{1}{t} \cdot C_{tot}(t)$.

Determine $\tilde{C}_r := \min_{t} \{ \tilde{C}_t \}$.

Result: After the r-th period <u>all</u> elements should be replaced; individual defective elements are replaced at once. The average cost per period is $\tilde{C}_r$ m.u.

Example
See 10.2.1 , in addition let the rate of interest be 8%, i.e.
i = 0.08.

t	P(t)	W(t)	q^{-t}	$W(t) \cdot q^{-t}$	$\sum_{\tau=1}^{t} W(\tau) \cdot q^{-\tau}$	$C_{tot}(t)$	$\tilde{C}_t$	
1	0.1	200	0.9259	185.18	185.18	2,407.34	2,407.34	
2	0.15	320	0.8573	274.34	459.52	3,093.15	1,546.58	
3	0.2	462	0.7938	366.74	826.26	4,066,38	1,355.46	← min
4	0.3	734.2	0.7350	539.64	1,365.90	5,567.70	1,391.93	
5	0.25	766.72	0.6806	521.83	1,887.73	7,024.39	1,404.88	

$\tilde{C}_r = \tilde{C}_3$ → after 3 month all tubes are replaced, individual defective tubes are replaced at once. The average cost per month with respect to the rate of interest is $ 1,355.46.

11. Inventory Models

Hypotheses

Given an enterprise, for which an optimal inventory policy must be determined; the demand per period, as well as the ordering and restocking costs, and the inventory costs are known. No time elapses between the time an order is made and the restocking is completed. Inventory sales are approximated by a continuous function; the demand must be fully satisfied. Find the optimal number per shipment to order; the number of orders per period and the corresponding cost.

Definitions

- x : number of ordered units
- L : average inventory (on-hand) balance; $L := x/2$
- D : demand per period
- b : number of orders per period; $b := D/x$
- c_o : ordering and restocking costs
- c_i : inventory costs (incl. interest charges, labour costs, etc.)
- C_{tot} : total costs; $C_{tot} := c_o \cdot b + c_i \cdot L$.

11.0.1 The Classical Inventory Model (Andler)

Procedure

Determine the optimal number $\bar{x}$ to order, so that

$$\bar{x} := \sqrt{\frac{2 \cdot c_o \cdot D}{c_i}} \quad ;$$

334 *Inventory Models*

the optimal number of orders $\bar{b}$ (times to order), so that

$$\bar{b} := \sqrt{\frac{D \cdot c_i}{2 \cdot c_o}} \quad ;$$

the corresponding total cost $\bar{C}_{tot}$, so that

$$\bar{C}_{tot} := \sqrt{2 \cdot c_o \cdot c_i \cdot D} \quad .$$

Example
An enterprise has a demand for 10,000 tape recorders yearly. The costs for one order and the refillment of the store are $400. Assume the interest charge per tape recorder is $2/yr. How many orders should be placed per year? How many tape recorders should be ordered each time? What are the total costs per year?

$D = 10,000; \quad c_o = 400; \quad c_i = 2;$

$$\bar{b} = \sqrt{\frac{10,000 \cdot 2}{2 \cdot 400}} = 5 \quad ; \qquad \bar{x} = \sqrt{\frac{2 \cdot 400 \cdot 10,000}{2}} = 2,000 \quad ;$$

$$\bar{C}_{tot} = \sqrt{2 \cdot 400 \cdot 2 \cdot 10,000} = \$\ 4000 \quad .$$

11.0.2 An Inventory Model with Penalties for Undersupplied Demands

Hypotheses
Given the Classical Inventory Model (11.0.1) in which the demand need not be completely satisfied. If the demand is undersupplied then each missing unit originates penalties of c_p m.u.

Procedure
Determine the optimal number $\bar{x}$ to order, so that

$$\bar{x} := \sqrt{\frac{2 \cdot c_o \cdot D}{c_i}} \cdot \sqrt{\frac{c_p}{c_i + c_p}} \quad ;$$

the optimal number of orders $\bar{b}$, so that

$$\bar{b} := \sqrt{\frac{D \cdot c_i}{2 \cdot c_o}} \cdot \sqrt{\frac{c_p}{c_i + c_p}} \quad ;$$

the corresponding total costs $\bar{C}_{tot}$, so that

$$\bar{C}_{tot} := \sqrt{\frac{2 \cdot c_o \cdot c_i \cdot c_p \cdot D}{c_i + c_p}} \quad .$$

Example
A tradesman, dealing in shirts, has a monthly demand for 4,000 shirts. The ordering cost is $ 3 per order and the inventory cost is $ 15 per shirt and month. If a customer is not supplied, a penalty of $ 1 per shirt is incurred. What is the optimal inventory policy for the tradesman ?

$D = 4{,}000; \quad c_o = 3; \quad c_i = 15; \quad c_p = 1 ;$

$$\bar{x} = \sqrt{\frac{2 \cdot 3 \cdot 4{,}000}{15}} \cdot \sqrt{\frac{1}{15+1}} = 10;$$

$$\bar{b} = \sqrt{\frac{4{,}000 \cdot 15}{2 \cdot 3}} \cdot \sqrt{\frac{1}{15+1}} = 25;$$

$$\bar{C}_{tot} = \sqrt{\frac{2 \cdot 3 \cdot 15 \cdot 1 \cdot 4{,}000}{15+1}} = \$\ 150,$$

i.e. only $\bar{x} \cdot \bar{b} = 250$ shirts per month are ordered and sold because of the extremely high inventory cost c_i.

11.0.3 An Inventory Model with Terms for Delivery

Hypotheses
Given the Classical Inventory Model (11.0.1) in which the time between the placement of the order and the restocking of the store is not zero.

336 *Inventory Models*

Definitions

t_d : delivery and restocking time

$D(t_d)$: number of items sold during the delivery and restocking time;
$$D(t_d) := t_d \cdot D .$$

Description

Step 1: Determine the optimal number $\bar{x}$ to order without considering the delivery and restocking time (according to 11.0.1).

Step 2: Is $\bar{x} \geq D(t_d)$?
If yes: Stop, $\bar{x}$; $\bar{b}$; $\bar{C}_{tot}$ (according to 11.0.1) is an optimal solution.
If no : Go to step 3.

Step 3: Stop, $\tilde{x}$; $= D(t_d)$ is a feasible solution, which is as near to the optimum (without terms of delivery) as possible. The corresponding number of orders is $\tilde{b}$: $= D/D(t_d)$, the total cost is

$$\tilde{C}_{tot} := \frac{D \cdot c_o}{D(t_d)} + \frac{c_i \cdot D(t_d)}{2} .$$

Example 1

Given the example of (11.0.1) with the delivery time of 18 days (assuming 360 days per year).

$t_d = 18/360 = 1/20;\quad D(t_d) = (1/20) \cdot 10,000 = 500;$

$\bar{x} = 2,000 \overset{!}{\geq} 500 \to$ Stop, $\bar{x} = 2,000;\quad \bar{b} = 5;$

$\bar{C}_{tot} = \$ 4,000$ is the optimal inventory policy.

Example 2

Given the example of (11.0.1) with the delivery time of 90 days (assuming 360 days per year).

$t_d = 90/360 = 1/4;\quad D(t_d) = (1/4) \cdot 10,000 = 2,500;$

$\bar{x} = 2,000 < 2,500 \to$ Stop, $\tilde{x} = 2,500;$

$\tilde{b} = 10,000/2,500 = 4$; $\tilde{C}_{tot} = \frac{10,000 \cdot 400}{2,500} + \frac{2 \cdot 2,500}{2} = \$\ 4,100$

is the optimal inventory policy.

11.0.4 An Inventory Model with Damage to Stock

Hypotheses
Given the Classical Inventory Model (11.0.1) in which damages to the stock are considered.

Definitions
$\tilde{D}$: number of items bought per period
ΔD : number of items damaged per period
p : cost per item .
Let $D: = \tilde{D} - \Delta D$; $b: = \tilde{D}/x$.

Procedure
Determine the optimal number $\bar{x}$ to order, so that

$$\bar{x}: = \sqrt{\frac{p^2 \cdot \Delta D^2}{c_i^2} + \frac{2 \cdot D \cdot c_o}{c_i} - \frac{p \cdot \Delta D}{c_i}}\ ;$$

the optimal number of orders $\bar{b}$, so that

$$\bar{b}: = \tilde{D} \cdot \sqrt{\frac{c_i \cdot \bar{x} + 2 \cdot p \cdot \Delta D}{2 \cdot c_o \cdot D \cdot \bar{x}}}\ ;$$

the corresponding total cost $\bar{C}_{tot}$, so that

$$\bar{C}_{tot}: = \bar{b} \cdot c_o + c_i \cdot \bar{x}/2 + p \cdot \Delta D\ .$$

Example
A pocket computer company will sell 500 items per month. Because of a lack of qualified help 50 items will be damaged beyond repair in a month. The costs per order and restocking of the store are

$ 3506.25; the inventory costs are $ 10 per item and month; the cost for one pocket computer is $ 100. What is the optimal number of pocket computers to order and how many orders should be placed per month ? What is the minimum price for one pocket computer, if the company does not wish to incur any loss for the damaged items ?

$D = 500;\quad \Delta D = 50;\quad \tilde{D} = 550;\quad c_o = 3506.25;\quad c_i = 10;\quad p = 100;$

$$\bar{x} = \sqrt{\frac{100^2 \cdot 50^2}{10^2} + \frac{2 \cdot 500 \cdot 3506.25}{10}} - \frac{100 \cdot 50}{10} = 275;$$

$$\bar{b} = 550 \cdot \sqrt{\frac{10 \cdot 275 + 2 \cdot 100 \cdot 50}{2 \cdot 3506.25 \cdot 500 \cdot 275}} \approx 2;$$

$\bar{C}_{tot} = 2 \cdot 3506.25 + 10 \cdot 275/2 + 100 \cdot 50 = \$\ 13{,}387.5;$

$$\frac{\bar{C}_{tot}}{D} = \frac{13{,}387.5}{500} = 26.775 \rightarrow \text{the minimum price for one pocket}$$

computer is $ 100 + 26.775 = $ 126.775 .

11.0.5 An Inventory Model with Rebates (different price intervals)

Hypotheses
Given the Classical Inventory Model (11.0.1) in which rebates are considered. A rebate is granted on the price when an order exceeds a specified amount.

Definitions
r_j : minimal selling quantity at the j-th price quotation;
$$r_1 < r_2 < \ldots < r_n$$
ε_j : price reduction per item at the j-th price quotation;
$$\varepsilon_1 < \varepsilon_2 < \ldots < \varepsilon_n$$

c_{i_j} : inventory cost per item corresponding to the j-th price quotation; $c_{i_1} > c_{i_2} > \ldots > c_{i_n}$.

Description

Step 1: Determine

$$C_{tot}^{(o)} := \sqrt{2 \cdot c_o \cdot c_i \cdot D}$$

$$C_{tot}^{(j)} := \frac{c_o \cdot D}{r_j} + \frac{c_{i_j} \cdot r_j}{2} - \varepsilon_j \cdot D \quad \forall \; j=1,\ldots,n \;.$$

Step 2: Determine

$$C_{tot}^{(s)} := \underset{j \in [0;n]}{\text{minimum}} \; \{C_{tot}^{(j)}\} \;.$$

Step 3: Is s = 0 ?

If yes: Go to step 4.

If no : Go to step 5.

Step 4: Stop, the optimal number $\bar{x}$ to order is

$$\bar{x} := \sqrt{\frac{2 \cdot c_o \cdot D}{c_i}} \;;$$

the optimal number of orders $\bar{b}$ is

$$\bar{b} = \sqrt{\frac{D \cdot c_i}{2 \cdot c_o}} \;;$$

the corresponding total cost $\bar{C}_{tot}$ is

$$\bar{C}_{tot} := C_{tot}^{(o)} \;.$$

Step 5: Stop, the optimal number $\bar{x}$ to order is $\bar{x} := r_s$;

the optimal number of orders $\bar{b}$ is $\bar{b} := D/r_s$;

the corresponding total cost $\bar{C}_{tot} := C_{tot}^{(s)}$.

Example

A wine bottler requires 30,000 bottles every 3 month. The supplier offers the following rebates:

340 *Inventory Models*

on orders of 4,000 bottles $ 0.25 off;
on orders of 5,000 bottles $ 0.5 off;
on orders of 10,000 bottles $ 0.$\bar{6}$ off.

The cost per order is $ 1,000. The inventory costs are $ 10 without rebate, $ 8, 6 or 5 with increasing number of bottles ordered. How many orders should be placed and what is the total cost ?

$D = 30,000;\quad c_o = 1,000;\quad c_i = 10;\quad c_{i_1} = 8;\quad c_{i_2} = 6;\quad c_{i_3} = 5;$

$r_1 = 4,000;\qquad r_2 = 5,000;\qquad r_3 = 10,000;$

$\varepsilon_1 = 1/4;\qquad \varepsilon_2 = 1/2;\qquad \varepsilon_3 = 2/3;$

$c_{tot}^{(o)} = \sqrt{2 \cdot 1,000 \cdot 10 \cdot 30,000} \approx \$\ 24,494.9\ ;$

$c_{tot}^{(1)} = \dfrac{1,000 \cdot 30,000}{4,000} + \dfrac{8 \cdot 4,000}{2} - (1/4) \cdot 30,000 = \$\ 16,000\ ;$

$c_{tot}^{(2)} = \dfrac{1,000 \cdot 30,000}{5,000} + \dfrac{6 \cdot 5,000}{2} - (1/2) \cdot 30,000 = \$\ 6,000\ ;$

$c_{tot}^{(3)} = \dfrac{1,000 \cdot 30,000}{10,000} + \dfrac{5 \cdot 10,000}{2} - (2/3) \cdot 30,000 = \$\ 8,000\ ;$

$c_{tot}^{(s)} = c_{tot}^{(2)} \rightarrow \bar{x} = r_2 = 5,000;\quad \bar{b} = b;\quad \bar{c}_{tot} = \$\ 6,000\ .$

11.0.6 An Inventory Model with Respect to Transportation Capacity

<u>Hypotheses</u>

Given the Classical Inventory Model (11.0.1) in which the capacity of the conveyance (e.g. truck, airplane, etc.) is considered. That is because of limitations in the transporting capacity more than one vehicle, or more than one trip may be necessary to deliver one order.

Definitions

x : capacity of the vehicle
p : number of required trips per order
C_f : fixed transportations cost per order
c_v : variable transportation cost per trip
$\tilde{C}$: transportation cost per order;

$$\tilde{C} := C_f + c_v \cdot p, \text{ if } (p-1) \cdot x < x \leq p \cdot x .$$

Description

Step 1: Calculate
$$p_1 := \frac{D \cdot c_v + \sqrt{D^2 \cdot c_v^2 + 2 \cdot c_i \cdot x^2 \cdot C_f \cdot D}}{c_i \cdot x^2}$$

and set $p^{(1)} := \langle p_1 \rangle$.

Step 2: Calculate $p_2 := \sqrt{\dfrac{2 \cdot C_f \cdot D}{c_i \cdot x^2}}$

and set $p^{(2)} := [p_2]$; $p^{(3)} := \langle p_2 \rangle$.

Step 3: Calculate
$$c^{(1)} := \sqrt{2 \cdot c_i \cdot D \cdot (C_f + c_v \cdot p^{(1)})} ;$$

$$c^{(j)} := \frac{C_f \cdot D}{x \cdot p^{(j)}} + \frac{c_v \cdot D}{x} + \frac{c_i \cdot x \cdot p^{(j)}}{2} \quad \forall \; j=2,3 .$$

Step 4: Determine $c^{(s)} := \underset{j=1,2,3}{\text{minimum}} \{c^{(j)}\}$

and set $\bar{C}_{tot} := c^{(s)}$.

Step 5: Determine the optimal number $\bar{x}$ to order, so that

$$\bar{x} := x \cdot p^{(j)}, \text{ if } \bar{C}_{tot} = c^{(j)} \quad \forall \; j=1,2,3;$$

the optimal number of orders is $\bar{b} := D/\bar{x}$; per order $p^{(s)}$ trips are required.

Note: In general $\bar{b} \notin \mathbb{N}$ (rounding off is necessary).

Example
A furniture company sells 400 wall cabinets yearly. The firm owns only one type of truck which can transport only 5 cabinets per trip to the company warehouse. The inventory cost per item and year is $ 60; the transportation cost function is : $\tilde{C} = 1,800 + 250 \cdot p$. How many orders should be placed ? How many trips are necessary ? What is the total cost ?

$D = 400; \quad x = 5; \quad c_i = 60; \quad C_f = 1,800; \quad c_v = 250;$

$$p_1 = \frac{400 \cdot 250 + \sqrt{400^2 \cdot 250^2 + 2 \cdot 60 \cdot 5^2 \cdot 1,800 \cdot 400}}{60 \cdot 5^2} \approx 140.18;$$

$$p_2 = \sqrt{\frac{2 \cdot 1,800 \cdot 400}{60 \cdot 5^2}} \approx 30.98; \quad p^{(1)} = 141; \quad p^{(2)} = 30; \quad p^{(3)} = 31;$$

$$C^{(1)} = \sqrt{2 \cdot 60 \cdot 400 \cdot (1,800 + 250 \cdot 141)} \approx \$ 42,171 ;$$

$$C^{(2)} = \frac{1,800 \cdot 400}{5 \cdot 30} + \frac{250 \cdot 400}{5} + \frac{60 \cdot 5 \cdot 30}{2} = \$ 29,300;$$

$$C^{(3)} = \frac{1,800 \cdot 400}{5 \cdot 31} + \frac{250 \cdot 400}{5} + \frac{60 \cdot 5 \cdot 31}{2} \approx \$ 29,295.16 ;$$

$\bar{C}_{tot} = C^{(s)} = C^{(3)}; \quad \bar{x} = 5 \cdot 31 = 155; \quad \bar{b} = 400/155 \approx 2.58 ;$

as $\bar{b} \notin \mathbb{N}$ it is necessary to round off:

(1) $\bar{b} = 2; \rightarrow \bar{x} = 200; p = 40;$ the total cost then is $ 29,600;

(2) $\bar{b} = 3; \rightarrow \bar{x}_1 = 135; \bar{x}_2 = 135; \bar{x}_3 = 130; \sum_{i=1}^{3} \bar{x}_i = 400;$
$p_1 = 27; p_2 = 27; p_3 = 26; p = \sum_{i=1}^{3} p_i = 80;$
the total cost then is $\approx$ $ 29,450.

12. Sequencing Models

12.0.1 JOHNSON's Algorithm for Two Machines

Hypotheses
Given a matrix $D : D_{[m \times 2]}$, where

d_{ij}: working time of order A_i on machine M_j ;

$d_{ij} \geq 0 \; \forall \; i=1,\ldots,m; \; j=1,2$;

under the scheduling rule that each order is placed on M_1 first and than on M_2, determine a planning schedule, so that the total duration for all orders is minimal.

Principle
In the algorithm a minimal element is determined in the matrix of "working time requirements" and the appropriate order is placed as close as possible at the beginning or at the end. Such orders are eliminated from the matrix and the next order is positioned in the same manner. The total time requirement is determined in a GANTT-diagram.

Description

Step 1: Define $S := \{d_{ij}\}$ and set the running-indices $p := 1$; $q := m$.

Step 2: Determine $d_{rs} := \min \{d_{ij} \mid d_{ij} \in S\}$.
Note: If more than one element d_{ij} is minimal, select one of them at random.

Step 3: Is $s = 1$?
If yes: Set $A_r^{(p)} := A_r$; $p := p + 1$. Go to step 4.

If no : Set $A_r^{(q)} := A_r$; q: = q - 1. Go to step 4.

Step 4: Define $S : S - \{d_{rj}\} \; \forall \; j=1,2$.

Step 5: Is $S = \emptyset$?
If yes: Go to step 6.
If no : Go to step 2.

Step 6: The sequence $(A_i^{(1)},\ldots,A_j^{(p)}, A_k^{(q)},\ldots,A_l^{(m)})$ is an optimal planning schedule. Determine (e.g. in a GANTT-diagram) the total duration for all orders.

Example
Given the following matrix D:

D :	M_1	M_2
A_1	2	4
A_2	7	5
A_3	6	4
A_4	9	8
A_5	8	2
A_6	7	1
A_7	3	9
A_8	5	1

$S = \{d_{ij}\} \; \forall \; i=1,\ldots,8; \; j=1,2;$
$p = 1; \; q = 8; \; d_{rs} = d_{82}; \; s = 2;$
$A_8^{(8)} = A_8; \; q = 7;$
$S = \{d_{ij}\} \; \forall \; i=1,\ldots,7; \; j=1,2;$
$d_{rs} = d_{62}; \; s = 2; \; A_6^{(7)} = A_6; \; q = 6;$
$S = \{d_{ij}\} \; \forall \; i=1,2,3,4,5,7; \quad j=1,2;$
$d_{rs} = d_{52}; \; s = 2; \; A_5^{(6)} = A_5; \; q = 5;$
$S = \{d_{ij}\} \; \forall \; i=1,2,3,4,7; \; j=1,2;$

$d_{rs} = d_{11}; \; s = 1; \; A_1^{(1)} = A_1; \; p = 2;$
$S = \{d_{ij}\} \; \forall \; i=2,3,4,7; \; j=1,2; \quad d_{rs} = d_{71}; \; s = 1; \; A_7^{(2)} = A_7; \; p = 3;$
$S = \{d_{ij}\} \; \forall \; i=2,3,4; \; j = 1,2; \quad d_{rs} = d_{32}; \; s = 2; \; A_3^{(5)} = A_3; \; q = 4;$
$S = \{d_{ij}\} \; \forall \; i=2,4; \; j = 1,2; \quad d_{rs} = d_{22}; \; s = 2; \; A_2^{(4)} = A_2; \; q = 3;$
$S = \{d_{41},d_{42}\} \; ; \; d_{rs} = d_{42}; \; s = 2; \; A_4^{(3)} = A_4; \; q = 2;$
$S \stackrel{!}{=} \emptyset \to$ the optimal sequence of the orders is :
$(A_1^{(1)}, A_7^{(2)}, A_4^{(3)}, A_2^{(4)}, A_3^{(5)}, A_5^{(6)}, A_6^{(7)}, A_8^{(8)});$

the corresponding GANTT-diagram :

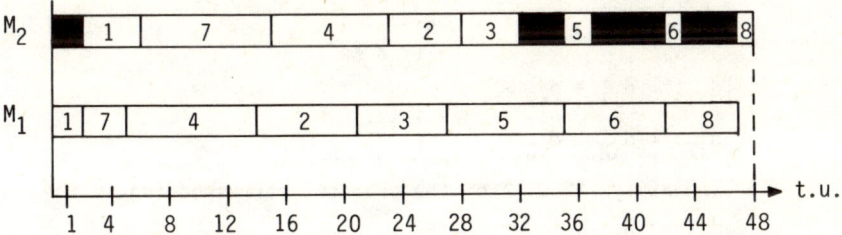

the total duration for all orders is 48 t.u., machine M_2 has an idle-time of total 14 t.u.

12.0.2 JOHNSON's Algorithm for Three Machines (special case)

Hypotheses

Given a matrix $D : D_{[m \times 3]}$, where

d_{ij}: working time of order A_i on machine M_j;
    $d_{ij} \geq 0 \quad \forall\ i=1,\ldots,m;\ j=1,2,3$;

each order requires all machines, the sequence is given with M_1, M_2, M_3. Determine a planning schedule, so that the total duration for all orders is minimal.

Principle

If the maximum time requirement on the second machine is less than or equal to the minimum time requirements on the first and third machines, then the time requirements of the first and second as well as those of the second and third machine are combined and the problem is handled as in 12.0.1 .

Description

Step 1: Is $(\max_i\{d_{i2}\} \leq \min_i\{d_{i1}\}) \wedge (\max_i\{d_{i2}\} \leq \min_i\{d_{i3}\})$?

If yes: Go to step 2.
If no : Stop, the algorithm can not determine an optimal

346   *Sequencing Models*

solution for this problem.

Step 2: Set up the matrix $\tilde{D} : \tilde{D}_{[m \times 2]}$, so that

$$\tilde{d}_{i1} := d_{i1} + d_{i2}$$
$$\tilde{d}_{i2} := d_{i2} + d_{i3}$$
$\forall\ i=1,\ldots,m$

and apply (12.0.1) on the problem, corresponding to the matrix $\tilde{D}$.

Result: An optimal sequence of orders $(A_i^{(1)},\ldots,A_i^{(m)})$; the total duration for all orders is determined in a GANTT-diagram, using the data from matrix D.

Example

Given the following matrix D:

D:	M_1	M_2	M_3
A_1	4	2	5
A_2	6	4	5
A_3	6	4	6
A_4	7	2	4
A_5	5	3	6
A_6	5	1	5

$\tilde{D}$:	$\tilde{M}_1$	$\tilde{M}_2$
A_1	6	7
A_2	10	9
A_3	10	10
A_4	9	6
A_5	8	9
A_6	6	6

(12.0.1) applied on the matrix $\tilde{D}$ yields the following optimal sequence of the orders: $(A_1^{(1)}, A_6^{(2)}, A_5^{(3)}, A_3^{(4)}, A_2^{(5)}, A_4^{(6)})$; the corresponding GANTT-diagram :

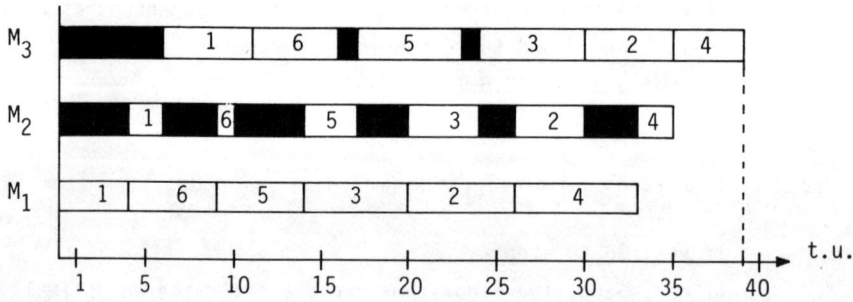

the total duration for all orders is 39 t.u., machine M_2 has an idle-time of total 19.t.u, M_3 has an idle-time of total 8 t.u.

12.0.3 A Heuristic Solution for a Sequencing Problem

Hypotheses
Given the matrix $D : D_{[m \times n]}$, where
d_{ij}: working time of order A_i on machine M_j;
$\quad d_{ij} \geq 0 \quad \forall\ i=1,\ldots,m;\ j=1,\ldots,n;\ n \geq 3;$
each order requires all machines, the sequence is given with M_1, $M_2,\ldots,M_n$. Determine a planning schedule, so that the total duration for all orders is as near to the minimum time as possible.

Principle
In this method the following rule is obeyed:
The order with the shortest working time requirement (on machine M_1) comes first, the one with the second shortest comes second etc.
If this rule does not lead to a unique sequence, then the time requirement on the next machine determines which order has priority.

Description
Step 1: Define $S := \{d_{ij}\}$; $Q := S$; set the running-index $p := 1$.

Step 2: Set the running-index $s := 1$.

Step 3: Determine $d_{rs} := \min\ \{d_{is} | (d_{is} \in S) \wedge (d_{is} \in Q)\}$.

Step 4: $\exists\ (d_{is} = d_{rs} | i \neq r)$?
      If yes: Go to step 5.
      If no : Go to step 7.

Step 5: Define $Q := Q - \{d_{kj} | d_{ks} > d_{rs}\}\ \forall\ j=1,\ldots,n$
      and set $s := s + 1$.

Step 6: Is $s \leq n$?

348 *Sequencing Models*

>If yes: Go to step 3.
>If no : Go to step 8.

Step 7: Set $A_r^{(p)} := A_r$; go to step 9.

Step 8: Set $A_r^{(p)} := A_r$, where $r := \{i | d_{is} \in Q\}$.

Step 9: Define $S := S - \{d_{rj}\}$ $\forall$ $j=1,\ldots,n$; $Q := S$.

Step 10: Is $S = \emptyset$?
>If yes: Go to step 11.
>If no : Set $p := p + 1$; go to step 2.

Step 11: The sequence $(A_i^{(1)},\ldots,A_i^{(m)})$ approximates the optimal schedule for the orders. Determine the total duration for all orders in a GANTT-diagram.

Note: If at some point two or more orders are waiting for the same machine, chose as next the order which has the shorter working time for that machine. To be exact, solve the remaining subproblem with one of the preceding algorithms.

Example

Given the following matrix D:

D:	M_1	M_2	M_3	M_4
A_1	6	6	4	4
A_2	1	3	5	1
A_3	3	3	2	6
A_4	2	4	6	3
A_5	3	1	2	6

$S = Q = \{d_{ij}\}$ $\forall$ $i=1,\ldots,5$; $j=1,\ldots,4$; $p = 1$; $s = 1$;

$d_{rs} = d_{21}$; $A_2^{(1)} = A_2$; $S = Q = \{d_{ij}\}$ $\forall$ $i=1,3,4,5$;

$j=1,\ldots,4$; $p = 2$; $d_{rs} = d_{41}$; $A_4^{(2)} = A_4$;

$S = Q = \{d_{ij}\}$ $\forall$ $i=1,3,5$; $j = 1,\ldots,4$; $p = 3$;

$d_{rs} = d_{31} = d_{51}$; $Q = \{d_{ij}\}$ $\forall$ i=3,5; j=1,...,4; s = 2;
$d_{rs} = d_{52}$; $A_5^{(3)} = A_5$; $S = Q = \{d_{ij}\}$ $\forall$ i=1,3; j=1,...,4; p = 4;
s = 1; $d_{rs} = d_{31}$; $A_3^{(4)} = A_3$; $S = Q = \{d_{1j}\}$ $\forall$ j=1,...,4; p = 5;
$d_{rs} = d_{11}$; $A_1^{(5)} = A_1$; $S \stackrel{!}{=} \emptyset$ → an approximate optimal sequence of orders is $(A_2^{(1)}, A_4^{(2)}, A_5^{(3)}, A_3^{(4)}, A_1^{(5)})$;

the corresponding GANTT-diagram:

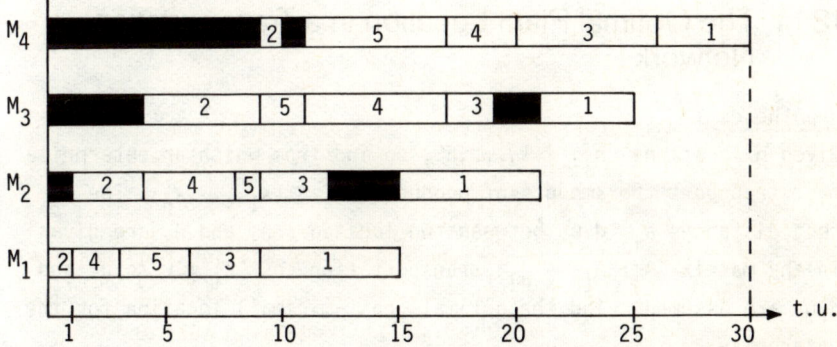

the total duration for all orders is 30 t.u., the following idle-times occur:
M_2 : 4 t.u.; M_3 : 6 t.u.; M_4 : 10 t.u. .

13. Plant Location Models

13.1 Exact Methods

13.1.1 The Optimal Plant Location in a Transportation Network I

Hypotheses

Given n locations n_j, $j=1,\ldots,n$, to and from which an enterprise must transport the amounts of products $x^T = (x_1,\ldots,x_n)$. The direct distances a_{ij} d.u. between two locations n_i and n_j are given in the matrix $A : A_{[n \times n]}$, transportation cost c_t m.u./q.u. and d.u. are assumed. Find the optimal (cost-minimal) location for the enterprise.

Principle

The cost for each possible station site is determined. The place with minimal cost is selected as the optimal station site.

Description

Step 1: Determine the total distance matrix $E : E_{[n \times n]}$ from the direct distance matrix A, using one of the algorithms (3.1.5; 3.1.6; 3.1.7).

Step 2: Calculate the vector $D : D_{[n \times 1]}$, so that
$D: = c_t \cdot E \cdot x$ and determine
$d_k: = \min_i \{d_i \in D\}$.

Result: The location n_k is optimal for the enterprise, the total transportation cost is d_k m.u.

Example

Given the following transportation network

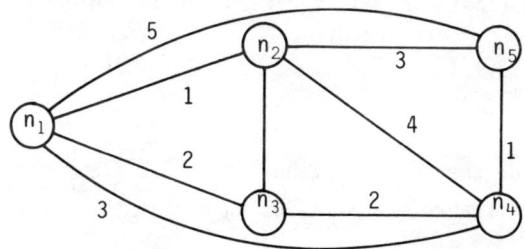

the quantity to be delivered is $x^T = (50; 75; 30; 100; 25)$; $c_t = 2$;

$$A = \begin{pmatrix} 0 & 1 & 2 & 3 & 5 \\ 1 & 0 & 2 & 4 & 3 \\ 2 & 2 & 0 & 2 & \infty \\ 3 & 4 & 2 & 0 & 1 \\ 5 & 3 & \infty & 1 & 0 \end{pmatrix} ; \quad E = \begin{pmatrix} 0 & 1 & 2 & 3 & 4 \\ 1 & 0 & 2 & 4 & 3 \\ 2 & 2 & 0 & 2 & 3 \\ 3 & 4 & 2 & 0 & 1 \\ 4 & 3 & 3 & 1 & 0 \end{pmatrix} ; \quad D = \begin{pmatrix} 1,070 \\ 1,170 \\ 1,050 \\ 1,070 \\ 1,230 \end{pmatrix} \leftarrow \min$$

$d_k = d_3 \rightarrow$ the location n_3 is optimal with a total transportation cost of 1,050 m.u.

13.1.2 The Optimal Plant Location in a Transportation Network II

Hypotheses

Given n locations n_j, $j=1,\ldots,n$, to and from which an enterprise must transport products. The direct distances a_{ij} d.u. between two locations n_i and n_j are given in the matrix $A : A_{[n \times n]}$.

Let $S : S_{[n \times p]}$ be the matrix of the sales of the products, so that

s_{ij}: number of the j-th product delivered to location n_i;

$F : F_{[n \times q]}$ be the matrix of the procurement of the components, so that

f_{ij}: number of the j-th component picked up from location n_i.

352 Plant Location Models

Transportation cost $c_t(s)$ m.u./q.u. and d.u. for the products and $c_t(f)$ m.u./q.u. and d.u. for the components are assumed.
Find the optimal (cost-minimal) location for the enterprise.

Principle See 13.1.1

Description

Step 1: Determine the total distance matrix $E : E_{[n \times n]}$ from the direct distance matrix A, using one of the algorithms (3.1.5; 3.1.6; 3.1.7).

Step 2: Calculate the vector $D : D_{[n \times 1]}$, so that

$D := E \cdot (S \cdot c_t(s) + F \cdot c_t(f))$ and determine

$d_k := \min_i \{d_i \in D\}$.

Result: The location n_k is optimal for the enterprise, the total transportation cost is d_k m.u.

Example
Given the following transportation network

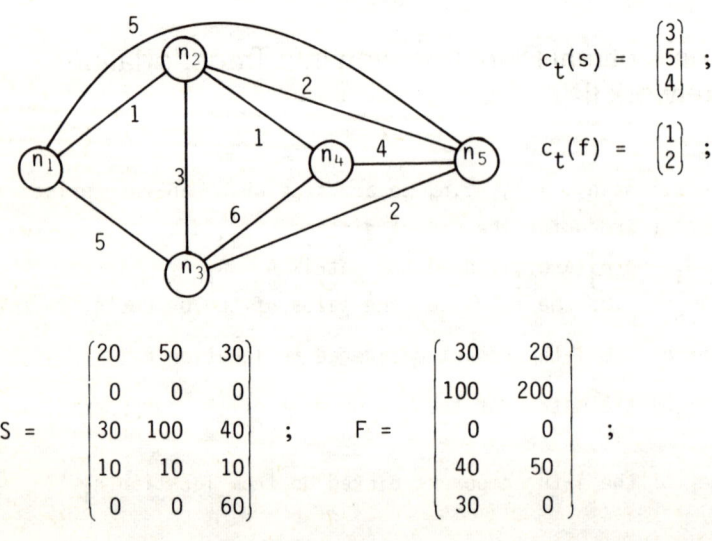

$$c_t(s) = \begin{pmatrix} 3 \\ 5 \\ 4 \end{pmatrix} ;$$

$$c_t(f) = \begin{pmatrix} 1 \\ 2 \end{pmatrix} ;$$

$$S = \begin{pmatrix} 20 & 50 & 30 \\ 0 & 0 & 0 \\ 30 & 100 & 40 \\ 10 & 10 & 10 \\ 0 & 0 & 60 \end{pmatrix} ; \quad F = \begin{pmatrix} 30 & 20 \\ 100 & 200 \\ 0 & 0 \\ 40 & 50 \\ 30 & 0 \end{pmatrix} ;$$

$$A = \begin{pmatrix} 0 & 1 & 5 & \infty & 5 \\ 1 & 0 & 3 & 1 & 2 \\ 5 & 3 & 0 & 6 & 2 \\ \infty & 1 & 6 & 0 & 4 \\ 5 & 2 & 2 & 4 & 0 \end{pmatrix} ; E = \begin{pmatrix} 0 & 1 & 4 & 2 & 3 \\ 1 & 0 & 3 & 1 & 2 \\ 4 & 3 & 0 & 4 & 2 \\ 2 & 1 & 4 & 0 & 3 \\ 3 & 2 & 2 & 3 & 0 \end{pmatrix} ; D = \begin{pmatrix} 4,830 \\ 3,550 \\ 5,080 \\ 5,310 \\ 4,780 \end{pmatrix} \leftarrow \min$$

$d_k = d_2 \rightarrow$ the location n_2 is optimal with a total transportation cost of 3,550 m.u.

13.1.3 The Optimal Plant Location on a Straight Line

Hypotheses

Given n stations n_j, j=1,...,n, lying approximately on a straight line; suppose these are the locations of the supplies and markets of an enterprise. The amount of the sale or procurement is given in the vector $x^T = (x_1,...,x_n)$; transportation cost c_t m.u./q.u. and d.u. are assumed. The distance between two neighbouring locations n_i and n_j is given by a_{ij} d.u., the locations are ordered with increasing indices. Find the optimal (cost-minimal) location for the enterprise.

Procedure

Determine $k := \min \{j | \sum_j x_j \geq T/2\}$, where $T := \sum_{j=1}^{n} x_j$.

Result: The location n_k is optimal. The distances between the stations have no influence on the solution.

The total transportation cost is $C_{tot} := c_t \cdot \sum_{j=1}^{n} a_{jk} \cdot x_j$ m.u.

Example

Given five locations, in which the supplies/ markets of an enterprise are located. The enterprise demands/supplies the following quantity in each location: $x^T(50; 30; 20; 60; 40)$.
Let the distances between the stations be :
$a_{12} = 15$; $a_{23} = 20$; $a_{34} = 10$; $a_{45} = 15$; and $c_t = 2.5$.

What is the optimal location for the enterprise ?

$T = \sum_{j=1}^{5} x_j = 200$; by $k = 3$ (or, when beginning from the other side, by $k = 4$) the half of the total quantity to be transported is exceeded; here two optimal locations exist, n_3 and n_4; the total transportation cost is

$C_{tot}(n_3) = 2.5 \cdot (35 \cdot 50 + 20 \cdot 30 + 0 \cdot 20 + 10 \cdot 60 + 25 \cdot 40) = 9,875$ m.u.
$C_{tot}(n_4) = 2.5 \cdot (45 \cdot 50 + 30 \cdot 30 + 10 \cdot 20 + 0 \cdot 60 + 15 \cdot 40) = 9,875$ m.u.

13.1.4 The Optimal Plant Location with Respect to Rectangular Transportation Movements

Hypotheses
Given n locations n_k, $k=1,\ldots,n$, in which the suppliers and customers of an enterprise are located. The locations are identified by their co-ordinates $(y_{1k}; y_{2k})$; to each of these locations corresponds a known quantity x_k of sales/procurement. Find the optimal location $(y_1^*; y_2^*)$ for the enterprise under the assumption that transportation movements are only possible parallel to the two co-ordinate axes. Transportation cost c_t m.u./q.u. and d.u. are assumed.

Principle
The method determines an optimum with respect to each co-ordinate. The point of intersection of both optima yields the optimal station site.

Description
Step 1: Determine the total quantity to be transported $T: = \sum_{k=1}^{n} x_k$.

Step 2: (changing the indices)
Assign the index i to the y_1-co-ordinate (abscissa) and the index j to the y_2-co-ordinate (ordinate);
the location n_k with the p-smallest y_1-value receives the index $i = p$, $\forall p = 1,\ldots,n$;

the location n_k with the q-smallest y_2-value receives the index $j = q$, $\forall\ q=1,\ldots,n$.

Result: The location n_k is now represented by n_{ij}, the corresponding transportation amount is now x_{ij}.

Step 3: Determine the index r, so that
$$r := \min \{i \mid \sum_i x_{ij} \geq T/2\}\ .$$

Step 4: Determine the index s, so that
$$s := \min \{j \mid \sum_j x_{ij} \geq T/2\}\ .$$

Result: The location n_{rs} with the co-ordinates $(y_1^*;y_2^*) = (y_{1r};y_{2s})$ is the optimal station site; the total transportation cost is
$$C_{tot} := c_t \cdot \sum_{k=1}^{n} (|y_1^* - y_{1k}| + |y_2^* - y_{2k}|) \cdot x_k \quad \text{m.u.}$$

Example

Given the following five locations n_k with the co-ordinates $(y_{1k};y_{2k})$ and the transportation quantity x_k:

n_k	y_{1k}	y_{2k}	x_k
n_1	2	6	80
n_2	5	5	20
n_3	8	3	40
n_4	3	2	50
n_5	6	9	60

$$c_t = 1;\quad T = \sum_{k=1}^{5} x_k = 250;$$

356 Plant Location Models

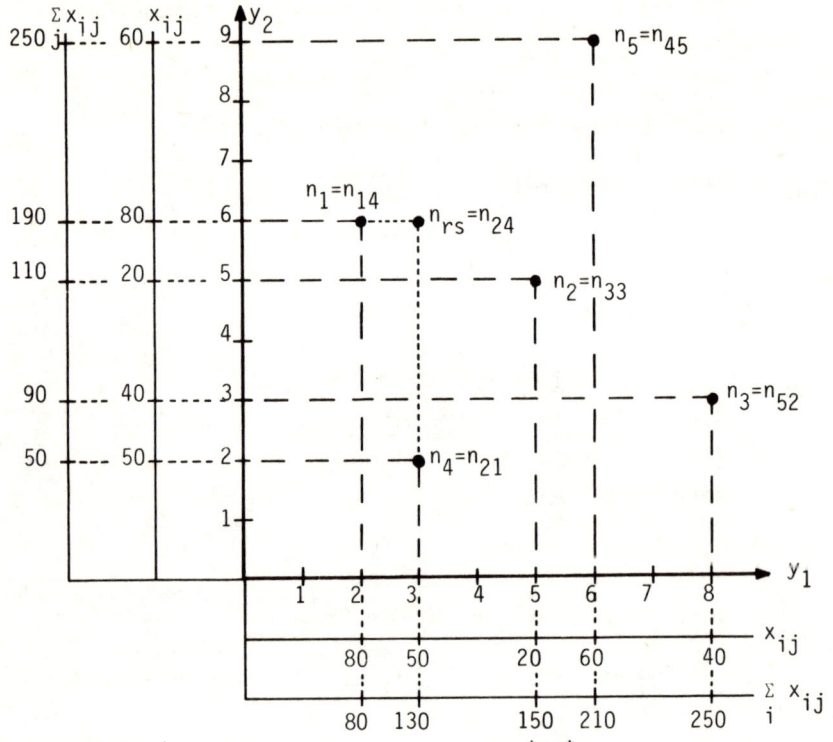

$r = 2;\ s = 4;\ n_{24}$ with the co-ordinates $(y_1^*;y_2^*) = (3;6)$ is the optimal location; the total transportation cost is

$$C_{tot} = [(|3-2|+|6-6|)80+(|3-5|+|6-5|)20+(|13-8|+|6-3|)40 \\ + (|3-3|+|6-2|)50+(|3-6|+|6-9|)60] = 1{,}020 \text{ m.u.}$$

13.2 Heuristic Methods

13.2.1 The Center of Gravity-Method

<u>Hypotheses</u>
Given n locations n_j, $j=1,\ldots,n$, in which the suppliers and customers of an enterprise are located. The locations are identified by their co-ordinates $(y_{1j};y_{2j})$; to each of these locations corresponds

a known quantity x_j of sales/procurement; a direct connection exists between all locations. Assuming transportation cost c_t m.u./q.u. and d.u., determine an optimal location $(y_1^*; y_2^*)$ for the enterprise.

Principle
This method determines the "center of gravity" of the n-sided polygon.

Procedure
Determine

$$y_1^* = \frac{\sum_{j=1}^{n} x_j^k \cdot y_{1j}}{\sum_{j=1}^{n} x_j^k} \quad ; \quad y_2^* = \frac{\sum_{j=1}^{n} x_j^k \cdot y_{2j}}{\sum_{j=1}^{n} x_j^k}, \text{ where } k \in \mathbb{N};$$

the total transportation cost is

$$C_{tot} = c_t \cdot \sum_{j=1}^{n} x_j \cdot \sqrt{(y_1^* - y_{1j})^2 + (y_2^* - y_{2j})^2} \quad \text{m.u.}$$

Example
Given the following five locations n_j with the co-ordinates $(y_{1j}; y_{2j})$ and the transportation quantity x_j:

n_j	y_{1j}	y_{2j}	x_j
n_1	2	6	80
n_2	5	5	20
n_3	8	3	40
n_4	3	2	50
n_5	6	9	60

$c_t = 2.5;$

(α) $k = 2$;

$$y_1^* = \frac{80^2 \cdot 2 + 20^2 \cdot 5 + 40^2 \cdot 8 + 50^2 \cdot 3 + 60^2 \cdot 6}{80^2 + 20^2 + 40^2 + 50^2 + 60^2} \approx 3.9103 \ ;$$

$$y_2^* = \frac{80^2 \cdot 6 + 20^2 \cdot 5 + 40^2 \cdot 3 + 50^2 \cdot 2 + 60^2 \cdot 9}{80^2 + 20^2 + 40^2 + 50^2 + 60^2} \approx 5.6966;$$

the location with the co-ordinates $(y_1^*; y_2^*) = (3.9103; 5.6966)$ is optimal;

$C_{tot} = 2.5 \cdot (80 \cdot 1.92 + 20 \cdot 1.3 + 40 \cdot 4.91 + 50 \cdot 3.81 + 60 \cdot 4.79) =$
$= 2,135.55$ m.u.;

(β) $k = 1$;

$$y_1^* = \frac{80 \cdot 2 + 20 \cdot 5 + 40 \cdot 8 + 50 \cdot 3 + 60 \cdot 6}{80 + 20 + 40 + 50 + 60} = 4.36 ;$$

$$y_2^* = \frac{80 \cdot 6 + 20 \cdot 5 + 40 \cdot 3 + 50 \cdot 2 + 60 \cdot 9}{80 + 20 + 40 + 50 + 60} = 5.36 ;$$

the location with the co-ordinates $(y_1^*; y_2^*) = (4.36; 5.36)$ is optimal;

$C_{tot} = 2.5 \cdot (80 \cdot 2.47 + 20 \cdot 0.72 + 40 \cdot 6.1 + 50 \cdot 3.61 + 60 \cdot 3.94) =$
$= 2,183.05$ m.u.

13.2.2 A Solution by Vector Summation

Hypotheses

See 13.2.1 . In addition, a (small) value ε is given, which denotes the tolerance distance for the solution location from the optimal location.

Principle

The method determines a sequence of resultants (whose length is monotone decreasing), in whose direction the current suboptimal solution is shifted. If the length of the resultant becomes less than or equal to the given value ε , the current suboptimum is considered as "almost optimal".

Description

Step 1: Assume the convex hull of the locations has been constructed through a certain connection of the locations. Select any arbitrary point $S = (y_1^*; y_2^*)$ inside of the convex hull for the initial solution.

Note: If there are any of the given locations inside the convex hull, it is convenient to select that location (inside of the convex hull) with the greatest transportation quantity.

Step 2: Determine $\tan \alpha^{(j)} := \dfrac{y_{2j} - y_2^*}{y_{1j} - y_1^*} \quad \forall\; j = 1, \ldots, n$

and define $\delta_j := \begin{cases} +1, & \text{if } y_{1j} \geq y_1^* \\ -1, & \text{if } y_{1j} < y_1^* \end{cases} \quad \forall\; j = 1, \ldots, n$

Step 3: Determine the points $S_j = (\tilde{y}_{1j}; \tilde{y}_{2j})$, so that

$$\tilde{y}_{1j} := y_1^* + \delta_j \cdot \sqrt{\dfrac{\lambda^2 \cdot x_j^2}{1 + (\tan \alpha^{(j)})^2}} \quad \forall\; j = 1, \ldots, n,$$

$$\tilde{y}_{2j} := y_2^* + \tan \alpha^{(j)} \cdot (\tilde{y}_{1j} - y_1^*)$$

where $\lambda \in (0; 1]$.

Step 4: Determine the point $P = (\hat{y}_1; \hat{y}_2)$, so that

$$\hat{y}_1 := \tilde{y}_{11} + \sum_{j=2}^{n} (\tilde{y}_{1j} - y_1^*) \;;\; \hat{y}_2 := \tilde{y}_{21} + \sum_{j=2}^{n} (\tilde{y}_{2j} - y_2^*).$$

Step 5: Is $\gamma := \sqrt{(y_1^* - \hat{y}_1)^2 + (y_2^* - \hat{y}_2)^2} \leq \varepsilon$?

If yes: Stop, $S = (y_1^*; y_2^*)$ is optimal with respect to the given value ε; the total transportation cost is

$$C_{tot} := c_t \cdot \sum_{j=1}^{n} x_j \cdot \sqrt{(y_1^* - y_{1j})^2 + (y_2^* - y_{2j})^2} \quad \text{m.u.}$$

If no : Go to step 6.

Step 6: Determine $\tan \beta := \dfrac{\hat{y}_2 - y_2^*}{\hat{y}_1 - y_1^*}$

and define $\delta := \begin{cases} +1, & \text{if } \hat{y}_1 \geq y_1^* \\ -1, & \text{if } \hat{y}_1 < y_1^* \end{cases}$.

Step 7: Determine the point $S' = (y_1'; y_2')$, so that

$$y_1' := y_1^* + \delta \cdot \sqrt{\dfrac{1/16 \cdot \gamma^2}{1 + (\tan \beta)^2}} \; ;$$

$$y_2' := y_2^* + \tan \beta \cdot (y_1' - y_1^*),$$

and the new initial solution $S = (y_1^*; y_2^*)$, where $y_1^* := y_1'$; $y_2^* := y_2' + 0.1$. Go to step 2.

Note 1: Instead of completing step 6 and step 7, a new initial solution can be determined by selecting any point between S and P.

Note 2: The vector summation can also be done graphically. This procedure is convenient, when solving smaller problems by hand.

Example

Given the following five locations n_j with the co-ordinates $(y_{1j}; y_{2j})$ and the transportation quantity x_j:

n_j	y_{1j}	y_{2j}	x_j
n_1	2	6	80
n_2	5	5	20
n_3	8	3	40
n_4	3	2	50
n_5	6	9	60

$c_t = 2.5$;
$\varepsilon = 0.3$;
$\lambda = 0.1$;

$S = n_2 = (5; 5)$;

$\tan \alpha^{(1)} = -1/3$; $\delta_1 = -1$; $S_1 = (-2.59; 7.53)$;
$\tan \alpha^{(2)} = 0$; $\delta_2 = 1$; $S_2 = (5; 5)$;
$\tan \alpha^{(3)} = -2/3$; $\delta_3 = 1$; $S_3 = (8.33; 2.78)$;
$\tan \alpha^{(4)} = 3/2$; $\delta_4 = -1$; $S_4 = (2.23; 0.84)$;
$\tan \alpha^{(5)} = 4$; $\delta_5 = 1$; $S_5 = (6.46; 10.82)$;

$P = (-0.57; 6.97)$; $\gamma = 5.908 > \epsilon$; $\tan \beta = -0.358$; $\delta = -1$;
$S' = (3.61; 5.5)$; $S = (3.61; 5.6)$;

$\tan \alpha^{(1)} = -0.248$; $\delta_1 = -1$; $S_1 = (-4.154; 7.525)$;
$\tan \alpha^{(2)} = -0.432$; $\delta_2 = 1$; $S_2 = (5.446; 4.807)$;
$\tan \alpha^{(3)} = -0.592$; $\delta_3 = 1$; $S_3 = (7.052; 3.563)$;
$\tan \alpha^{(4)} = 5.92$; $\delta_4 = -1$; $S_4 = (2.775; 0.655)$;
$\tan \alpha^{(5)} = 1.423$; $\delta_5 = 1$; $S_5 = (7.06; 10.509)$;

$P = (3.739; 4.659)$; $\gamma = 0.95 > \epsilon$; $\tan \beta = -7.295$; $\delta = 1$;
$S' = (3.64; 5.36)$; $S = (3.64; 5.46)$;

$\tan \alpha^{(1)} = -0.329$; $\delta_1 = -1$; $S_1 = (-3.959; 7.96)$;
$\tan \alpha^{(2)} = -0.338$; $\delta_2 = 1$; $S_2 = (5.535; 4.82)$;
$\tan \alpha^{(3)} = -0.564$; $\delta_3 = 1$; $S_3 = (7.124; 3.495)$;
$\tan \alpha^{(4)} = 5.406$; $\delta_4 = -1$; $S_4 = (2.731; 0.544)$;
$\tan \alpha^{(5)} = 1.5$; $\delta_5 = 1$; $S_5 = (6.968; 10.452)$;

$P = (3.839; 5.431)$; $\gamma = 0.201 \overset{!}{<} 0.3 = \epsilon \rightarrow$ Stop, $S = (3.64; 5.46)$ is an optimal location for $\epsilon = 0.3$;

$C_{tot} = 2.5 \cdot (80 \cdot 1.73 + 20 \cdot 1.44 + 40 \cdot 5.01 + 50 \cdot 3.52 + 60 \cdot 4.25) =$
$\phantom{C_{tot}} = 1{,}996.5$ m.u.

362 *Plant Location Models*

Graphical solution: (Note: 10 q.u. are equivalent to 1 d.u.)

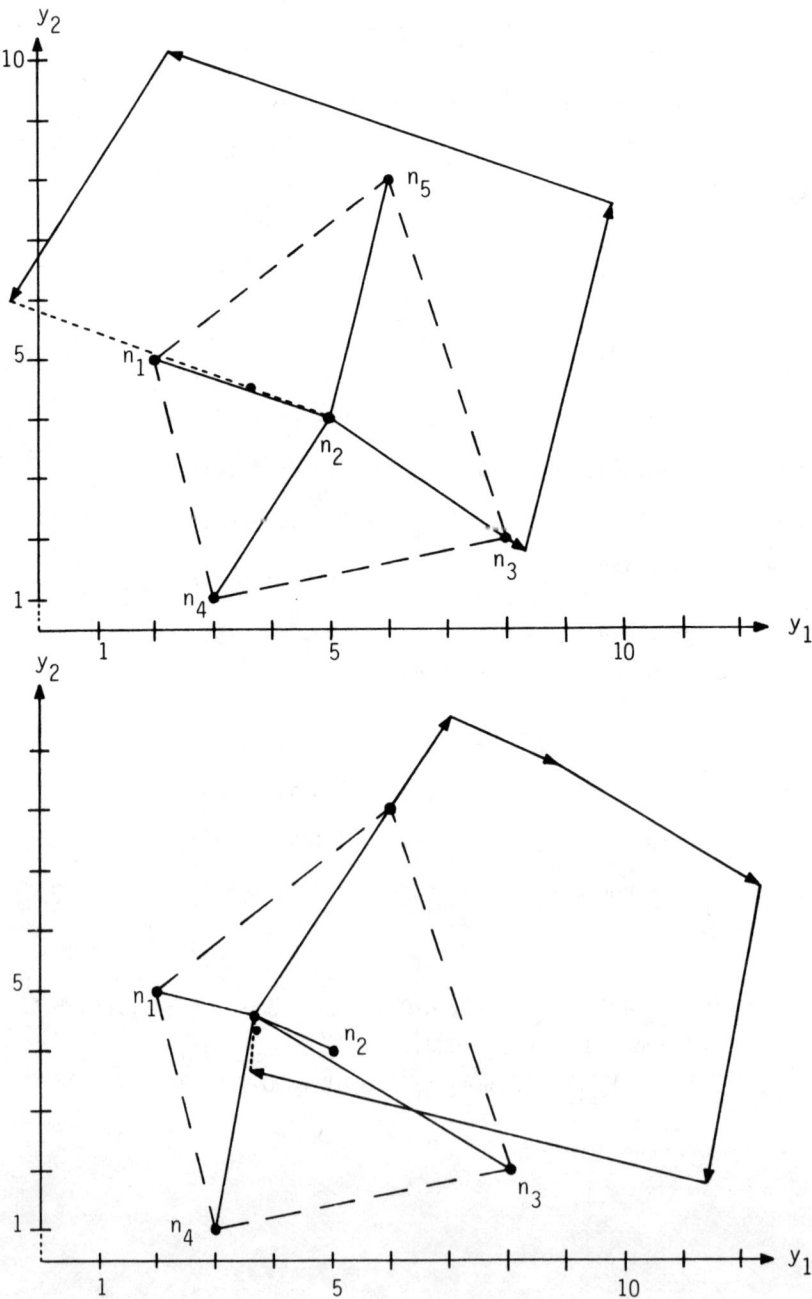

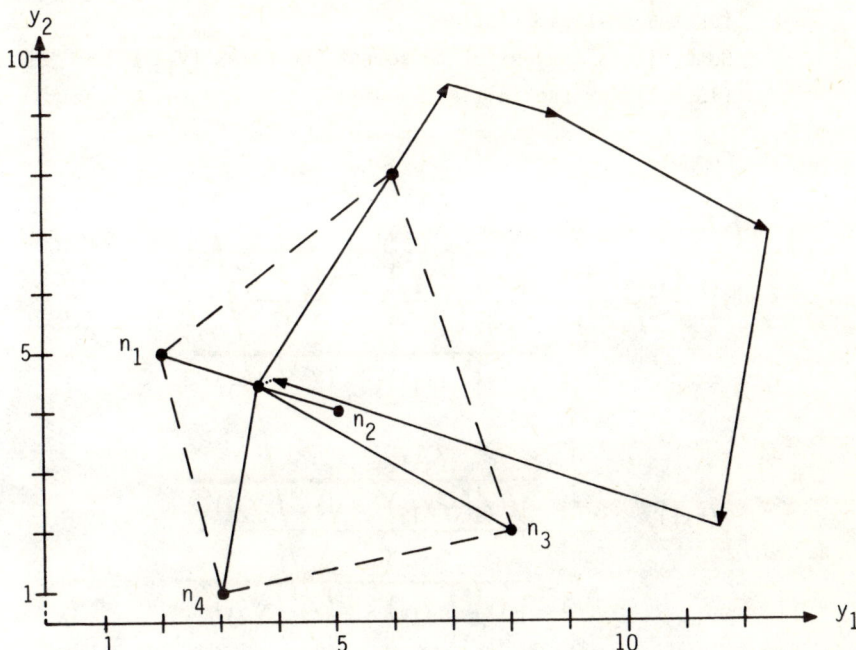

13.2.3 An Iterative Method

Hypotheses
See 13.2.1 . In addition, a (small) value ε is given, which denotes the minimal shift of a solution $P^{(i-1)}$ to $P^{(i)}$. Set the running-index $k := 0$.

Principle
The method determines a sequence of points in the plane, until the value of the straight line between two consecutive points becomes less than or equal to the given value ε. The last generated co-ordinates are considered as "almost optimal".

Description
Step 1: Select an arbitrary point $P^{(0)} = (y_1^{(0)}; y_2^{(0)})$

364 *Plant Location Models*

for the initial solution.
Note: It is convenient to select the point $(y_1^*; y_2^*)$ of (13.2.1) for the initial solution.

Step 2: Compute

$$y_1^{(k-1)} := \frac{\sum_{j=1}^{n} \frac{x_j \cdot y_{1j}}{\sqrt{(y_1^{(k)}-y_{1j})^2 + (y_2^{(k)}-y_{2j})^2}}}{\sum_{j=1}^{n} \frac{x_j}{\sqrt{(y_1^{(k)}-y_{1j})^2 + (y_2^{(k)}-y_{2j})^2}}}$$

$$y_2^{(k+1)} := \frac{\sum_{j=1}^{n} \frac{x_j \cdot y_{2j}}{\sqrt{(y_1^{(k)}-y_{1j})^2 + (y_2^{(k)}-y_{2j})^2}}}{\sum_{j=1}^{n} \frac{x_j}{\sqrt{(y_1^{(k)}-y_{1j})^2 + (y_2^{(k)}-y_{2j})^2}}}$$

Step 3: Is $\gamma_{k+1} := \sqrt{(y_1^{(k+1)} - y_1^{(k)})^2 + (y_2^{(k+1)} - y_2^{(k)})^2} \leq \varepsilon$?

If yes: Stop, the point $P^{(k+1)} = (y_1^{(k+1)}; y_2^{(k+1)})$ is optimal with respect to the given value ε. The total transportation cost is

$$C_{tot} := c_t \cdot \sum_{j=1}^{n} x_j \cdot \sqrt{(y_1^{(k+1)}-y_{1j})^2+(y_2^{(k+1)}-y_{2j})^2} \; m.u.$$

If no : Set $k := k + 1$; go to step 2.

Example

Given the following five locations n_j with the co-ordinates $(y_{1j}; y_{2j})$ and the transportation quantity x_j:

n_j	y_{1j}	y_{2j}	x_j
n_1	2	6	80
n_2	5	5	20
n_3	8	3	40
n_4	3	2	50
n_5	6	9	60

$c_t = 2.5$;

$\varepsilon = 1/20$;

$P^{(o)} = (y_1^{(o)}; y_2^{(o)}) = (4; 3)$;

$$y_1^{(1)} = \frac{\dfrac{80 \cdot 2}{\sqrt{(4-2)^2+(3-6)^2}} + \dfrac{20 \cdot 5}{\sqrt{(4-5)^2+(3-5)^2}} + \dfrac{40 \cdot 8}{\sqrt{(4-8)^2+(3-3)^2}} +}{\dfrac{80}{\sqrt{(4-2)^2+(3-6)^2}} + \dfrac{20}{\sqrt{(4-5)^2+(3-5)^2}} + \dfrac{40}{\sqrt{(4-8)^2+(3-3)^2}} +}$$

$$\frac{\dfrac{50 \cdot 3}{\sqrt{(4-3)^2+(3-2)^2}} + \dfrac{60 \cdot 6}{\sqrt{(4-6)^2+(3-9)^2}}}{\dfrac{50}{\sqrt{(4-3)^2+(3-2)^2}} + \dfrac{60}{\sqrt{(4-6)^2+(3-9)^2}}} = 4.03$$

$$y_2^{(1)} = \frac{\dfrac{80 \cdot 6}{\sqrt{(4-2)^2+(3-6)^2}} + \dfrac{20 \cdot 5}{\sqrt{(4-5)^2+(3-5)^2}} + \dfrac{40 \cdot 3}{\sqrt{(4-8)^2+(3-3)^2}} +}{\dfrac{80}{\sqrt{(4-2)^2+(3-6)^2}} + \dfrac{20}{\sqrt{(4-5)^2+(3-5)^2}} + \dfrac{40}{\sqrt{(4-8)^2+(3-3)^2}} +}$$

$$\frac{\dfrac{50 \cdot 2}{\sqrt{(4-3)^2+(3-2)^2}} + \dfrac{60 \cdot 9}{\sqrt{(4-6)^2+(3-9)^2}}}{\dfrac{50}{\sqrt{(4-3)^2+(3-2)^2}} + \dfrac{60}{\sqrt{(4-6)^2+(3-9)^2}}} = 4.6$$

to simplify collect the results in a table as follows:

$p^{(k)}$	$y_1^{(k)}$	$y_2^{(k)}$	γ_k
$p^{(0)}$	4	3	–
$p^{(1)}$	4.03	4.6	$\approx 8/5$
$p^{(2)}$	3.98	5.11	$\approx 1/2$
$p^{(3)}$	3.93	5.3	$\approx 1/5$
$p^{(4)}$	3.87	5.37	$\approx 9/100$
$p^{(5)}$	3.83	5.4	$1/20$

$\gamma_5 = \varepsilon = 1/20 \rightarrow$ Stop, $p^{(5)} = (3.83; 5.4)$ is an optimal location for $\varepsilon = 1/20$;

$C_{tot} = 2.5 \cdot (80 \cdot 1.92 + 20 \cdot 1.24 + 4 \cdot 4.81 + 50 \cdot 3.5 + 60 \cdot 4.2) = 1,994.5$ m.u.

Appendix

Table 1: $q^{(k)} = (1 + i)^k$

i \ k	1	2	3	4	5	6	7	8	9	10
0.01	0.9901	0.9803	0.9706	0.9610	0.9515	0.9421	0.9327	0.9234	0.9143	0.9053
0.02	0.9804	0.9612	0.9423	0.9239	0.9057	0.8879	0.8705	0.8535	0.8368	0.8203
0.03	0.9709	0.9426	0.9152	0.8885	0.8626	0.8375	0.8131	0.7894	0.7664	0.7441
0.04	0.9615	0.9246	0.8890	0.8548	0.8219	0.7903	0.7599	0.7307	0.7026	0.6756
0.05	0.9526	0.9070	0.8639	0.8227	0.7835	0.7462	0.7107	0.6768	0.6446	0.6139
0.06	0.9434	0.8900	0.8396	0.7921	0.7473	0.7050	0.6651	0.6274	0.5919	0.5584
0.07	0.9346	0.8734	0.8163	0.7629	0.7130	0.6664	0.6227	0.5820	0.5439	0.5083
0.08	0.9259	0.8573	0.7938	0.7350	0.6806	0.6302	0.5835	0.5403	0.5003	0.4632
0.09	0.9174	0.8417	0.7722	0.7084	0.6499	0.5963	0.5470	0.5019	0.4604	0.4224
0.10	0.9091	0.8264	0.7513	0.6830	0.6209	0.5645	0.5132	0.4665	0.4241	0.3855
0.11	0.9009	0.8116	0.7312	0.6587	0.5934	0.5346	0.4816	0.4339	0.3909	0.3522
0.12	0.8929	0.7972	0.7118	0.6355	0.5658	0.5066	0.4523	0.4039	0.3606	0.3220
0.13	0.8850	0.7831	0.6930	0.6133	0.5428	0.4803	0.4251	0.3762	0.3329	0.2946
0.14	0.8772	0.7695	0.6750	0.5921	0.5194	0.4556	0.3996	0.3506	0.3075	0.2697
0.15	0.8696	0.7561	0.6575	0.5718	0.4972	0.4323	0.3759	0.3269	0.2843	0.2472

Table 2: $q^{-k} = (1 + i)^{-k}$

i \ k	1	2	3	4	5	6	7	8	9	10
0.01	1.0100	1.0201	1.0303	1.0406	1.0510	1.0615	1.0721	1.0829	1.0937	1.1046
0.02	1.0200	1.0404	1.0612	1.0824	1.1041	1.1262	1.1487	1.1717	1.1951	1.2190
0.03	1.0300	1.0609	1.0927	1.1255	1.1593	1.1941	1.2299	1.2668	1.3048	1.3439
0.04	1.0400	1.0816	1.1249	1.1699	1.2167	1.2653	1.3159	1.3686	1.4233	1.4802
0.05	1.0500	1,1025	1.1576	1.2155	1.2763	1.3401	1.4071	1.4775	1.5513	1.6289
0.06	1.0600	1.1236	1.1910	1.2625	1.3382	1.4185	1.5036	1.5938	1.6895	1.7908
0.07	1.0700	1.1449	1.2250	1.3108	1.4026	1.5007	1.6058	1.7182	1.8385	1.9672
0.08	1.0800	1.1664	1.2597	1.3605	1.4693	1.5869	1.7138	1.8509	1.9990	2.1589
0.09	1.0900	1.1881	1.2950	1.4116	1.5386	1.6771	1.8280	1.9926	2.1719	2.3674
0.10	1.1000	1.2100	1.3310	1.4641	1.6105	1.7716	1.9487	2.1436	2.3579	2.5937
0.11	1.1100	1.2321	1.3676	1.5181	1.6851	1.8704	2.0762	2.3045	2.5580	2.8394
0.12	1.1200	1.2544	1.4049	1.5735	1.7673	1.9738	2.2107	2.4760	2.7731	3.1058
0.13	1.1300	1.2769	1.4429	1.6305	1.8424	2.0820	2.3526	2.6584	3.0040	3.3946
0.14	1.1400	1.2996	1.4815	1.6890	1.9254	2.1950	2.5023	2.8526	3.2519	3.7072
0.15	1.1500	1.3225	1.5209	1.7490	2.0114	2.3131	2.6600	3.0591	3.5179	4.0456

Table 3: e^{-k}

k	e^{-k}	k	e^{-k}	k	e^{-k}	k	e^{-k}
.01	.9900	.26	.7711	.51	.6005	.76	.4677
.02	.9802	.27	.7634	.52	.5945	.77	.4630
.03	.9704	.28	.7558	.53	.5886	.78	.4584
.04	.9608	.29	.7483	.54	.5827	.79	.4538
.05	.9512	.30	.7408	.55	.5769	.80	.4493
.06	.9418	.31	.7334	.56	.5712	.81	.4449
.07	.9324	.32	.7261	.57	.5655	.82	.4404
.08	.9231	.33	.7189	.58	.5599	.83	.4360
.09	.9139	.34	.7118	.59	.5543	.84	.4317
.10	.9048	.35	.7047	.60	.5488	.85	.4274
.11	.8958	.36	.6977	.61	.5434	.86	.4232
.12	.8869	.37	.6907	.62	.5379	.87	.4190
.13	.8781	.38	.6839	.63	.5326	.88	.4148
.14	.8694	.39	.6771	.64	.5273	.89	.4107
.15	.8607	.40	.6703	.65	.5220	.90	.4066
.16	.8521	.41	.6637	.66	.5169	.91	.4025
.17	.8437	.42	.6570	.67	.5117	.92	.3985
.18	.8353	.43	.6505	.68	.5066	.93	.3946
.19	.8270	.44	.6440	.69	.5016	.94	.3906
.20	.8187	.45	.6376	.70	.4966	.95	.3867
.21	.8106	.46	.6313	.71	.4916	.96	.3829
.22	.8025	.47	.6250	.72	.4868	.97	.3791
.23	.7945	.48	.6188	.73	.4819	.98	.3753
.24	.7866	.49	.6126	.74	.4771	.99	.3716
.25	.7788	.50	.6065	.75	.4724	1.00	.3679

k	e^{-k}	k	e^{-k}
1.1	.3329	3.1	.04505
1.2	.3012	3.2	.04076
1.3	.2725	3.3	.03688
1.4	.2426	3.4	.03337
1.5	.2231	3.5	.03020
1.6	.2019	3.6	.02732
1.7	.1827	3.7	.02472
1.8	.1653	3.8	.02237
1.9	.1496	3.9	.02024
2.0	.1353	4.0	.01832
2.1	.1225	4.1	.01657
2.2	.1108	4.2	.01500
2.3	.1003	4.3	.01357
2.4	.09072	4.4	.01228
2.5	.08208	4.5	.01111
2.6	.07427	4.6	.01005
2.7	.06721	4.7	.009095
2.8	.06081	4.8	.008230
2.9	.05502	4.9	.007447
3.0	.04979	5.0	.006738

Table 4 : Random numbers with an equal distribution

2049	9135	6601	5112	5266	6728	2188	3846	3734	4017
7087	2825	8667	8831	1617	7239	9622	1622	0409	5822
6187	0189	5748	0380	8820	3606	7316	4297	2160	8973
4178	3758	0191	5361	9605	9605	1526	8370	4969	0463
3844	4187	2145	8365	5964	0501	9196	4456	6573	9751
4932	7312	3579	7867	4000	1056	8830	1662	0869	2580
7032	7117	8283	0767	1074	5571	8464	8057	1698	7343
9058	0455	2463	1812	7991	6371	6881	5540	2774	2163
8476	5802	9722	0325	1001	1337	2042	9695	5242	8808
5749	8807	5926	7406	0537	2700	4704	8674	8016	9034
4829	5817	7136	8337	9009	9304	4010	5817	7126	9175
0634	5168	4614	4769	3246	7559	8221	8030	6603	4735
2829	7794	9246	6385	4330	3750	1255	2157	9415	4161
6851	0071	7014	9558	9956	2151	9359	6002	6458	8926
4262	8457	9217	8257	5462	2705	9735	2004	2800	6913
8660	7450	8436	9606	9996	9287	0972	4065	4140	9403
1962	4877	2789	9528	4654	2356	6478	7438	6292	1122
3100	1200	2648	0649	2406	4507	4102	2498	7771	6545
4658	6444	6375	6053	4427	1425	1938	9075	2627	4471
9335	9644	5599	6565	8249	6899	8688	7347	9876	0079
5688	9609	0763	1792	4400	3490	8251	6512	1296	5359
9981	6570	9188	3064	5262	7601	4111	7835	1855	4210
2041	9584	5237	8034	5720	4701	9016	8033	5941	4710
7470	7973	3232	5128	8315	7261	5645	1775	1428	1486

Appendix 371

9053	1349	4317	1495	9279	0474	5898	7622	0344	2412
3358	7831	2562	1115	1940	8754	4528	0335	0755	3294
3768	2114	2675	4256	9672	9264	3236	5791	8289	0402

7082	0519	5926	7306	2429	0199	3925	7661	6604	4570
7176	1045	9291	1734	5984	3088	0943	9323	7545	8128
8817	4047	7333	7390	2280	7320	5015	7812	6053	4372

8572	8448	9060	0079	5633	0388	9623	1694	6614	2802
7245	8673	9770	8346	9333	9368	4390	5368	8324	6634
0787	2616	6460	9258	4275	9127	7982	4834	4933	7102

5476	8770	7390	2335	2677	4597	7797	8760	5522	0374
7715	3563	4950	3707	8933	3102	1587	7336	7943	2301
3454	5165	5122	7100	5089	1244	5316	2230	3731	4669

5173	2842	5529	0841	7762	4943	5279	4453	6010	7884
6982	3868	0176	8023	7819	4782	5676	7465	8792	7513
0130	3536	0034	6191	0704	7602	3990	2271	8877	6844

1198	3035	9335	9699	4403	3048	8234	1416	3706	9143
4999	4950	4053	6294	0680	3117	4294	2768	1003	1568
3922	9964	3487	8903	6533	5209	2952	5523	0274	9608

0974	3689	7763	5119	6602	4891	5275	5181	2128	5327
4153	8232	9981	9184	2291	5232	6985	4320	2048	9300
3392	6048	5311	1391	8125	9314	5933	6146	7525	2079

3621	5593	7559	8211	6141	8419	3933	7992	6591	6890
7087	2714	8663	8057	1587	7347	9831	0485	7876	3919
9456	8382	2860	2270	9033	5050	5825	5589	8277	1817

Table 5 : Area under the standardized normal distribution function

$$\Phi(x) = \int_{-\infty}^{x} \frac{1}{\sqrt{2\cdot\pi}} \cdot e^{-\frac{x^2}{2}} \quad ; \quad \Phi(-x) = 1 - \Phi(x)$$

x	0	1	2	3	4	5	6	7	8	9
0.0	.5000	.5040	.5080	.5120	.5160	.5199	.5239	.5279	.5319	.5359
0.1	.5398	.5438	.5478	.5517	.5557	.5596	.5636	.5675	.5714	.5754
0.2	.5793	.5832	.5871	.5910	.5948	.5987	.6026	.6064	.6103	.6141
0.3	.6179	.6217	.6255	.6293	.6331	.6368	.6406	.6443	.6480	.6517
0.4	.6554	.6591	.6628	.6664.	6700	.6736	.6772	.6808	.6844	.6879
0.5	.6915	.6950	.6985	.7019	.7054	.7088	.7123	.7157	.7190	.7224
0.6	.7258	.7291	.7324	.7357	.7389	.7422	.7454	.7486	.7518	.7549
0.7	.7580	.7612	.7642	.7673	.7704	.7734	.7764	.7794	.7823	.7852
0.8	.7881	.7910	.7939	.7967	.7996	.8023	.8051	.8078	.8106	.8133
0.9	.8159	.8186	.8212	.8238	.8264	.8289	.8315	.8340	.8365	.8389
1.0	.8413	.8438	.8461	.8485	.8508	.8531	.8554	.8577	.8599	.8621
1.1	.8643	.8665	.8686	.8708	.8729	.8749	.8770	.8790	.8810	.8830
1.2	.8849	.8869	.8888	.8907	.8925	.8944	.8962	.8980	.8997	.9015
1.3	.9032	.9049	.9066	.9082	.9099	.9115	.9131.	9147	.9162	.9177
1.4	.9192	.9207	.9222	.9236	.9251	.9265	.9279	.9292	.9306	.9319
1.5	.9332	.9345	.9357	.9370	.9382	.9394	.9406	.9418	.9429	.9441
1.6	.9452	.9463	.9474	.9484	.9495	.9505	.9515	.9525	.9535	.9545
1.7	.9554	.9564	.9573	.9582	.9591	.9599	.9608	.9618	.9625	.9633
1.8	.9146	.9649	.9656	.9664	.9671	.9678	.9686	.9693	.9699	.9706
1.9	.9713	.9719	.9726	.9732	.9738	.9744	.9750	.9756	.9761	.9767
2.0	.9772	.9778	.9783	.9788	.9793	.9798	.9803	.9808	.9812	.9817
2.1	.9821	.9826	.9830	.9834	.9838	.9842	.9846	.9850	.9854	.9857
2.2	.9861	.9864	.9868	.9871	.9875	.9878	.9881	.9884	.9887	.9890
2.3	.9893	.9896	.9898	.9901	.9904	.9906	.9909	.9911	.9913	.9916.
2.4	.9918	.9920	.9922	.9925	.9927	.9929	.9931	.9932	.9934	.9936
2.5	.9938	.9940	.9941	.9943	.9945	.9946	.9948	.9949	.9951	.9952
2.6	.9953	.9955	.9956	.9957	.9959	.9960	.9961	.9962	.9963	.9964
2.7	.9965	.9966	.9967	.9968	.9969	.9970	.9971	.9972	.9973	.9974
2.8	.9974	.9975	.9976	.9977	.9977	.9978	.9979	.9979	.9980	.9981
2.9	.9981	.9982	.9982	.9983	.9984	.9984	.9985	.9985	.9986	.9986

Bibliography

A tripel [a,b,c] shall precede each reference in the bibliography, so that:
- a: chapter
- b: B, if it is a book
  A, if it is an article or other special research
- c: running-index in alphabetical order according to author.

The books and articles cited in the general bibliography encompass the broad aspects of Operations Research and as such are not listed according to the chapters of this handbook.

The most important journals and their abbreviations are:

CACM	: Communications of the Association for Computing Machinery
JACM	: Journal of the Association for Computing Machinery
MS	: Management Science
NRLQ	: Naval Research Logistics Quarterly
ORQ	: Operational Research Quarterly
ORSA	: Journal of the Operations Research Society of America
Op.Res.Verf.	: Operations Research Verfahren
SIAM	: Journal of the Society for Industrial applied Mathematics
Ufo	: Unternehmensforschung
ZOR	: Zeitschrift für Operations Research

Chapter 0

[0.B.1.] Ackoff, R.L., Sasieni, M.W.; Operations Research, Verlag Kunst und Wissen, Stuttgart 1970

[0.B.2.] Angermann, A.; Entscheidungsmodelle, F. Nowack Verlag, Frankfurt/Main, 1963

[0.B.3.] Brockhoff, K.; Unternehmensforschung, W. de Gruyter, Berlin - New York 1973

[O.B.4.] Churchman, C.W.; Ackoff, R.L.; Arnoff, E.L.; Operations Research, R. Oldenbourg Verlag, Wien - München 1961

[O.B.5.] Desbazeille, G.; Unternehmensforschung, Berliner Union/ Kohlhammer, Stuttgart 1970

[O.B.6.] Dück, W.; Bliefernich, M.; (eds.) Operationsforschung, Deutscher Verlag der Wissenschaften, Berlin 1972

[O.B.7.] Henn, R.; Künzi, H.P.; Einführung in die Unternehmensforschung, Band I und II, Heidelberger Taschenbücher, Bde. 38 und 39, Springer Verlag, Berlin - Heidelberg - New York 1968

[O.B.8.] Kaufmann, A.; Faure, R.; Methoden des Operations Research, W. de Gruyter, Berlin - New York 1974

[O.B.9.] Körth, H.; Otto, C.; Runge, W.; Schoch, M.; (Hrsg.), Lehrbuch der Mathematik für Wirtschaftswissenschaften, Westdeutscher Verlag Opladen 1973

[O.B.10] Kulhavy, E.; Operations Research, Betriebswirtschaftlicher Verlag Dr. Gabler, Wiesbaden 1963

[O.B.11] Kwak, N.K.; Mathematical Programming with Business Applications, Mc Graw Hill, New York 1973

[O.B.12] Mc Millan, Jr. C.; Mathematical Programming, J. Wiley & Sons, New York 1970

[O.B.13.] Mitchell, G.H.; Operational Research: Techniques and Examples, The English Universities Press Ltd., London 1972

[O.B.14.] Müller-Merbach, H.; Operations Research, Verlag F. Vahlen, München 1973

[O.B.15.] Neumann, K.; Operations Research A - F (working papers), Karlsruhe 1972

[O.B.16.] Sasieni, M.W.; Yaspan, A.; Friedman, L.; Operations Research: Methods and Problems, J. Wiley & Sons Inc., New York 1959

[O.B.17.] Schneeweiß, H.; Entscheidungskriterien bei Risiko, in: Ökonometrie und Unternehmensforschung Band VI, Springer Verlag, Berlin - Heidelberg - New York 1967

[O.B.18.] Stahlknecht, P.; Operations Research, in: Schriften zur Datenverarbeitung, Band 3, F. Vieweg & Sohn, Braunschweig 1970
[O.B.19.] Theil, H.; Boot, J.C.G.; Kloek, T.; Operations Research and Quantitative Economics, Mc Graw - Hill, New York 1965
[O.B.20.] Weber, H.H.; Einführung ins Operations Research, Akademische Verlagsgesellschaft, Frankfurt/Main 1972
[O.B.21.] Autorenkollektiv, Mathematische Standardmodelle der Operationsforschung, Verlag Die Wirtschaft, Berlin 1972
[O.A.1.] Stahlknecht, P.; OR - Ein Leitfaden für Praktiker, Teil 1: Input - Output - Modelle/Optimierungsverfahren, in: Elektronische Datenverarbeitung, Beiheft 6, F. Vieweg & Sohn, Braunschweig 1965
[O.A.2.] Stahlknecht,P.; OR - Ein Leitfaden für Praktiker, Teil 2: Simulationsmethoden/Ablauf - und Terminplanung, in: Elektronische Datenverarbeitung, Beiheft 7, F. Vieweg & Sohn, Braunschweig 1966

Chapter 1

[1.B.1.] Arrow, K.J.; Hurwicz, L.; Uzawa, H.; Studies in Linear and Nonlinear Programming, Stanford University Press, Stanford (Calif.) 1958
[1.B.2.] Berge, C.; Ghouila-Houri, A.; Programming, Games and Transportation Networks, J. Wiley & Sons Inc. New York 1965
[1.B.3.] Bloech, J.; Lineare Optimierung für Wirtschaftswissenschaftler, Westdeutscher Verlag Opladen 1974
[1.B.4.] Charnes, A; Cooper, W.W.; Henderson, A.; An Introduction to Linear Programming, J. Wiley & Sons, New York 1953
[1.B.5.] Collatz, L.; Wetterling, W.; Optimierungsaufgaben, Heidelberger Taschenbücher, Band 15, Springer Verlag, Berlin - Heidelberg - New York 1966
[1.B.6.] Cooper,L.; Steinberg, D.; Introduction to Methods of Optimization, W.B. Saunders Co., Philadelphia-London-Toronto 1970

[1.B.7.] Dantzig, G.B.; Linear Programming and Extensions, Princetown University Press, Princetown N.J. 1962

[1.B.8.] Dinkelbach, W.; Sensitivitätsanalysen und parametrische Programmierung, in: Ökonometrie und Unternehmensforschung, Band XII, Springer Verlag, Berlin - Heidelber - New York 1969

[1.B.9.] Ferguson, R.D.; Sargent, L.F.; Linear Programming: Fundamentals and Applications, Mc Graw - Hill Comp., New York 1958

[1.B.10.] Gal, T.; Betriebliche Entscheidungen und parametrische Programmierung, W. de Gruyter, Berlin - New York 1972

[1.B.11.] Gale, D.; The Theory of Linear Economic Models, Mc Graw - Hill Book Comp., New York - Toronto - London, 1960

[1.B.12.] Gass, S.I.; Linear Programming, Mc Graw - Hill Book Comp., New York 1969

[1.B.13.] Glicksman, A.M.; An Introduction to Linear Programming and the Theory of Games, J. Wiley & Sons Inc., New York - London 1963

[1.B.14.] Hadley, G.; Linear Programming, Addison Wesley Publ.Comp., Reasing (Mass.) 1969

[1.B.15.] Judin, D.B.; Golstein, E.G.; Lineare Optimierung I, Akademie - Verlag, Berlin 1968

[1.B.16.] Judin, D.B.; Golstein, E.G.; Lineare Optimierung II, Akademie - Verlag, Berlin 1970

[1.B.17.] Karlin, S.; Mathematical Methods and Theory of Games, in: Programming and Economics, Vol. II: The Theory of Infinite Games, Addison Wesley, Reading (Mass.) 1959

[1.B.18.] Kaufmann, A.; Methods and Models of Operations Research, Prentice Hall Inc., Englewood Cliffs N.J. 1963

[1.B.19.] Krekó, B.; Lehrbuch der linearen Optimierung, Deutscher Verlag der Wissenschaften, Berlin 1964

[1.B.20.] Kromphard, W.; Henn, R.; Förstner, K.; Lineare Entscheidungsmodelle, Springer Verlag, Berlin - Göttingen - Heidelberg 1962

[1.B.21.] Kuhn, H.W.; Tucker, A.W.; (eds.), Linear Inequalities and Related Systems, Annals of Mathematics Study No. 38, Princetown University Press, Princetown N.J. 1956

[1.B.22.] Luenberger, D.G.; Introduction to Linear and Nonlinear Programming, Addison Wesley Publ. Comp., Reading (Mass.) 1973

[1.B.23.] Naylor, H.; Byrne, E.T.; Linear Programming, Wadsworth Publ. Comp. Inc., Belmont (Calif.) 1963

[1.B.24.] Piehler, J.; Einführung in die lineare Optimierung, Verlag Harri Deutsch, Zürich - Frankfurt/Main 1969

[1.B.25.] Saaty, T.L.; Mathematical Methods of Operations Research, Mc Graw - Hill Book Comp., New York 1959

[1.B.26.] Sakarovitch, M.; Notes on Linear Programming, Van Nostrand Reinhold Comp., New York 1972

[1.B.27.] Simonnard, M.; Linear Programming, Prentice Hall Inc., Englewood Cliffs N.J. 1966

[1.B.28.] Vajda, S.; Mathematical Programming, Addison Wesley Publ. Comp., Reading (Mass.) 1961

[1.B.29.] Vajda, S.; The Theory of Games and Linear Programming, Methuen & Comp. Ltd. and Science Paperbacks, London 1970

[1.B.30.] Vazsonyi, A.; Scientific Programming in Business and Industry, J. Wiley & Sons, New York 1958

[1.B.31.] Zimmermann, H.-J.; Zielinski, J.; Lineare Programmierung, W. de Gruyter, Berlin - New York 1971

[1.A.1.] Beale, E.M.L.; "An Alternate Method of Linear Programming" Proc. Cambridge Phil. Soc. 50/4; 1954

[1.A.2.] Beale, E.M.L.; "Cycling in the Dual Simplex Algorithm", in: NRLQ 2; 1955

[1.A.3.] Dantzig, G.B.; Ford Jr., L.R.; Fulkerson, D.R.; "A Primal - Dual Algorithm for Linear Programs" in: [1.B.21.]

[1.A.4.] Dantzig, G.B.; Wolfe, P.; "Decomposition Principle for Linear Programs", ORSA 8/1, 1960

[1.A.5.] Eaves, B.C.; "The Linear Complementary Problem, in: MS 17/9, 1971
[1.A.6.] Frisch, R.; "The Multiplex Method for Linear Programming", in: Memorandum fra Universitetets Socialøkonomiske Institutt, Oslo 1955
[1.A.7.] Gal, T.; Nedoma, J.; "Methode zur Lösung mehrparametrischer linearer Programme", in: Op.Res.Verf. XII, 1972
[1.A.8.] Habr, J.; "Die Frequenzmethode zur Lösung des Transportproblems und verwandter linearer Programmierungsprobleme", in: Wissenschaftliche Zeitschrift der Universität Dresden 10/5, 1961
[1.A.9.] Hadley, G.F.; Simmonard, W.A.; "A Simplified two phase technique for the Simplex Method", in: NRLQ 6/3, 1959
[1.A.10.] Kelley, Jr. J.E.; "Parametric Programming and the Primal-Dual Algorithm", in: ORSA 7/3, 1959
[1.A.11.] Kuhn, H.W.; "The Hungarian Method for the Assignment Problem", in: NRLQ 2/1; 1955
[1.A.12.] Künzi, H.P.; "Die Duoplex - Methode", in: Ufo 7/3, 1963
[1.A.13.] Künzi, H.P.; Kleibohm, K.; "Das Triplex - Verfahren", in: Ufo 12/3, 1963
[1.A.14.] Lemke, C.E.; "The Dual Method of Solving the Linear Programming Problem", in: NRLQ 1/1, 1954
[1.A.15.] Orchard - Hayes, W.; "Evolution of Linear Programming Computing Techniques", in: MS 4/2, 1958
[1.A.16.] Saaty, T.; Gass, S.; "The Computational Algorithm for the parametric objective function", in: NRLQ 2/1 & 2, **1955**
[1.A.17.] Wagner, H.M.; "A Two - Phase Method for the Simplex - Tableau", in: ORSA 4/4, 1956

Chapter 2

[2.B.1.] Brauer, K.M.; Binäre Optimierung, C. Heymanns Verlag, Köln 1969
[2.B.2.] Burkard, R.E.; Methoden der Ganzzahligen Optimierung, Springer Verlag, Wien - New York 1972

[2.B.3.] Cooper, L.; Steinberg, D.; [1.B.6.]
[2.B.4.] Greenberg, H.; Integer Programming, in: Mathematics in Science and Engineering, vol. 76, Academic Press, New York - London 1971
[2.B.5.] Hu, T.C.; Integer Programming and Network Flows; Addison Wesley Publ. Co. Reading (Mass.), 1970
[2.B.6.] Korbut, A.A.; Finkelstein, J.J.; Diskrete Optimierung, Akademie - Verlag, Berlin 1971
[2.B.7.] Saaty, T.L.; [1.B.25.]
[2.B.8.] Saaty, T.L.; Optimization in Integers and Related Extremal Problems, Mc. Graw - Hill Book Comp., New York 1970

[2.A.1.] Balas, E.; "An Additive Algorithm for Solving Linear Programs with Zero - One Variables", in: ORSA 13/4, 1965
[2.A.2.] Balas, E.; "Discrete Programming by the Filter Method", in: ORSA 15/5, 1967
[2.A.3.] Benders, J.F.; "Partitioning Procedures for Solving Mixed - Variables Programming Problems", in: Numerische Mathematik 4, 1962
[2.A.4.] Ben - Isreal, A.; Charnes, A.; "On some Problems on Diophantine Programming", in: Cahiers Centre Etud. Rech. Opérat. 4, 1962
[2.A.5.] Dakin, R.J.; "A Tree Search Algorithm for Mixed Integer Programming Problems", in: The Computer Journal 8/3, 1965
[2.A.6.] Driebeek, N.J.; "An Algorithm for the Solution of Mixed Integer Programming Problems", in: MS 12/7, 1966
[2.A.7.] Glover, F.; "A Multiphase Dual Algorithm for the Zero - One Integer Programming Problem", in: ORSA 13/6, 1965
[2.A.8.] Glover, F.; "A New Foundation for a Simplified Primal Integer Programming Algorithm", in: ORSA 16/4; 1968
[2.A.9.] Gomory, R.E.; "An all - integer programming Algorithm", in: Industrial Scheduling, Englewood Cliffs, N.J., Prentice Hall 1963
[2.A.10.] Gomory, R.E.; "Solving Linear Programming Problems in Integers", in: Proc.Sympos.Appl.Math. vol. X, 1960

[2.A.11.] Gomory, R.E.; "An Algorithm for Integer Solutions to Linear Programs", in: Graves, R.L.; Wolfe, P.(Hrsg.); Recent Advances in Mathematical Programming, Mc Graw - Hill Book Comp., New York 1963

[2.A.12.] Kolesar, P.J.; "A Branch and Bound Algorithm for the Knapsack Problem", in: MS 13/9, 1967

[2.A.13.] Korte, B.; "Ganzzahlige Programmierung - Ein Überblick", in: Unternehmensforschung heute, Lecture Notes in Operations Research and Mathematical Systems Bd. 50, Springer Verlag, Berlin - Heidelberg - New York, 1971

[2.A.14.] Korte, B.; Krelle, W.; Oberhofer, W.; "Ein lexikographischer Suchalgorithmus zur Lösung allgemeiner ganzzahliger Programmierungsaufgaben", in: Ufo 13, 1969 (2 articles)

[2.A.15.] Land, A.H.; Doig, A.G.; "An Automatic Method of Solving Discrete Programming Problems", in: Econometrica 28/3, 1960

[2.A.16.] Lemke, C.E.; Spielberg, K.; "Direct Search Algorithm for Zero - One and Mixed - Integer Programming", in: ORSA 15/5, 1967

[2.A.17.] Noltemeier, H.; "Ganzzahlige Programme und kürzeste Wege in Graphen", in: Op. Res. Verf. XIV, 1972

[2.A.18.] Noltemeier, H.; "Sensitivitätsanalyse bei diskreten linearen Optimierungsproblemen", in: Lecture Notes in Operations Research and Mathematical Systems vol. 30; Springer Verlag, Berlin - Heidelberg - New York, 1970

[2.A.19.] Noltemeier, H.; "Zum asymptotischen Algorithmus", in: Op. Res. Verf. XVII, 1973

[2.A.20.] Picard, J.-C.; Ratliff, H.D.; "A Graph - Theoretic Equivalence for Integer Programs", in: ORSA 21/1; 1973

[2.A.21.] Young, R.D.; "A Simplified Primal (All Integer) Integer Programming Algorithm", in: ORSA 16/4, 1968

Chapter 3

[3.B.1.] Bellman, R.; Cooke, K.L.; Lockett, J.A.; Algorithms, Graphs and Computers;in: Mathematics in Science and Engineering, vol. 62, Academic Press, New York - London 1970

[3.B.2.] Berge, C.; Ghouila - Houri, A., [1.B.2.]

[3.B.3.] Berge, C.; The Theory of Graphs and its Applications, J. Wiley & Sons Inc.,New York 1962

[3.B.4.] Busacker, R.G.; Saaty, T.L.; Finite Graphs and Networks, Mc Graw Hill, New York 1965

[3.B.5.] Domschke, W.; Kürzeste Wege in Graphen: Algorithmen, Verfahrensvergleiche; Mathematical Systems in Economics, Bd. 2, Hain Verlag, Anton Hain, Meisenheim/Glan 1972

[3.B.6.] Ford, Jr. L.R.; Fulkerson, D.R.; Flows in Networks, Princetown University Press, Princetown N.J. 1962

[3.B.7.] Harary, F.; Graph Theory; Addison Wesley Publ. Comp., Reading (Mass.) 1972

[3.B.8.] Hu, T.C.; [2.B.5.]

[3.B.9.] Kaufmann, A.; Einführung in die Graphentheorie; in: Orientierung und Entscheidung; R. Oldenbourg, München - Wien 1971

[3.B.10.] Knödel, W.; Graphentheoretische Methoden und ihre Anwendungen, in: Ökonometrie und Unternehmensforschung, Band XIII, Springer Verlag; Berlin - Heidelberg - New York 1969

[3.B.11.] König, D.; Theorie der endlichen und unendlichen Graphen, New York, Chelsea Publ. Comp. 1950; (reprint; Leipzig, 1936)

[3.B.12.] Marshall, C.W.; Applied Graph Theory; J. Wiley & Sons Inc., New York 1971

[3.B.13.] Noltemeier, H.; Graphentheorie, W. de Gruyter; Berlin - New York 1975

[3.B.14.] Ore, O.; Graphs and their Uses; Random House; The L.W. Singer Comp., New York 1963

[3.A.1.] Breyer, F.; "Ein Algorithmus zur Konstruktion kostenminimaler Zirkulationen in Netzen", in: Op. Res. Verf. XV, 1973

[3.A.2.] Busacker, R.G.; Gowen, P.J.; "A Procedure for Determining a Family of Minimal - Cost Network Flow Patterns"; ORO Technical Report 15, Operations Research Office, Johns Hopkins University, 1961

[3.A.3.] Dantzig, G.B.; "On the Shortest Route Through a Network", in: MS 6/2, 1960

[3.A.4.] Dantzig, G.B.; Fulkerson, D.R.; "Computation of Maximal Flows in Networks", in: NRLQ 2/4, 1955

[3.A.5.] Dijkstra, E.W.; "A Note on Two Problems in Connection with Graphs", in: Numerische Mathematik 1, 1959

[3.A.6.] Domschke, W.; "Ein neues Verfahren zur Bestimmung kostenminimaler Flüsse in Kapazitätendigraphen", in: Op. Res. Verf. XVI, 1973

[3.A.7.] Edmonds, J.; Karp, R.M.; "Theoretical Improvements in Algoritmic Efficiency for Network Flow Problems", in: JACM 19/2, 1972

[3.A.8.] Elmaghraby, S.E.; "Some Network Models in Management Science", in: Lecture Notes in Operations Research and Mahtematical Systems, vol. 29, Springer Verlag, Berlin - Heidelberg - New York 1970

[3.A.9.] Farbey, B.A.; Land, A.H.; Murchland, J.D.; "The Cascade Algorithm for Finding all Shortest Distances in a Directed Graph", in: MS 14/1, 1967

[3.A.10.] Flood, M.M.; "The Traveling Salesman Problem", in: ORSA 4/1, 1956

[3.A.11.] Floyd, R.W.; "Algorithm 97: Shortest Path", in: CACM 5/6, 1962

[3.A.12.] Ford, Jr. L.R.; Fulkerson, D.R.; "A Simple Algorithm for Finding Maximal Network Flows and An Application to the Hitchcook Problem", in: Canadian Journal of Mathematics 9/2, 1957

[3.A.13.] Fulkerson, D.R.; "An out - of - kilter Method for Minimal Cost Flow Problems", in: SIAM 9/1, 1961
[3.A.14.] Hasse, M.; "über die Behandlung graphentheoretischer Probleme unter Verwendung der Matrizenrechnung", in: Wissenschaftliche Zeitschrift der Technischen Universität Dresden 10, 1961
[3.A.15.] Hässig, K.; "Theorie verallgemeinerter Flüsse und Potentiale", in: Op. Res. Verf. XXI, 1975
[3.A.16.] Hoffman, A.J.; "A Generalization of Max Flow - Min Cut", in: Math. Progr. 6/3, 1974
[3.A.17.] Klein, M.; "A Primal Method for Minimal Cost Flows with Applications to the Assignment and Transportation Problems", in: MS 14/3, 1967
[3.A.18.] Kruskal Jr., J.B.; "On the Shortest Spanning Tree of a Graph and the Traveling Salesman Problem", in: Proc. Am. Math. Soc. 7, 1956
[3.A.19.] Land, A.H.; Stairs, S.W.; "The Extension of the Cascade Algorithm to Larger Graphs", in: MS 14/1, 1967
[3.A.20.] Little, J.D.C.; Murty, K.G.; Sweeny, D.W.; Karel, C.; "An Algorithm for the Traveling Salesman Problem", in: ORSA 11/5, 1963
[3.A.21.] Noltemeier, H.; [2.A.17.]
[3.A.22.] Noltemeier, H.; [2.A.19.]
[3.A.23.] Noltemeier, H.; "The Distribution of Minimal Total Duration in Networks with high 'Arc Density'", in: Op. Res. Verf. V, 1968
[3.A.24.] Noltemeier, H.; "An Algorithm for the Determination of Longest Distances in a Graph", in: Math. Progr. 9, 1975
[3.A.25.] Picard, J.-C.; Ratliff, H.D.;[2.A.20.]
[3.A.26.] Steckhan, H.; "Güterströme in Netzen"; in: Lecture Notes in Operations Research and Mathematical Systems, vol.88, Springer Verlag, Berlin - Heidelberg - New York 1973

Chapter 4

[4.B.1.] Berbig, R.; Franke, R.; Netzplantechnik, Verlag für Bauwesen, Berlin 1973

[4.B.2.] Federal Electric Corporation; A Programmed Introduction to PERT, J. Wiley & Sons Inc., New York 1965

[4.B.3.] Gewald, K.; Kasper, K.; Schelle, H.; Netzplantechnik; Band 2: Kapazitätsoptimierung, R. Oldenbourg, München - Wien 1972

[4.B.4.] Götzke, H.; Netzplantechnik, Fachbuchverlag Leipzig, 1969

[4.B.5.] Küpper, W.; Lüder, K.; Streitferdt, L.; Netzplantechnik, Physica Verlag, Würzburg - Wien 1975

[4.B.6.] Miller, R.W.; PERT, R.v. Decker's Verlag G. Schenk, Hamburg - Berlin 1965

[4.B.7.] Riester, W.F.; Schwinn, R.; Projektplanungsmodelle, Physica Verlag, Würzburg - Wien 1970

[4.B.8.] Thumb, N.; Grundlagen und Praxis der Netzplantechnik, Verlag Moderne Industrie, München 1968

[4.B.9.] Voigt, J.-P.; Fünf Wege der Netzplantechnik, Verlagsgesellschaft R. Müller, Köln - Braunsfeld 1971

[4.B.10.] Wille, H.; Gewald, K.; Weber, H.D.; Netzplantechnik, Band I: Zeitplanung; R. Oldenbourg, München - Wien 1967

[4.B.11.] Zimmermann, H.J.; Netzplantechnik, Sammlung Göschen, Bd. 4011, W. de Gruyter, Berlin - New York 1971

[4.A.1.] Clingen, C.T.; "A Modification of Fulkerson's PERT Algorithm", in: ORSA 12/4, 1964

[4.A.2.] Kelley, Jr. J.E.; "Critical Path Planning and Scheduling: Mathematical Basis", in: ORSA 9/3, 1961

[4.A.3.] Lambourn, S.; "Ressource Allocation and Multiproject Scheduling (RAMPS) - A New Tool on Planning and Control", in: The Computer Journal 5, 1963

[4.A.4.] Martin, J.J.; "Distribution of the Time Through a Directed Acyclic Network", in: ORSA 13/1, 1965

[4.A.5.] Meyer, H.; "Stochastische Netzwerke vom Typ GERT", in: Op. Res. Verf. XVII, 1973
[4.A.6.] Neumann, K.; "Entscheidungsnetzpläne", in: Op. Res. Verf. XIII, 1972
[4.A.7.] Noltemeier, H.; "Netzplantechnik"; in: Management Lexikon, Gabler Verlag, Wiesbaden 1972
[4.A.8.] Pritsker, A.A.B.; "GERT Networks", in: The Production Engineer 1968
[4.A.9.] van Slyke, R.W.; "Monte Carlo Methods and the PERT Problem", in: ORSA 11/5, 1963
[4.A.10.] Welsh, D.J.A.; "Super - critical Arcs of PERT - Networks", in: ORSA 14/1, 1966

Chapter 5

[5.B.1.] Abt, C.C.; Serious Games, The Viking Press, New York 1970
[5.B.2.] Berge, C.; Ghouila - Houri, A.; [1.B.2.]
[5.B.3.] Burger, E.; Einführung in die Theorie der Spiele, W. de Gruyter, Berlin 1966
[5.B.4.] Collatz, L.; Wetterling, W., [1.B.5.]
[5.B.5.] Davis, M.D.; Game Theory, Basic Book Inc. New York - London 1970
[5.B.6.] Dresher, M.; Strategische Spiele, Verlag Industrielle Organisation, Zürich 1961
[5.B.7.] Glicksman, A.M.; [1.B.13.]
[5.B.8.] Karlin, S.; [1.B.17.]
[5.B.9.] Kromphard, W.; Henn, R.; Förstner, K.; [1.B.20.]
[5.B.10.] Luce, R.D.; Raiffa, H.; Games and Decisions, J. Wiley & Sons Inc., London - Sydney 1957
[5.B.11.] Mc Kinsey, J.; Introduction to the Theory of Games, Mc Graw - Hill, New York 1952
[5.B.12.] Morgenstern, O.; Spieltheorie und Wirtschaftswissenschaft, R. Oldenbourg, Wien - München 1963
[5.B.13.] v. Neumann, J.; Morgenstern, O.; Theory of Games and Economic Behaviour, Princetown University Press, Princetown N.J. 1953

[5.B.14.] Owen, G.; Game Theory, W.B. Saunders Comp., Philadelphia 1968
[5.B.15.] Vajda, S.; [1.B.29.]

[5.A.1.] Bohnenblust, H.F.; "The Theory of Games", in: Modern Mathematics for the Engineer (ed. Beckenbach) New York 1970
[5.A.2.] Brown, R.H.; "The Solution of a Certain Two - Person Zero-Sum Game", in: ORSA 5/1, 1957
[5.A.3.] Lemke, C.E.; "Bimatrix Equilibrium Points and Mathematical Programming", in: MS 11, 1965 (ser. A)
[5.A.4.] Nash, J.F.; "Equilibrium Points in N - Person Games", in: Proc. Nat. Ac. Sc. 36; 1950
[5.A.5.] Nash, J.F.; "Non - cooperative games"; in: Annals of Mathematics, 54, 1951
[5.A.6.] Neumann, K.; "Kooperative und nichtkooperative Zweipersonenspiele"; in: Op. Res. Verf. IX, 1971
[5.A.7.] Rosenmüller, J.; "On a Generalization of the Lemke - Howson - Algorithm to noncooperative n - Person - Games"; in: SIAM 21/1, 1971
[5.A.8.] Rosenmüller, J.; "Kooperative Spiele und Märkte", in: Lecture Notes in Economics and Mathematical Systems, vol. 53, 1971

Chapter 6

[6.B.1.] Beckmann, M.J.; Dynamic Programming of Economic Decisions, Springer Verlag; Berlin - Heidelberg - New York 1968
[6.B.2.] Bellman, R.; Dynamic Programming; Princetown University Press; Princetown N.J. 1962
[6.B.3.] Bellman, R.; Dreyfus, S.; Applied Dynamic Programming; Princetown 1962
[6.B.4.] Hadley, G.; Non - Linear and Dynamic Programming; Addison Wesley Publ. Comp. Reading (Mass.) 1964

[6.B.5.] Howard, R.A.; Dynamic Programming and Markov Processes; J. Wiley & Sons Inc.; New York 1960
[6.B.6.] Jacobs, O.L.R.; An Introduction to Dynamic Programming; The Theory of Multistage Decision Process; London 1967
[6.B.7.] Nemhauser, G.L.; Einführung in die Theorie der dynamischen Programmierung; R. Oldenbourg; München - Wien 1969
[6.B.8.] Neumann, K.; Dynamische Optimierung; BI Hochschultaschenbücher Bd. 714a; Mannheim 1969
[6.B.9.] Schneeweiß, C.; Dynamisches Programmieren; Physica Verlag; Würzburg - Wien 1974
[6.B.10.] Vazsonyi, A.; 1.B.30.

[6.A.1.] Bellman, R.; "Notes in the Theory of Dynamic Programming III; Equipment Replacement Policy"; RAND Report P 632; 1955
[6.A.2.] Bellman, R.; "Dynamic Programming and the Numerical Solution of Variational Problems"; in: ORSA 5/2; 1957
[6.A.3.] Bellman, R.; Dreyfus, S.; "Dynamic Programming and the Reliability of Multicomponent Devices"; in: ORSA 6/2; 1958
[6.A.4.] Dreyfus, S.E.; "Computational Aspects of Dynamic Programming", in: ORSA 5/3, 1957
[6.A.5.] Ferschl, F.; "Grundzüge des Dynamic Programming", in: Ufo 3, 1959
[6.A.6.] Held, M.; Karp, R.M.; "A Dynamic Programming Approach to Sequencing Problems", in: SIAM 10, 1962
[6.A.7.] Künzi, H.P.; Müller, O.; Nievergelt, E.; "Einführungskurs in die dynamische Programmierung", in: Lecture Note in Operations Research and Mathematical Systems, vol. 6, Springer Verlag, Berlin - Heidelberg - New York, 1968
[6.A.8.] Mitten, L.G.; Nemhauser, G.L.; "Multistage Optimization", in: Chemical Engineering Progress 59, 1963

Chapter 7

[7.B.1.] Feller, W.; An Introduction to Probability Theory and its

Applications I., J. Wiley & Sons Inc., New York 1957
[7.B.2.] Ferschl, F.; Zufallsabhängige Wirtschaftsprozesse, Physica Verlag, Würzburg - Wien 1964
[7.B.3.] Kaufmann, A.; [1.B.18.]
[7.B.4.] Khintchine, A.Y.; Mathematical Methods in the Theory of Queueing, in: Griffin's Statistical Monographs and Courses; No. 7
[7.B.5.] Lee, A.M.; Applied Queueing Theory, Mc Millan & Co. Ltd., London 1966
[7.B.6.] Morse, P.M.; Queues, Inventories and Maintenance, J. Wiley & Sons Inc., New York 1967
[7.B.7.] Newell, G.F.; Applications of Queueing Theory, London 1971
[7.B.8.] Page, E.; Queueing Theory in OR, London, Butterworth 1972
[7.B.9.] Prabhu, N.U.; Queues and Inventories, J. Wiley & Sons Inc., New York 1965
[7.B.10.] Ruiz - Pala, E.; Avila - Beloso, K.; Wartezeit und Warteschlange, Verlag Anton Hain, Meisenheim/Glan 1967
[7.B.11.] Saaty, T.L.; Elements of Queueing Theory, Mc Graw - Hill Book Comp., New York 1961
[7.B.12.] Schassberger, R.; Warteschlangen, Springer Verlag, Wien - New York 1973
[7.B.13.] Takács, L.; Introduction to the Theory of Queues, Oxford University Press, New York 1962

[7.A.1.] Doig, A.; "A Bibliography of the Theory of Queues", in: Biometrika 44, 1957
[7.A.2.] Erlang, A.K.; "Solution of some problems in the Theory of Probabilities of Significance in Automatic Telephone Exchanges", in: The Post Office Electrical Engineer's Journal 10, 1917/18
[7.A.3.] Förstner, K.; "Über das Auftreten und die Berechnung von Warteschlangen", in: Op. Res. Verf. I, 1963
[7.A.4.] Luchak, G.; "The Solution of the Single - Channel Queueing Equations Characterized by a Time - Dependent Poisson -

Distributed Arrival Rate and a General Class of Holding - Times", in: ORSA 4/6, 1956
[7.A.5.] Luchak, G.; "The Distribution of the Time Required to Reduce to Some Preassigned Level a Single - Channel Queue Characterized by a Time - Dependent Poisson - Distributed Arrival Rate and a General Class of Holding Times", in: ORSA 5/2, 1957
[7.A.6.] Meisling, T.; "Discrete - Time Queueing Theory", in: ORSA 6/1, 1958
[7.A.7.] Phipps, Jr. T.E.; "Machine Repair as a Priority Waiting - Line Problem", in: ORSA 4/1, 1956
[7.A.8.] Saaty, T.L.; "Résumé of Useful Formulas in Queueing Theory", in: ORSA 5/2, 1957
[7.A.9.] Scott, M.; Ulmer, M.B.; "Some Results for a Simple Queue with Limited Waiting Room", in: ZOR 16, ser. A, 1972
[7.A.10.] Stange, K.; "Zwei in Reihe geordnete Schalter (mit einer Warteschlange dazwischen) bei Exponentialverteilung der Ankünfte und Abgänge", in: Ufo 6, 1962
[7.A.11.] Takács, L.; "Priority Queues", in: ORSA 12/1, 1964

Chapter 8

[8.B.1.] Abadie, J.M. (ed.); Non - Linear Programming, North-Holland Publ. Comp. Amsterdam 1967
[8.B.2.] Arrow, K.J.; Hurwicz, L.; Uzawa, H.; [1.B.1.]
[8.B.3.] Collatz, L.; Wetterling, W.; [1.B.5.]
[8.B.4.] Cooper, L.; Steinberg, D.; [1.B.6.]
[8.B.5.] Karlin, S.; [1.B.17.]
[8.B.6.] Künzi, H.P.; Krelle, W.; Nichtlineare Programmierung, Heidelberger Taschenbücher Band 172, Springer Verlag, Berlin - Heidelberg - New York 1975 (reprint; 1962)
[8.B.7.] Luenberger, D.G.; [1.B.22.]
[8.B.8.] Mangasarian, O.L.; Nonlinear Programming, Mc Graw - Hill Book Comp., New York 1969

[8.B.9.] Zangwill, W.I.; Nonlinear Programming: A Unified Approach, Prentice Hall Inc., Englewood Cliffs, N.J. 1969

[8.B.10.] Zoutendijk, G.; Methods of feasible Directions, Elsevier Publ. Comp., Amsterdam 1960

[8.A.1.] Beale, E.M.L.; "On Quadratic Programming", in: NRLQ 6/3, 1959

[8.A.2.] Beale, E.M.L.; "On Minimizing a Convex Function Subject to Linear Inequalities", in: J. Roy. Stat. Soc. 17 B, 1955

[8.A.3.] Czap, H.; Dualität bei nichtlinearen Optimierungsproblemen und exakten Penalty - Funktionen", in: Op. Res. Verf. XXI, 1975

[8.A.4.] Fiacco, A.V.; Mc Cormick, G.P.; "The Sequential Unconstrained Minimization Technique for Nonlinear Programming, a Primal Dual Method", in: MS 10/2, 1964

[8.A.5.] Frank, M.; Wolfe, P.;"An Algorithm for Quadratic Programming", in NRLQ 3/1 & 2, 1956

[8.A.6.] Kuhn, H.W.; Tucker, A.W.; "Non - Linear Programming", in: Proc. of the Second Berkely Symp. on Math. Stat. and Probab., Berkeley (Calif.), University Press 1950

[8.A.7.] Lemke, C.E.; "A Method of Solution for Quadratic Programs", in: MS 8/4, 1962

[8.A.8.] Markovitz, H.M.; "The Optimization of a Quadratic Function subject to Linear Constraints", in: NRLQ 3/1 & 2, 1956

[8.A.9.] Moeschlin, O.; "Zur Stützpunktwahl beim Schnittverfahren der nichtlinearen Programmierung, Teil I und II; in: Ufo 14, 1970

[8.A.10.] Rosen, J.B.; "Gradient Projection Method for Non - Linear Programming: Part I, Linear Constraints", in: SIAM 8/1, 1960

[8.A.11.] Rosen, J.B.; "The Gradient Projection Method for Non - Linear Programming: Part II", in: SIAM 9/4, 1961

[8.A.12.] Wolfe, P.; "Methods of Non - Linear Programming", in: Graves, R.L.; Wolfe, P. (eds.); Recent Advances in Mathematical Programming, Mc Graw - Hill Book Comp., New York 1963

[8.A.13.] Wolfe, P.; "The Simplex - Method for Quadratic Programming", in: Econometrica 27, 1959

[8.A.14.] Zangwill, W.I.; "Non - Linear Programming via Penalty Functions", in: MS 13/5, 1967

Chapter 9

[9.B.1.] Buxton, J.N. (ed.); Simulation Programming Languages, in: Proceedings of the IFIP Working Conference on Simulation Programming Languages, North Holland Publ. Comp., Amsterdam,1968

[9.B.2.] Hammersley, J.M.; Handscomb, D.C.; Monte Carlo Methods, Methuen,London, 1964

[9.B.3.] Koxholt, R.; Die Simulation - ein Hilfsmittel der Unternehmensforschung, R. Oldenbourg, München - Wien 1967

[9.B.4.] Krüger, S.; Simulation, W. de Gruyter, Berlin - New York 1975

[9.B.5.] Meyer, R.C.; Newell, W.T.; Pazer, H.L.; Simulation in Business and Economics, Prentice Hall Inc.,Englewood Cliffs, N.J. 1969

[9.B.6.] Mertens, P.; Simulation, Poeschel Verlag, Stuttgart 1969

[9.B.7.] Mihram, G.A.; Simulation: Statistical Foundations and Methodology, in: Mathematics in Science and Engineering, vol. 92, Academic Press, New York - London 1972

[9.B.8.] Naylor, T.H.; Balintfy, J.L.; Burdick, D.S.; Chu, K.; Computer Simulation Techniques, J. Wiley & Sons Inc., New York 1966

[9.B.9.] Niemeyer, G.; Die Simulation von Systemabläufen mit Hilfe von Fortran IV, W. de Gruyter, Berlin - New York 1972

[9.B.10.] Niemeyer, G.; Systemsimulation, Akademische Verlagsgesellschaft, Frankfurt/Main 1973

[9.B.11.] Tocher, K.D.; The Art of Simulation, The English Universities Press Ltd., London 1967

[9.A.1.] Buxton, J.N.; Laski, J.G.; "Control and Simulation Language", in: The Computer Journal 5/3, 1963
[9.A.2.] Clark, C.E.; "The Utility of Statistics of Random Numbers", in: ORSA 8/2, 1960
[9.A.3.] Clark, C.E.; "Importance Sampling in Monte Carlo Analysis", in: ORSA 9/5, 1961
[9.A.4.] Conway, R.W.; "Some Tactical Problems in Digital Simulation", in: MS 10/1, 1963
[9.A.5.] Dimsdale, B.; Markowitz, H.M.; "A Description of the SIMSCRIPT Language", in: IBM Systems Journal 3/1, 1964
[9.A.6.] Harling, J.; "Simulation Techniques in Operations Research: A Review", in: ORSA 6/3, 1958
[9.A.7.] Jacoby, J.H.; Harrison, S.; "Multi - Variable Experimentation and Simulation Models", in: NRLQ 9/2, 1962
[9.A.8.] Namnech, P.; "Vergleich von Zufallszahlen-Generatoren", in: Elektronische Rechenanlagen 8, 1966
[9.A.9.] Schneeweiß, H.; "Simulationstechnik", in: AKOR - Schrift Nr. 5, Berlin - Köln - Frankfurt/Main 1971
[9.A.10.] van Slyke, R.M.; [4.A.9.]

Chapter 10

[10.B.1.] Barlow, R.E.; Pochan, F.; Mathematical Theory of Reliability, J. Wiley & Sons Inc., New York 1965
[10.B.2.] Bussmann, K.F.; Mertens, P.; Operations Research und Datenverarbeitung bei der Instandhaltungsplanung, Poeschel-Verlag, Stuttgart 1968
[10.B.3.] Cox, D.R.; Renewal Theory, London 1962
[10.B.4.] Lloyd, D.K.; Lepow, M.; Reliability, Management Methods and Mathematics, Prentice Hall Inc., Englewood Cliffs 1962
[10.B.5.] Proschan, F.; Pólya Type Distributions in Renewal Theory with an Application to an Inventory Problem, Prentice Hall Inc., Englewood Cliffs N.J. 1960

[10.B.6.] Roberts, N.H.; "Mathematical Methods in Reliability Engineering, New York 1962

[10.A.1.] Bellman, R.; [6.A.1.]
[10.A.2.] Chung - Kai Lai; Polland, H.; "An Extension of Renewal Theory", in: Proc. Am. Math. Soc. 3, 1952
[10.A.3.] Derman, C.; Sachs, J.; "Replacement of Periodically inspected Equipment", in: NRLQ 7/4, 1960
[10.A.4.] Dreyfus, S.E.; "A Generalized Equipment Replacement Study", in: Rand Report 1957
[10.A.5.] Ehrenfeld, S.; "Interpolation of the Renewal Function", in: ORSA 14/1, 1966
[10.A.6.] Marathe, V.P.; Nair, K.P.K.; "Multistage Planned Replacement Strategies", in: ORSA 14/5, 1966
[10.A.7.] Pyke, R.; "Markov Renewal Processes with Finitely many States", in: Ann. Math. Statist. 32/4, 1961
[10.A.8.] Sasieni, N.W.; "A Markov Chain Process in Industrial Replacement", in: ORQ 7/1, 1956
[10.A.9.] Smith, W.L.;"Renewal Theory and its Ramifications", in: J. Roy. Statist. Soc. Ser. B. 20/2, 1958
[10.A.10.] Vergin, R.C.; "Optimal Renewal Policies for Complex Systems", in: NRLQ 15/4, 1968

Chapter 11

[11.B.1.] Andler, K.; Rationalisierung der Fabrikation und optimale Losgröße, Oldenbourg, München - Berlin 1929
[11.B.2.] Baffa, E.S.; Taubert, W.H.; Production - Inventory Systems: Planning and Control, Richard D. Irwin Inc., 1972
[11.B.3.] Kaufmann, A.; [1.B.18.]
[11.B.4.] Müller, E.; Simultane Lagerdisposition und Fertigungsablaufplanung bei mehrstufiger Mehrproduktfertigung, W. de Gruyter, Berlin - New York 1972
[11.B.5.] Naddor, E.; Lagerhaltungssysteme, Verlag Harri Deutsch,

Frankfurt/Main - Zürich 1971
[11.B.6.] Proschan, R.; [10.B.5.]
[11.B.7.] Vazsonyi, A.; [1.B.30.]
[11.B.8.] Whitin, T.M.; The Theory of Inventory Management, Princetown University Press, Princetown N.J. 1953

[11.A.1.] Beckmann, M.J.; "An Inventory Model for Arbitrary Interval and Quality Distributions of Demand", MS 8/1, 1961
[11.A.2.] Beckmann, M.J.; "An Inventory Policy for Repair Parts", in: NRLQ 6/3, 1959
[11.A.3.] Bishop, J.L. Jr.; "Experience with a Successful System for Forecasting and Inventory Control", in: ORSA 22/6, 1974
[11.A.4.] Fromovitz, S.; "A Class of One - Period Inventory Models", in: ORSA 13/5, 1965
[11.A.5.] Haehling von Lanzenauer, Ch.; "Optimale Lagerhaltung bei merhstufigen Produktionsprozessen", in: Ufo 11, 1967
[11.A.6.] Henn, R.;"Fließbandfertigung und Lagerhaltung bei mehreren Gütern", in: Ufo 9, 1965
[11.A.7.] Kao, E.P.C.; "A Discrete Time Inventory Model with Arbitrary Interval and Quantity Distributions of Demand", in: ORSA 23/6, 1975
[11.A.8.] Morse, P.M.; "Solutions of a Class of Discrete - Time Inventory Problems", in: ORSA 7/1, 1959
[11.A.9.] Ohse, D.; "Zur Bestimmung der wirtschaftlichen Bestellmenge bei Preisstaffeln", in: Ufo 14/3, 1970
[11.A.10.] Popp, E.; "Einführung in die Theorie der Lagerhaltung", in: Lecture Notes in Economics and Mathematical Systems, vol. 7, Springer Verlag, Berlin - Heidelberg - New York 1968
[11.A.11.] Riepl, R.J.; "Ein deterministisches Mehrprodukt - Lagerhaltungsmodell", in: Op. Res. Verf. XVII, 1973
[11.A.12.] Scarf, H.; "The Optimality of (S,s) Policies in the Dynamic Inventory Problem", in: Mathematical Methods in the Social Sciences; Arrow, K.; Karlin, S.; Suppes, P.; (eds.)

Stanford University Press, Stanford (Calif.) 1960

[11.A.13.] Sivazlian, B.D.; "A Continous - review (s,S) Inventory System with Arbitrary Interarrival Distribution between Unit Demand", in: ORSA 22/1, 1974

[11.A.14.] Veinott, Jr. A.F.; "The Optimal Inventory Policy for Batch Ordering", in: ORSA 13/3, 1965

Chapter 12

[12.B.1.] Baker, K.R.; Introduction to Sequencing and Scheduling, J. Wiley & Sons Inc., New York 1974

[12.B.2.] Conway, R.W.; Maxwell, W.L.; Miller, L.W.; Theory of Scheduling, Addison Wesley Publ. Comp., Reading - Palo Alto - London 1967

[12.B.3.] Kern, W.; Optimierungsverfahren in der Ablauforganisation, Girardet Verlag, Essen 1967

[12.B.4.] Mensch, G.; Ablaufplanung, Westdeutscher Verlag, Köln und Opladen 1968

[12.B.5.] Müller, E.; [11.B.4.]

[12.B.6.] Müller - Merbach, H.; Optimale Reihenfolgen, in: Ökonometrie und Unternehmensforschung, Band XV, Springer Verlag, Berlin - Heidelberg - New York 1970

[12.B.7.] Muth, J.P.; Thompson, G.L.; Industrial Scheduling, Prentice Hall Inc., Englewood Cliffs 1963

[12.B.8.] Roy, B.; Ablaufplanung, Oldenbourg Verlag, München - Wien 1968

[12.B.9.] Siegel, T.; Optimale Maschinenbelegungsplanung, E. Schmidt Verlag, Berlin 1974

[12.A.1.] Bowman, E.H.; "The Schedule Sequencing Problem", in: ORSA 7/5, 1959

[12.A.2.] Elmaghraby, S.E.; "The Machine Sequencing Problem", in: NRLQ 15/2, 1968

[12.A.3.] Giffler, D., Thompson, G.L., "Algorithm for Solving Production - Scheduling Problems", in: ORSA 8/4, 1960

[12.A.4.] Johnson, S.M.; "Discussion: Sequencing in Jobs on Two Machines with Arbitrary Time Lags", in: MS 5/3, 1959
[12.A.5.] Johnson, S.M.; "Optimal Two and Three - Stage Production Schedules with Set Up Times Included", in: NRLQ 1/1,1954
[12.A.6.] Korte, B.; Oberhofer, W.; "Zwei Algorithmen zur Lösung eines komplexen Reihenfolgeproblems", in: Ufo 12/4, 1968
[12.A.7.] Lomnicki, Z.A.; "A 'branch and bound' Algorithm for the exact Solution of the Three - Machine Scheduling Problem", in: OPQ 16/1, 1965
[12.A.8.] Manne, A.S.; "On the Job Shop Scheduling Problem", in: ORSA 8/2, 1960
[12.A.9.] Noltemeier, H.; "Produktionsablaufplanung - ausgewählte Probleme und Methoden", in: Informationssysteme im Produktionsbereich, Oldenbourg, München - Wien 1975
[12.A.10.] Szwarc, W.; "Mathematical Aspects of the 3 x n Job - Shop - Sequencing Problem", in: NRLQ 21/1, 1974
[12.A.11.] Szwarc, W.; "Optimal Elimination Methods in the m x n Flow Shop Scheduling Problem", in: ORSA 6/6, 1973

Chapter 13

[13.B.1.] Behrens, K.C; Allgemeine Standortbestimmungslehre, Westdeutscher Verlag Köln und Opladen, 1961
[13.B.2.] Bloech, J.; Optimale Industriestandorte, Physica Verlag, Würzburg - Wien 1970
[13.B.3.] Bloech, J.; Ihde, G.-B.; Betriebliche Distributionsplanung, Physica Verlag, Würzburg - Wien 1972
[13.B.4.] Christaller, W.; Die zentralen Orte in Süddeutschland, Jena 1933
[13.B.5.] Grundmann, W.; et. al.; Mathematische Methoden zur Standortbestimmung, Verlag Die Wirtschaft, Berlin 1968
[13.B.6.] Lösch, A.; Die räumliche Ordnung der Wirtschaft, Jena 1944
[13.B.7.] Thünen, J.H.v.; Der isolirte Staat in Beziehung auf Landwirtschaft und Nationalökonomie, part 1, 3rd printing,

Schumacher - Zarchlin (ed.), Berlin 1875

[13.B.8.] Weber, A.; Über den Standort der Industrien, Part 1, Reine Theorie des Standorts, Tübingen 1922

[13.A.1.] Beckmann, M.J.; "Über den optimalen Standort eines Verkehrsnetzes", in: Zeitschrift für Betriebswirtschaft 35, 1965

[13.A.2.] Böventer, E.v.; "Die Struktur der Landschaft, Versuch einer Synthese und Weiterentwicklung der Modelle J.H. von Thünens, W. Christallers und A. Löschs, optimales Wachstum und optimale Standortverteilung", in: Schneider, E. (ed.), Schriften des Vereins für Socialpolitik, N.F., Bd. 27, Berlin 1962

[13.A.3.] Davis, P.L.; Ray, T.L.; "A Branch-Bound Algorithm for the Capacitated Facilities Location Problem", in: NRLQ 16/3, 1969

[13.A.4.] Domschke, W.; "Modelle und Verfahren zur Bestimmung betribelicher und innerbetrieblicher Standorte - Ein Überblick", in: Discussion Paper Nr. 24, Institut für Wirtschaftstheorie und Operations Research, Universität Karlsruhe 1973 and in ZOR 19/2, 1975 ser. B.

[13.A.5.] Efroymson, M.A.; Ray, T.L.; "A Branch-Bound Algorithm for Plant Location", in: ORSA 14/3, 1966

[13.A.6.] Eiselt, H.A.; von Frajer, H.; "Der optimale Industriestandort - Ein Verfahren mit den Kriterien Transportkosten, Arbeitsmarkt, Boden", in: Neues Archiv für Niedersachsen 24/4, 1975

[13.A.7.] Feldmann, E.; Lehrer, F.A.; Ray, T.L.; "Warehouse Location under Continous Economies of Scale", in: MS 12, 1966

[13.A.8.] Khumawala, B.M.; "An Efficient Branch and Bound Algorithm for the Warehouse Location Problem", in: MS 18, 1972

[13.A.9.] Kuehn, A.A.; Hamburger, M.J.; "A Heuristic Program for Locating Warehouses", in: MS 9, 1963

[13.A.10.] Launhardt, W.; "Die Bestimmung des zweckmäßigsten Standortes einer gewerblichen Anlage", in: Zeitschrift des Vereines Deutscher Ingenieure 26, 1882

[13.A.11.] Manne, A.S.; "Plant Location under Economies of Scale - Decentralization and Computation", in: MS 11, 1964

[13.A.12.] Müller - Merbach, H.; "Modelle der Stadtortbestimmung", in: Kosiol (ed.), Handwörterbuch des Rechnungswesens, Stuttgart 1970

[13.A.13.] Spielberg, K.; "Algorithms for the Simple Plant-Location Problem with Some Side Conditions", in: ORSA 17, 1969

[13.A.14.] Spielberg, K.; "Plant Location with Generalized Search Origin", in: MS 16, 1962

[13.A.15.] Warszawski, A.; "Pseudo - Boolean Solutions to Multidimensional Location Problems", in: ORSA 22/5, 1974